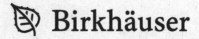

Modern Birkhäuser Classics

Many of the original research and survey monographs in pure and applied mathematics published by Birkhäuser in recent decades have been groundbreaking and have come to be regarded as foundational to the subject. Through the MBC Series, a select number of these modern classics, entirely uncorrected, are being re-released in paperback (and as eBooks) to ensure that these treasures remain accessible to new generations of students, scholars, and researchers.

Vladimir P. Vizgin

Unified Field Theories

in the first third of the 20th century

Reprint of the 1994 Edition

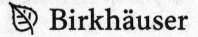 Birkhäuser

Vladimir P. Vizgin
Institute for the History
of Science and Technology
Russian Academy of Sciences
Staropansky per. 1/5
109012 Moscow
Russia
vlvizgin@gmail.com

2010 Mathematics Subject Classification: 53-03, 83-03, 01A60, 83-02, 78-03

ISBN 978-3-0348-0173-7   e-ISBN 978-3-0348-0174-4
DOI 10.1007/978-3-0348-0174-4

Library of Congress Control Number: 2011930253

Cover design: deblik, Berlin

Printed on acid-free paper

Springer Basel AG is part of Springer Science+Business Media

www.birkhauser-science.com

In memory of my parents

Vizgin Pavel Aleksandrovich and *Vizgina Ekaterina Mikhailovna*

Contents

Preface

Despite the rapidly expanding ambit of physical research and the continual appearance of new branches of physics, the main thrust in its development was and is the attempt at a theoretical synthesis of the entire body of physical knowledge. The main triumphs in physical science were, as a rule, associated with the various phases of this synthesis. The most radical expression of this tendency is the program of construction of a unified physical theory. After Maxwellian electrodynamics had unified the phenomena of electricity, magnetism, and optics in a single theoretical scheme on the basis of the concept of the electromagnetic field, the hope arose that the field concept would become the precise foundation of a new unified theory of the physical world. The limitations of an electromagnetic-field conception of physics, however, already had become clear in the first decade of the 20th century.

The concept of a classical field was developed significantly in the general theory of relativity, which arose in the elaboration of a relativistic theory of gravitation. It was found that the gravitational field possesses, in addition to the properties inherent in the electromagnetic field, the important feature that it expresses the metric structure of the space-time continuum. This resulted in the following generalization of the program of a field synthesis of physics: The unified field representing gravitation and electromagnetism must also describe the geometry of space-time. In light of this program, the main task became the search for geometrical schemes more general than Riemannian geometry that would lead to equations of the unified field sufficiently nontrivial to include among their solutions the corpuscular and quantum aspects of matter.

This book is an account of that program, grandiose in its inspiration, for a field geometrical synthesis of physics. It includes the program's first successes, the hopes that were associated with it at the start of the 1920s, its difficulties, failures, and paths of further development up to the beginning of the 1930s, when the successes of quantum theory and the discovery of new elementary particles and interactions between them pushed the uncompleted program to the background and put before it new and extremely serious problems.

In the 1970s and 1980s the problem of a unified field theory of the fundamental physical interactions again became very topical, but it must be said that the aim then was a theory of the four fundamental interactions: gravitational, electromagnetic, weak, and strong. In this modern form, the program is still essentially geometrical. Its geometrism is expressed by the so-called gauge structure of the fundamental fields, which makes it possible to use the geometry of fiber spaces. The gauge field concept, which had opened up a new resource for the field-theoretical synthesis of physics, arose historically from unified geometrized field theories of the 1920s. Indeed, the actual forms of field geometrization used in the unified theories of Weyl, Kaluza, and Einstein have not entirely lost their significance. The present program of field synthesis of physics is not in opposition to the quantum-theoretical program, as it was in the 1920s and 1930s, however. Quite the contrary, in fact—these two programs, for a long time in opposition, have now found a way to unify their efforts in the attempt to solve the fundamental problem of modern physics—the construction of a consistent unified theory of the elementary particles and their interactions.

I thank G.E. Gorelik, G.M. Idlis, A.B. Kozhevnikov, M.I. Monastyrskiĭ, and L.S. Polak, who read a draft of the book and made many helpful comments. Discussions of the draft or fragments of it in the department of the history of physics and mechanics of the Institute of the History of Natural Sciences and Technology of the USSR Academy of Sciences were very stimulating. I am grateful to A.T. Grigor'yan, the leader of the department, for his constant interest in this work. Finally, I thank Dr. Julian B. Barbour for his expert translation.

Introduction

In 1979, the Nobel Prize for physics was awarded to the American physicists S. Weinberg and S. Glashow and the Pakistani physicist A. Salam for the creation of a unified gauge theory of weak and electromagnetic interactions. As Salam noted in his Nobel address, this event occurred on the centenary of the death of Maxwell, who created the first field-theoretical synthesis of electricity, magnetism, and light, and also on another centenary, the birth of Einstein, "the man who gave us the vision of an ultimate unification of *all* the forces" ([1], p. 526).

The aim of natural scientists to construct a unified scientific picture of the world, and of physicists to construct a unified physical theory, goes back to ancient natural philosophy, which strove to find the key to a unified description of the Universe either through the discovery of a single material first principle of the phenomena of nature, or, like the Pythagoreans, through a unified mathematical structure of the world, or on the basis of the atomistic conception.

For more than two centuries, beginning in the 17th century, the epoch of Galileo, Descartes, and Newton, classical mechanics played the part of the unifying principle behind the entire diversity of natural phenomena. The "mechanistic philosophy," the "mechanistic picture of the world," and the "mechanistic explanation of nature" also reigned in the 19th century. Adopting positions close to the methodology of scientific research programs or Kuhn's paradigm approach, we could speak of the classical mechanical program or the classical mechanical paradigm in the natural sciences (and, *a fortiori*, in the physics) of the 17th to 19th centuries.

The creation during the 1860s and 1870s of electromagnetic field theory, principally by Maxwell, led finally to the formulation in the middle of the 1890s of the electromagnetic field program and the electromagnetic picture of the world in physics. The successes of Maxwell's field synthesis of electric, magnetic, and optical phenomena and the subsequent successes of Lorentz's electron theory advanced the electromagnetic field program to a dominant position. Its leaders were E. Wiechert, M. Abraham, P. Drude, W. Kaufmann, W. Wien, A. Sommerfeld, J. Larmor, O. Lodge, and to a large degree J.J. Thomson, O. Heaviside, G. Fitzgerald, H.A. Lorentz, H. Poincaré, and P. Langevin. The first stages of the scientific revolution in physics that occurred at the turn of the 19th and 20th centuries took place largely under the motto of the transition of physics to the position of the electromagnetic field program.

In his study of the crisis of the philosophical foundations of physics at the beginning of the 20th century, Lenin was forced to analyze the scientific revolution in physics. He quite correctly identified the essence of this revolution, which, in his opinion, consisted of the destruction of the mechanistic philosophy and the transition of physics to the position of the electromagnetic field program.[1]

Considered in its entirety, the permanent achievement of Maxwell's theory and the electromagnetic field program was the creation of the field concept and the idea of basing a synthesis of physics on the field concept. However, this brilliantly conceived and apparently so-promising program was already seriously undermined at the beginning of the first decade of the 20th century. The creation in 1905 of the special theory of relativity, the appearance of the quantum theory of radiation, and, in the next decade, the development of the general theory of relativity and a quantum theory of the atom showed most clearly the limits of the electromagnetic field program. It was replaced in macroscopic physics and, in particular, in gravitational theory by the relativistic program; in microscopic physics it was replaced by the quantum theoretical program. In many branches of physics, for a long time these two programs either had no contact or else complemented each other. The first impressive theoretical success was achieved by the relativistic program—in the field of gravitational theory. The general theory of relativity, in essence a relativistic theory of the gravitational field, was created in 1915–1916. The significance of this theory went far beyond gravitational physics, however, since it was simultaneously a new theory of space and time that radically transformed the foundations of the space-time notions of classical physics and even the special theory of relativity. The basic feature of general relativity that distinguished

---

[1] See V.P. Vizgin, "Lenin's analysis of the state of physics at the turn of the 19th and 20th centuries" in: *Lenin's Philosophical Heritage and Modern Physics* [in Russian], Nauka, Moscow (1981), pp. 222–262.

it sharply from all other physical theories, including the first quantum theories, was the inherent idea of the geometrization of a physical interaction (the gravitational interaction). The interpretation of the gravitational field as the manifestation of space-time curvature and the new concept of space-time, which was actually identified with a physical field whose equations are determined by the distribution of matter (both matter possessing rest mass and the electromagnetic field), was a departure from the traditional theories of physics. The fundamental nature of general relativity, and the depth and originality of its basic ideas, advanced it to the status of a program theory. Indeed, Hilbert (1915) and Weyl (1918) developed the first unified theories of electromagnetic and gravitational fields on the basis of general relativity, and these led to the formulation of the program of unified geometrized field theories at the beginning of the 1920s.

This program is related to the electromagnetic field program, since in both a fundamental role is played by the concept of a field, namely, a classical field described by a system of partial differential equations. Another common feature is that, in their most radical forms, they attempt to reduce the concept of particles and the quantum behavior of fields and particles to specific manifestations of classical fields.

It is usually assumed that Einstein was the originator of the geometrized field program. This is not entirely true, since the programmatic outlines of the new approach were formulated, as we have noted, by Hilbert and Weyl, and Einstein was initially extremely skeptical about this program. On the other hand, it *is* true to the extent that the foundation and core of this program was supplied by the general theory of relativity, and without a doubt Einstein made the decisive contribution to its creation. Nevertheless, some years after Weyl's pioneering work (and we note in passing that it was in Weyl's and not Hilbert's form that the core of the geometrized field program was developed in its final form) Einstein did indeed become the leader of this program; moreover, he remained its leader to the end of his life.

At the beginning of the 1920s, the program inspired great hopes, although few physicists decided to be guided by it in their research. The failures of the global conceptions of Hilbert and Weyl, above all regarding the acquisition of new physical results and the establishment of connections with quantum theory, frightened away physicists, who mainly preferred to work on the basis of the quantum theoretical program, which did produce real physical results even if they were not always outstanding. In contrast, the mathematical depth of the geometrized field program and its connection with the latest advances of differential geometry attracted mathematicians, who played an important part in its unfolding.

Besides Hilbert and Weyl, we must mention Cartan, Schouten, Veblen, Levi-Civita, and Eisenhart. Physicists such as Schrödinger, Pauli, Eddington, Fock, Tamm, and others, who made important and sometimes decisive contributions to the development of the quantum program, were also interested in the geometrized field program and attempted to use it. The creation during 1925–1927 of quantum mechanics and its triumphs in subsequent years led to a pronounced loss of authority of the unified geometrized field program at the beginning of the 1930s, although from time to time some leaders of the quantum program, such as Pauli (at the beginning of the 1930s) or Schrödinger (in the 1940s), returned to unified field theories. Against a background of the outstanding successes of the quantum program, the (at times) elegant mathematical schemes of the unified geometrized field theories appeared fruitless, especially as the years and decades passed and the unified theories that successively supplanted their predecessors did not advance beyond global conceptions and abstract mathematical structures.

A fundamental crisis in the development of the unified theories occurred at the beginning of the 1930s, especially after 1932, which was a landmark in the history of elementary particles. These years saw the discovery of the neutron and positron, the development of the foundations of quantum electrodynamics, the development of the proton–neutron model of nuclei, the notion of nuclear exchange forces, and the first theory of weak interactions. Thus, two new types of interaction—the strong and the weak—appeared in physics. This circumstance emphasized still more the limitations of the geometrized field program, which aimed to unify the gravitational and electromagnetic fields and was still in essence classical. The beginning of the 1930s was indeed the time after which the program gradually lost its significance. It was probably only the authority of Einstein (and perhaps, Schrödinger, who in the 1940s and 1950s continued to work on unified theories) and the mathematical depth of some of these theories that attracted theoreticians to this field.

Although in the 1940s and 1950s there were occasional attempts to unify not only the electromagnetic and gravitational fields but also the meson field on the basis of the geometrization program, the overwhelming majority of physicists regarded these attempts as having no promise. The main efforts of theoreticians at that time were directed to the development of a consistent quantum electrodynamics and the extension of the methods of quantum field theory to the complete set of elementary particles. The first task was brilliantly completed by Tomonaga, Schwinger, and Feynman, who developed renormalization theory. The solution of the second problem was hindered by the unceasing discovery of more and more new elementary particles and properties of them. It was at the end of the 1940s and the beginning of the

1950s that there began a period of new discoveries in the physics of elementary particles. The hyperons and K mesons, and also the so-called resonances, were discovered. In view of this abundance of different types of elementary particles and the interactions between them, the approach based on the unified geometrized field program appeared extremely abstract and devoid of a real empirical basis. The discovery of semi-empirical laws, such as the conservation of strangeness, which gave certain hopes for the construction of a sensible theoretical scheme in the foreseeable future, was regarded as a great success in particle physics. The discovery of parity nonconservation in weak interactions in the second half of the 1950s complicated the situation in particle physics still further.

By this time, Einstein had already died (1955), and almost none of the leading physicists took the program of geometric unification seriously. In 1958, the position of the overwhelming majority of physicists was clearly expressed by Pauli in the English edition of his famous encyclopedia article on the theory of relativity, to which he wrote a special supplement devoted to unified geometrized field theories after 1920:

> Most physicists, including the author, agree with the analysis of Bohr and Heisenberg in their judgment of the epistemological situation produced by these developments [i.e., the uncertainty principle and complementarity], and therefore hold a complete solution of the open problems of physics through a return to the classical field concepts to be impossible. ([2], p. 224)

Nevertheless, Pauli considered the main lines of development of the program, emphasizing once more the existence of fundamental difficulties of a physical nature in the path of its realization, difficulties that had not been overcome during the almost forty years that had elapsed since the appearance of the first unified geometrized field theories.

A few years after this, an even harsher estimate of the geometrized field program was given by the well-known Soviet general relativist A.Z. Petrov:

> None of the existing "unified theories" has yet gone beyond abstract theoretical constructions or led to significant discoveries or consequences capable of experimental verification.... They have not played ... a heuristic role with respect to other branches of modern physics. ([3], p. 7)

The number of such comments could easily be multiplied. Incidentally, in the second half of the 1950s the idea of a unified field theory again came to the fore, though not on the basis of the earlier program. We are referring to Heisenberg's unified nonlinear spinor quantum field theory [4, 5]. It should be said here that the hopes for Heisenberg's theory were not justified, and the direction associated with the gauge field concept was more promising, leading ultimately to the now famous Weinberg–Salam theory. It is interesting to see, however, how Heisenberg, who received the baton of unified field theories from Einstein, estimated the geometrized field program. At the start of the 1960s, in a paper entitled "Comments on Einstein's outline of a unified field theory," he wrote of the program:

> This attempt, founded on a magnificent basis, at first appeared to have failed. At the time when Einstein was occupied with the problem of a unified field theory, new elementary particles were continually being discovered, and new fields were associated with them. As a consequence, a firm empirical basis did not yet exist for implementation of Einstein's program, and his attempt did not lead to any convincing results. However, the disaster that befell Einstein's program also had deeper reasons than merely uncertainty about empirical facts; the reasons are to be sought in the relationship of Einstein's field-theoretical notions to quantum theory. ([6], p. 63)

Without dwelling on the reasons for the failure of Heisenberg's conception, we mention that, while it retained Einstein's idea of nonlinearity of the field equations, it abandoned the idea of geometrization and general covariance. In contrast, the gauge concept of field unification, which led to the Weinberg–Salam theory and to realistic projects for the unification of not only the electromagnetic and weak but also the strong and even gravitational interactions, is, in the geometric respect, a certain generalization of the geometrized field program, taking into account both Einstein's idea of geometrizing physical interactions and the idea of general covariance [7]. The great successes of the gauge concept in the 1970s gradually changed attitudes to the unified geometrized field program and the unprecedented efforts of Einstein aimed at its realization. Yang, one of the creators of the concept of gauge fields, and Salam, Weinberg, and Glashow, whom we have already mentioned, emphasized that the concept went back to the early ideas that Einstein and Weyl developed on the basis of general relativity and the geometrized field program [1, 8]:

> On the long and difficult path to the understanding of nature, we again and again find ideas that are derived from Einstein. ([9], p. 8)

Thus, in connection with the present revival of the field unification idea, attitudes to the geometrized field program have changed, and there is now an interest in the history of unified geometrized field theories. Moreover, even a preliminary study of the formation of the basic ideas of the gauge concept indicates an intimate relationship between them and unified geometrized field theories. Thus, investigation of the by no means smooth path of development of these theories has now acquired particular importance.

Even if this renaissance had not occurred, and the geometrized field program had been finally written off as a dead end in the development of theoretical physics, study of the history of this program would still be an extremely important and interesting task for the historian of modern physics. The birth of the program at the turn of the century and during its first two decades; the attempts to use unified theories to master the mysteries of quantum theory; the adherence (at times transient, at times very long) to this program of many eminent physicists and mathematicians of this century such as Einstein, Hilbert, Weyl, Eddington, Schrödinger, Pauli, and Cartan; the gradual decrease of its authority after the creation of quantum mechanics and especially after the epoch of elementary particle physics had begun (in the early 1930s); the major contribution of this idea to the development of the most modern differential geometry; and, finally, the unparalleled doggedness of Einstein—these are all unquestionable realities of the physics of the 20th century, and, moreover, fundamental theoretical physics at that. Therefore, one can hardly understand the development of physics in our century if one ignores this line of development of theoretical physics that even recently was regarded as secondary or even a blind alley.

Although the program remained unrealized, and the paths of its realization by no means lay along the principal directions of the development of physics, study of this program is very instructive and makes it possible to investigate the steps in theoretical thought through which we pass when studying the evolution of programs (for example, that of quantum theory) that led to the creation of generally accepted scientific theories. Moreover, we shall see that many aspects of the actual historical process in its principal directions can be understood only when the attempts made on the lateral paths are taken into account. Of course, the marginal nature of programs like the one we are considering does not by any means become clear all at once but only in the process of the development of the program, which is accompanied by failures, especially when set against the background of successes of competing programs.

The investigation of unified geometrized field theories also provides interesting material from the methodological point of view. Problems of the structure of scientific theory, the functioning of methodological principles of physics, the interrelations between the physical aspects and the mathematical formalism, the axiomatics of physics, the heuristic role of theoretical schemes, and alternatives to the main program—all these fundamental problems of the philosophy and the methodology of scientific epistemology are at the center of attention in an analysis of the paths along which the unified geometrized field theories were developed.

Thus, the topicality and methodological value to the history of science of study of the unified geometrized field program are rather obvious. Nevertheless, corresponding studies by historians of science hardly exist. We do not pretend to fill this gap entirely. We state in advance the restrictions we have adopted. First, as prehistory and preconditions of the program, we consider only the electromagnetic field program, and also some characteristic features of general relativity as the core of the program of geometrized unification. Thus, the earlier history of the problem of the synthesis of physical knowledge associated, for example, with the classical-mechanical program (18th–19th centuries) is outside the ambit of our study.

Second, we restrict our investigation to about one and a half decades, the years after the first unified field theory characteristic of the program had been advanced (Weyl's theory in 1918). The years 1932–1933 are the upper chronological limit of our systematic analysis of the development of the geometrized field program. Why? The fact is that by this time extensive material on unified theories had been accumulated, and all the main forms of such theories had been created. In addition, in those years the advantages of the quantum-theoretical program for the solution of the problem of the structure of matter had become entirely obvious to the majority of physicists. Finally, as we have already noted, it was precisely at this time that, alongside the electromagnetic and gravitational fields, the existence of other fields associated with the newly discovered elementary particles became clear. It is at the beginning of the 1930s that we see the first theoretical schemes for the strong and weak interactions and the theory of the electromagnetic field being transformed into quantum electrodynamics. This does not mean that further unified geometrized field theories did not appear in subsequent years, but the propositions of the program were so undermined that even adherents like Weyl, in a certain sense its creator, recognized the defeat, or at least the lack of promise, of the strategy previously adopted by them.

Third, we have not, unfortunately, been able to consider all the forms of unified field theories that arose during the 1920s and 1930s. It has been necessary to concentrate mainly on the theoretical schemes that were regarded as the most promising and were developed with the greatest vigor. Since the indisputable leaders of the unification program were Weyl, Einstein, and Eddington, it is their studies, and also the studies that attracted their attention, that have been most fully investigated.

Finally, in this complete history we identify certain turning points and central themes. We mention as an example Weyl's theory and its subsequent developments, because it was the first that could be truly called a program theory and, precisely for this reason, had the greatest heuristic significance in the genesis of fundamental aspects of quantum theory. An example of a turning point in the development of the program was in 1921, when the theories of Eddington and Kaluza were advanced and when Einstein definitely agreed with the position of this program. Some of the central themes that are considered in our work are the evolution of Einstein's attitude to the program—from complete rejection to a complete switching over to it (of course, Einstein is a special figure in this history, and he successively tried all or almost all the geometrical field unification schemes that existed at the time); the heuristic role of unified theories in the genesis and subsequent development of quantum mechanics and quantum field theory (in particular, we investigate in detail the history of the development of the concept of gauge symmetry and gauge fields, since, on the one hand, unified field theories played a very important part of this process and, on the other hand, it was on the basis of this concept that, in the second half of the century, a real project of a new unified field theory arose); the clear regression (and its recognition) of the unified geometrized field program at the beginning of the 1930s, and so on. Therefore, our work is to a large degree an outline study.

We have already mentioned the almost complete absence of work devoted to unified geometrized field theories by historians of science. Nevertheless, our work has not been based solely on primary sources. In the first place, we have made important use of some reviews of unified geometrized field theories written by Tonnelat, Pauli, Bergmann, Eddington, Weyl, Beck, Landé, Jordan, Schrödinger, Schmutzer, Rumer, Treder, Petrov, and others [10–23]. We have also used scientific biographical material relating to the lives and work of Einstein, Hilbert, Weyl, Eddington, Pauli, Schrödinger, and other physicists and mathematicians of the 1920s and 1930s, correspondence and memoir material (here, Einstein's correspondence was particularly valuable), and studies by historians of science of related themes, in particular the history of quantum mechanics by Jammer, M. Klein, Mehra, Forman, Raman and Forman, Hund, and others [24].

For the discussion of the electromagnetic field program the studies of R. McCormmach, Hirosige, Goldberg, Pyenson, and others [24] were particularly important. Methodologically, we have based our work on Lakatos's concept of scientific research programs (although not in complete measure and with allowance for the criticism of it in Soviet literature) [25, 26].

CHAPTER 1

The Electromagnetic Program for the Synthesis of Physics

INTRODUCTORY COMMENTS

The development of physics, especially during periods of radical adjustment, demonstrates the fruitfulness of the methodological idealization usually associated with Lakatos's concept of scientific research programs (methodology of scientific research programs, MSRP). This methodology helps in the evaluation of many facts in the development of scientific knowledge. Moreover, extensive scientific-historical material confirms the real existence of entities, organized and functioning, such as scientific research programs. A program has a "core" structure, namely, a *hard core* consisting of one or several fundamental theoretical schemes, some additional propositions of a methodological (and sometimes even philosophical) nature, and a *defensive shield* of auxiliary hypothesis and structures[1] that makes it possible to identify problems for further investigation and to predict anomalies and eliminate them, or rather, transform them into confirmatory examples. Thus, the defensive shield of a program is its active organ, which develops in the process of the functioning of the program and determines the procedure of *positive heuristic, whereas the hard core is responsible for the strategy of the program and, as a rule, for the contribution that the program makes to the scientific (or more narrowly, physical) picture of the world.*

---

[1] The development of equivalent mathematical formalisms of core theories, for example, the analytic mechanics of Lagrange, Hamilton, Jacobi, and others, is an example of work in the domain of the defensive shield of, in the given case, the classical mechanical program, which contains as its basis the core theory of Newtonian mechanics.

The concept of a scientific research program is used generally in a quite wide sense; in particular, in the history of physics it can be applied to special branches of the subject and even to individual problems, for example, in the analysis of electrical conduction of metals. In these cases, the core of the program includes not only fundamental theories but also very special theoretical schemes. However, since the emergence of science, and in particular physics, a pioneering role in its development has been played by programs that relate to physics as a whole or to its main theoretical constructions [27]. We shall call such programs *global*. The global scientific research programs have the longest life, touching the very foundations of physical knowledge (if we are speaking of physics): The regression of certain global programs and the clear progression of others lead ultimately to the former being supplanted by the latter, which is naturally related to the phenomenon of a scientific revolution. Scientific pictures of the world are deeply related to global programs. Moreover, since the concept of the scientific picture of the world is almost universally accepted and has long been used by historians and methodologists of science [28–35], it is in fact often used when the concept of a global scientific research program is more effective and appropriate. The fact is that the concept of a picture of the world possesses certain shortcomings compared with that of a scientific research program. A picture of the world is usually associated with a transparent picture of the world from the point of view of fundamental scientific theories. It is not included in any effectively working methodological scheme like the MSRP, its structural outlines are less clear than in a scientific research program, and so on. A global program is not necessarily required to be transparent; the concept is more operative and methodological than a picture of the world, although there is more than a little in common between these two concepts, for both are global in nature, are based on fundamental theories, and often contain propositions of a philosophical nature. The mechanical, electromagnetic, and quantum relativistic pictures of the world (in the last case with certain reservations and restrictions), which characterize the 19th and 20th centuries, can be compared to corresponding global scientific research programs.

The problem of the unity of physical knowledge is intimately related to the investigation of global scientific research programs. The basic subject of our analysis will be unified geometrized field theories (mainly of the gravitational and electromagnetic fields), which strived in their radical forms to reduce the fundamental material elementary particles to a field. Thus, the frameworks of these theories have much in common with those of global research programs. It is natural, therefore, to speak of a global program based on two basic classical field theories—the electromagnetic and gravitational, the basic program function being fulfilled in fact by the theory of the gravitational field, i.e., the general theory of relativity, since the method of unifying the fields in all (or almost all) unified geometrized field programs is based on the characteristic feature of general relativity, the geometrization of physical interactions. The concept of a physical picture of the world based on these theories appears particularly unfortunate and actually was never used. This is explained not only by the fact that the geometrized field theories are not transparent but also by the fact that none of the theories became widely accepted. The various unified field theories, following one after another, remained rather at the level of a fundamental but unrealized program and did not lead to a picture of the world recognized by the scientific community of physicists. It was different in the case of the electromagnetic field program. Of course, it was more transparent than the geometrized field program. The important thing, however, was that at the turn of the century the majority of physicists adhered to the electromagnetic program (in one form or another), and many of them (and subsequently many historians of science as well) used the concept of the electromagnetic picture of the world, electromagnetic view of nature, etc. [29, 34, 35]. Even in this case, the concept of a global program is entirely acceptable and even appears more appropriate than the concept of a picture of the world.

In periods of scientific revolution there are, as a rule, several competing programs, namely, global scientific research programs. For example, after the creation of the special theory of relativity, the classical mechanical, electromagnetic, and relativistic (or relativistic field) programs really did compete in the solution of fundamental problems of gravitation and the structure of matter. In addition, the programs themselves can be divided into several subprograms, which differ not only in the defensive shield but also in some details, often important, of the core structure. Despite the sometimes sharp difference between the program frameworks, there is a certain affinity between the hard cores of competing programs. First, the fundamental theories that provide the basis of the program hard cores can be related by some form of correspondence principle (for example, the Einstein–Poincaré special principle of relativity in the relativistic program and the theory of space and time in the classical mechanical program). Second—and this is now more important for us—the sets of methodological presuppositions relating to the hard cores of different programs have, as a rule, some elements in common. These common elements belong to the norms of scientific investigation and methods of construction of scientific theories that have become established in the given field of science. Although different scientists may adhere to different global programs, they belong mainly to a single scientific community (in the given case, we have in mind the contemporary scientists). W. Thomson, a supporter of the classical mechanical program, Abraham and Mie, supporters and even leaders of the electromagnetic program, and Einstein, the leader of the relativistic program, undoubtedly belonged to a single scientific community. The assumptions and principles of methodological nature common to a given scientific community form the not very clearly delineated system of so-called methodological principles of physics, which only comparatively recently became the subject of detailed investigation as a whole [36]. During a time in which widely accepted programs are regressing and new theories are being created in a period of revolutionary transformation of scientific knowledge, the system of methodological principles ensures the stability and continuity necessary and also plays the part of one of the arbiters in the competition among the scientific research programs. The principles that form this system of methodological principles of physics include, for example, the principles of symmetry, conservation, causality, simplicity, observability, correspondence, and unity of physical knowledge. This system, like the very notion of a scientific research program, is an idealization whose validity is confirmed by scientific-historical material relating to the physics of the second half of the 19th century and the 20th century, and the expediency of its introduction is justified by the fact that it does, as we believe, augment the methodology of scientific research programs, making it more realistic and flexible [37, 38].

In this chapter, we shall consider some attempts to construct a unified field theory based on the electromagnetic program. They are of interest to us primarily because this program was the first field program for the synthesis of physics based on the concept of a classical—in the given case, electromagnetic—field described by a system of second-order partial differential equations. Further, one of the main aims of the electromagnetic program, like the geometrized field program, was the idea that it would be possible to reduce to the field all the remaining "matter," in the first place the simplest charged particles of matter, and atoms (understood as field concentrations, specific vortex formations, or as certain singularities). Finally, some specific projects for realization of the electromagnetic program, for example, the theories of Einstein (1908–1910) and Mie (1912–1913), were based on the idea of a nonlinear generalization of Maxwell's equations of the electromagnetic field, just as the unified geometrized field program was associated with generalization of the nonlinear equations of general relativity.

Despite such a strong affinity, there are deep differences between the electromagnetic and unified geometrized field programs. In the electromagnetic program, it was assumed that the gravitational field would be completely reduced to the electromagnetic field (though in some moderate forms of the program an independent existence of the gravitational field was allowed). The electromagnetic field, the basis of everything that exists, was assumed to be continuously distributed in Euclidean space, while after the discovery of special relativity, the space-time continuum in which the field was embedded was equipped with a four-dimensional pseudo-Euclidean structure, which was not subject to an influence of the field. Thus, in the electromagnetic program, the methodological idea of the synthesis was strongly reductionist; it was assumed that gravitation and matter, and the quantum manifestations of radiation discovered at the beginning of the century and, *a fortiori*, the complicated effects of solid-state physics, gases, liquids, and so on would be reduced to an electromagnetic field described either by Maxwell's equations or some modification of them. Anticipating, we note that the methodological idea of synthesis in the unified geometrized field program was essentially different, though in this program too there were elements of reductionism. Above all, unification was to be achieved on a geometrical basis. Both fields known at the beginning of the 1920s—the electromagnetic and the gravitational—were assumed to be manifestations of a single geometrical structure of space-time, just as in general relativity the gravitational field was regarded as a manifestation of the curvature of a four-dimensional pseudo-Riemannian space and described by metric tensor and corresponding Christoffel symbols (affine connection coefficients). In particular, in some forms electromagnetism was associated with the torsion of a suitably generalized pseudo-Riemannian space; in others, it was associated with the introduction of a fifth dimension; in yet others with the introduction of an asymmetric metric tensor of four-dimensional space, and so on. As a result, invariance principles, geometrical aspects, and the relativistic approach were in the forefront in the geometrized field programs, as in general relativity. The situation in the unification program with regard to the problem of reducing particles to a unified field was more obscure. The particles were assumed to be described either as certain singular solutions of the fundamental equations or, in contrast, by means of everywhere-regular solutions. Here, one can see a much greater affinity with the electromagnetic program, especially in the case when nonlinear generalizations of Maxwell's equations were used.

Historically, the program of the electromagnetic field synthesis of physics directly preceded the unified field program and, undoubtedly, influenced its formation in exactly the same way as the field concept that had arisen in Maxwell's theory was undoubtedly the key to the creation of general relativity. Moreover, some leaders of the geometrized field program, above all Einstein and Hilbert, had themselves taken part, before the creation of general relativity, in the development of unified field theories based on the electromagnetic program.

Thus, study of the electromagnetic program and the specific forms of unified theories based on it had a double interest for us—in logical respects as genuine precursors of the geometrized field program and the corresponding theories, and in historical respects as direct preconditions and factors in the formation of the geometrized field program.

THE DEVELOPMENT OF THE ELECTROMAGNETIC PROGRAM

The development of the electromagnetic program was quite complicated, and close examination of it does not have a place in the present work. There are, in fact, several deep and detailed studies that are actually devoted to this problem by McCormmach, Jammer, Hirosige, Illy, and others [33–35, 39–41]. In the majority of these studies, in particular in the very detailed paper of McCormmach, the electromagnetic picture of the world rather than the electromagnetic program appears as the topic. As we have already noted, however, we prefer to use the concept of a global scientific research program, in this case the electromagnetic field program.

Previously we mentioned the existence of several different forms of realization of the electromagnetic program (or even several forms of the program). We shall first describe the general features of all these forms. The basis of the hard core of this program was first supplied by Maxwell's theory of the electromagnetic field, which was then developed to the level of Lorentz's electron theory (Maxwell created the theory of the electromagnetic field by the middle of the 1860s and the first systematic exposition of the electron theory was given by Lorentz in 1892). It was also assumed that primary physical reality was constituted precisely *by* the electromagnetic field. It must be said, however, that by no means did all physicists use the field concept; the ether concept was more widespread and often replaced it. In addition, some moderate forms of the program did not require that the charged particles should be reducible to the field, although in the majority of programs it was assumed that the mass of the charged particles, especially the electrons, was entirely due to the electromagnetic field.

The electromagnetic field program, whose first formulations were advanced in the middle of the 1890s by Wiechert in Germany and Larmor in Britain and which became almost universally accepted during the period 1900–1905, was preceded by two more special programs: the Faraday program of short-range (local) interaction, which played a pioneering part in the genesis of Maxwell's theory, and the corpuscular electric program of Weber, which was most strongly developed by him in the period when the foundations of

Maxwell's theory had already been created, i.e., in the 1870s, and which had a strong influence on Lorentz. Nevertheless, the dominant global program in the 19th century was still the classical mechanical program, which in the most fundamental questions (the primary status of mass, the structure of the ether, the notions of space and time, and causality) was supported by the overwhelming majority of physicists, including both supporters of Faraday and Maxwell and supporters of Weber.

Lorentz greatly valued Weber's program, according to which everything existing was to be reduced to electric particles of two signs interacting in accordance with Weber's electrodynamic law, and in his electron theory Lorentz relied equally on the field concept and Maxwell's theory, on the one hand, and on the concept of charged particles and the corresponding parts of Weber's program, on the other.[2]

The radical form of the electromagnetic program advanced by Wiechert in 1894 and somewhat later by Larmor declared the ether to be primary reality; its excited states gave the charged particles (electrons), and the origin of their mass was explained on the basis of the concept of electromagnetic mass developed in the 1880s and 1890s, primarily by British scientists (J.J. Thomson, Heaviside, and also, somewhat later, Searle and Morton). It was assumed that the laws of Newtonian mechanics could be deduced from the equations of the electromagnetic field. A few words should be said about the concept of electromagnetic mass. In 1881, studying the motion of a small charged sphere, J.J. Thomson found that, because of the interaction of the charge of this sphere with the field produced by it, a force that resists the motion of the sphere arises. This could be interpreted as though the sphere possessed an additional mass. "In other words, it [i.e., the resistance experienced by the sphere] must be equivalent to an increase in the mass of the charged moving sphere," wrote Thomson ([41], p. 144 of the Russian translation). According

---

[2] McCormmach notes some important features of Weber's program that represent clear departures from the classical mechanical program and are fully developed either in the framework of the electron theory and the electrodynamics of moving bodies, which are essentially related to the electromagnetic program, or in the special theory of relativity. These are the replacement of Newtonian instantaneous action at a distance by the propagation of an electric force with finite velocity, the violation of Newton's third law for electrodynamic forces, the presence of an upper limit for the relative velocities of particles, the possibility of symmetric use of the coordinates of space and time in the description of electrodynamic phenomena, the idea that the apparent masses of the electric particles depend on their velocities, and a new formulation of the law of conservation of energy in a form suited to electrodynamics.

to Thomson's calculations, this additional mass was

$$\mu = \frac{4}{15}\frac{e^2}{ac^2},$$

where e is the charge of the sphere, a is its radius, and c is the speed of light.

This mass of electromagnetic origin, which is not associated with any definite amount of matter, was called *apparent* or *fictitious mass*. Having calculated the apparent mass of the earth under the assumption that it carries a certain electric charge, the greatest allowed by the geophysical ideas of that time, Thomson concluded that this apparent mass was exceptionally small compared with the ordinary mass of the earth and initially he was very far from the idea of interpreting the entire inertial mass of charged bodies as induced. Eight years later, Heaviside gave a more rigorous calculation of this effect. He obtained what was to become a classical result:[3]

$$\mu = \frac{2}{3}\frac{e^2}{ac^2}.$$

The Thomson–Heaviside effect suggested the idea of complete reduction of mechanical mass to the electromagnetic field. The question of the electromagnetic nature of mass became a central question for clarification of the viability of the electromagnetic program, which at the end of the 1890s acquired more and more support among physicists, especially in Germany and Britain. Although the concept of electromagnetic mass was the child of British scientists, their ideas about the ether were more mechanistic than those of the German physicists. Therefore, the British form of the electromagnetic program was more contradictory. Besides J.J. Thomson, Heaviside, Larmor, Fitzgerald, Lodge, and others in the British Isles were supporters of the electromagnetic program in one form or another during that time.

In contrast, Lorentz adopted a more careful position, holding back from global electromagnetism and preferring to work out specific problems of the electron theory and the electrodynamics of moving bodies, which ultimately led him to results of fundamental significance. In 1895 he was forced to admit the possibility of violation of the principle of the equality of action and reaction for the ether, and in 1899 he established a dependence of the mass of all bodies, not only charged particles, on their velocities. Actually, dependence of the

---

[3] Heaviside took a more literal view of his result, regarding the increase in the mass as a real physical effect.

electromagnetic mass of a charge on the velocity of its motion had already been established by J.J. Thomson in 1893, although this dependence differed from the one that was subsequently recognized as correct (in fact, it contained the characteristic relativistic factor and led to an infinite value of the mass for a velocity equal to the speed of light) [42]. The extension of dependence of mass on velocity to all bodies, including ones without charge, amounted to a clear recognition of the limitations of Newtonian mechanics and, thus, of the classical mechanical program as a whole. This result was a consequence of the introduction by Lorentz (and Fitzgerald) of the contraction hypothesis into the electron theory. It was a new and very strong stimulus for the development of non-Newtonian dynamics and the electromagnetic program.

Just a little earlier there had been an avalanche of experimental discoveries in the fields of x-rays, radioactivity, and electric charge in gases (1895–1896), which led to the discovery of the electron (J.J. Thomson and Wiechert, 1897). For the electron theory and the electromagnetic program as a whole this was an event of huge importance, not only because the fundamental particle expected by theoreticians had finally been discovered experimentally, but also because the experimentalists acquired a powerful tool for testing the new electromagnetic dynamics. The subsequent explanation of the Zeeman effect (Lorentz, 1897) and the electron theory of the electrical conduction of metals (Riecke, 1898) strengthened still further the position of the electron theory and the electromagnetic program.

In 1898, Des Coudres suggested that cathode rays could be used experimentally to test whether the mass of the electron could be reduced to an electromagnetic mass. The discovery of the electron, and also the demonstration by J.J. Thomson in 1900 of the identity of electrons obtained in different ways, created the prerequisites for the experimental resolution of this cardinal problem of the electromagnetic program. The discovery of radioactivity and the subsequent identification of beta rays with rapidly traveling electrons made an experimental verification of the dependence of the mass of electrons on the velocity of their motion entirely realistic.

In 1900, in a major program report at Leyden entitled "Electromagnetic theories of physical phenomena," Lorentz brilliantly described the successes of the electromagnetic program, paying in particular great attention to problems of the structure of matter, and he noted that as yet only gravitation could not be fitted into the framework of this program. In the same year, he attempted to fill this gap as well by developing a very promising scheme for the reduction of gravitation to electromagnetism based on a field generalization of the ideas of Weber, Zöllner, and Mossotti. In the same year, Wien developed an analogous theory of gravitation, in which he also postulated a completely electromagnetic origin of the mass of the charged particles that make up matter. Like Lorentz, Wien saw the main problem of the physics of the time as being its unification on the basis of the electromagnetic program, but, in contrast to Lorentz, he devoted particular attention to a systematic electromagnetic derivation of mechanics.

The electromagnetic program continued to strengthen its position. On the one hand, it was clearly progressive, transforming anomalies into confirmatory examples, and its theoretical growth in these years anticipated the also growing empirical material. On the other hand, theoreticians no longer succeeded fully in reconciling the electromagnetic program with the system of methodological principles. For example, the efforts of many theoreticians, in particular Poincaré, had the aim of reconciling the electron theory with the principle of conservation of momentum. To this end, Poincaré developed the concept of electromagnetic momentum, and the reconciliation was achieved. During 1902–1903, fundamental investigations of electron dynamics in the framework of the electromagnetic program were made by the young Göttingen theoretician Abraham, who became one of the leaders of the new program. His approach was more radical than Lorentz's. Whereas Lorentz allowed the existence of nonelectromagnetic forces precisely in order to explain the equilibrium of the model of a deformable electron, which he used in electron dynamics, Abraham developed a model of a nondeformable electron, which did not require this assumption. Although Lorentz continued to adhere to his model, which, naturally, was related to his electrodynamics of moving bodies, at the beginning of the century many physicists preferred Abraham's theory precisely because it was more thoroughly consistent with the electromagnetic program. As a result, the question of accurate experimental verification of the velocity dependence of the electron mass moved to the center of attention, since there was a difference between the corresponding formulas of Lorentz and Abraham.

At this stage, the Göttingen experimental physicist Kaufmann became one of the leaders of the electromagnetic program. In his review lecture at a congress of German natural scientists and doctors in 1901, he emphasized the successes of the electromagnetic program, which, like Lorentz, he regarded as a development of Weber's program, and he formulated five basic problems on the solution of which, in his view, the final triumph of the electromagnetic view of nature depended. His five problems were: (1) experimental proof that the entire electron mass could be reduced to electromagnetic mass; (2) the systematic reduction of mechanics to electromagnetism (in the spirit of Wien); (3) the experimental proof of the electron structure of matter (more precisely, that matter consisted solely of electrons); (4) the establishment of a connection between the properties of the periodic table of the elements and definite configurations of the electrons in atoms (here, great theoretical investigations were still needed); and (5) experimental verification of Wien's electromagnetic theory of gravitation. In 1902, Kaufmann confirmed Abraham's formula for the velocity dependence of the electron mass, and in 1904 Lorentz, in a lecture given to the Electrotechnical Society in Berlin, was inclined to recognize the successes and good prospects of the electromagnetic program in its radical form:

> On this basis [i.e., on the basis of Kaufmann's experiments] it must be recognized that negative electrons do not possess a true mass and have only an electron mass; they are, so to speak, only a charge without matter.... ([43], p. 27)[4]

Thus, by 1904–1905 the electromagnetic program had already reached its peak, though its supporters did not yet recognize that fact. Already at that time, after the foundations of the electrodynamics of moving bodies and the special theory of relativity had been laid, and the ideas of quanta had acquired an ever greater reality, the electromagnetic program began to exhibit signs of regression, and there gradually developed new competing programs associated with the extension of the special theory of relativity to the main branches of physics (the relativistic program) and the introduction of quantum ideas into the study of the structure of radiation and matter (quantum theoretical

---

[4] It is true that a bit later he spoke about this somewhat more carefully, in the subjunctive, regarding the results of Kaufmann's experiments as not yet sufficient for unambiguous recognition of the triumph of the Abraham–Kaufmann position:

> One could suppose that all and every ponderable matter consists of electrons and that all and every kinetic energy of moving bodies consists of the energy of electromagnetic fields. If this assumption were confirmed, then ultimately it would not be the case that electromagnetic phenomena were explained mechanically but, rather, mechanical phenomena would be explained by electromagnetic processes; in such a case, all and every technology would have as its basis electrotechnology. ([43], p. 32)

Incidentally, in preparing this lecture for publication in 1905 Lorentz already assumed that the "latest measurements of Kaufmann (1905) overturned the hypothesis of the deformable electron." Lorentz continued: "Thus, my attempt to derive from the electron theory complete independence of phenomena from the velocity of translational motion must be regarded as having failed." In other words, the results of Kaufmann's experiments led Lorentz, precisely in the year that the special theory of relativity was created, to abandon the electrodynamics of moving media that he had developed and which he had strongly related to his hypothesis of the deformable electron, and to give up the principle of relativity in the domain of electromagnetism.

program). The general opinion of physicists clearly registered a change in the situation only at the end of the first decade of the century.

In recognizing the reality of the electromagnetic program, we nevertheless wish to emphasize once more that this construction is a methodological idealization. A very wide spectrum of interpretations of it existed. Moreover, some of them, in which particular attention was devoted to the structure of the ether, were very mechanistic. At times concessions to the classical mechanical program were made even by Lorentz and Poincaré, for example, in discussing the concept of the momentum of the electromagnetic field. Even after the special theory of relativity had decisively broken the ether concept physicists sometimes returned to it. As we shall see, echoes of the ether concept can be found even in the physics of the 1920s [40].

DIFFICULTIES OF THE ELECTROMAGNETIC PROGRAM: SPECIAL RELATIVITY AND QUANTA

Investigations in the electron theory and the electrodynamics of moving bodies associated with the development of the electromagnetic program, especially the work of Lorentz and Poincaré, led ultimately to the special theory of relativity, which did not agree well with the basic propositions of the program. It was not merely that the formulas of relativistic dynamics agreed with those of Lorentz and not of Abraham, but above all the fact that special relativity cast doubt on the very existence of the ether, which was one of the main concepts of the electromagnetic program. In addition, the main program accent after the creation of special relativity was shifted from the field reductionism characteristic of the electromagnetic program to the heuristic associated with the requirement of Lorentz covariance and the relativistic methodological technique, the foundations of which were laid in Einstein's famous paper on the special theory of relativity (1905).

Special relativity was indeed a program theory; it formed the hard core of a new, relativistic program that led quickly to a relativistic rearrangement of most branches of physics and to the creation of the general theory of relativity, which significantly generalized the relativistic program itself.

The scientific community of physicists by no means immediately recognized the revolutionary conclusions of relativity theory or that it ran counter to the electromagnetic program. For more than five years after its creation, experiments of the Kaufmann type continued to be at the center of attention, and special relativity and Lorentz's theory gave the same result for the velocity dependence of the electron mass, in contrast to Abraham's theory associated with the radical form of the electromagnetic program. Therefore, for several years, most specialists still did not make a clear distinction between special relativity and Lorentz's theory, calling it most frequently the Einstein–Lorentz theory. Even in 1906 many physicists still adhered to the positions of the electromagnetic program. This is indicated, for example, by the discussions of Kaufmann's experiments at the congress of German natural scientists and doctors in 1906. In the opinion of Kaufmann himself and the majority of the participants, Abraham's theory agreed better with the results of these experiments, thus confirming their faith in the correctness of the electromagnetic program. Among the scientists present, the only one who defended the position of Lorentz and Einstein was Planck, who emphasized that Lorentz's result could be obtained in the framework of special relativity without any special assumptions about the form of the electron. The opponents of Planck—Abraham, Kaufmann, Bucherer, Sommerfeld, Runge, Gans and others—firmly supported the positions of the electromagnetic program, which were regarded as more progressive than the Lorentz–Einstein point of view (which they called *mechanico-relativistic conservatism*). We recall that after Kaufmann's experiments in 1905 Lorentz himself was prepared to give up (and in fact did give up) his theory, the conclusions of which agreed with those of special relativity, in favor of the maximalistic form of the electromagnetic program based on Abraham's theory.

Only after the work of Minkowski do we note a transition of physicists to the position of relativity theory; moreover, it was precisely in 1909–1911 that the experiments of the Kaufmann type very definitely began to give evidence in support of the special theory of relativity. It was in these years that physicists began to regard the relativity principle as a universal physical principle not dependent on the propositions of the electron theory and special relativity as a universal phenomenological theory only historically related to electrodynamics. The radical position of the electromagnetic program with regard to the electromagnetic nature of mass, in particular the electromagnetic interpretation of the velocity dependence of the electron mass, lost its significance to a very large degree, since special relativity explained the dependence without any special assumptions about the structure of electrons. From this period, i.e., 1909–1911, the deepest problems of the electromagnetic program, above all those associated with the electromagnetic concept of mass, which in 1900–1906 were at the center of attention of the scientific community, began gradually to lose their importance. Recognition of the incorrectness of Abraham's theory also amounted to recognition of the excessive claims of the electromagnetic program, which thus entered a stage of steady regression.

The limitations of the electromagnetic program were also manifested in another way. We are referring to the unsuccessful attempts to explain on the basis of this program, and indeed quite generally on the basis of any classical notions, the quantum properties of black body radiation, which were first discovered by Planck in 1900. Thus, already in 1903 Lorentz showed that the electron theory was capable of explaining only the long-wavelength limit of Planck's quantum formula. This was one of the first important indications of a limitation of the electromagnetic program—precisely in a period in which it was still in a progressive stage. After this, Lorentz repeatedly emphasized the incompatibility of quantum ideas with the electron theory in its classical form.

Although Planck himself was not a supporter of the electromagnetic program, he firmly believed in the correctness of Maxwell's classical field theory, and therefore he did not associate the quanta discovered by him with a real structure of radiation. An important part in the development of quantum ideas was played by Einstein's investigation during 1905–1909. Accepting a real existence of light quanta, he successfully explained a number of incomprehensible phenomena such as the photoelectric effect, fluorescence, and photoionization of gases. Then, applying Planck's formula to the vibrations of atoms in a solid, Einstein explained the deviation of the observed values of the specific heat from their values calculated in accordance with the formulas of classical physics. These studies of Einstein opened up new horizons for the application of quantum ideas and were taken up by Nernst (together with his collaborators), Sommerfeld, Debye, and others. By the end of the first decade of the century the idea of the universality of quantum laws began to take shape.

Although in those years almost all physicists either did not believe in the reality of quanta or regarded them in the perspective of being reducible to manifestations of a classical electromagnetic field or even an ether, the heuristic strength of the quantum ideas became more and more tangible, and their incompatibility with the electromagnetic program more and more obvious. A turning point in this process was the First Solvay Congress, which took place in 1911 in Brussels, at which leading physicists recognized the undoubted successes of the quantum theory and the inability of the classical physics, in particular the electromagnetic program, to fit the quantum theory into its framework.

In essence, the quantum theoretical program had already been established by this time; its first great success was to be Bohr's quantum theory of 1913. The problem of atomic structure proved to be beyond the electron theory and the electromagnetic program as a whole. The quantum program led to rapid progress in the solution of this problem, and this ended after a decade and a half with the creation of quantum mechanics.

Thus, the electromagnetic program, which had been born in the middle of the 1890s and had achieved its greatest successes and popularity in the first years of the 20th century, had suffered a marked reverse by the end of its first decade and entered a stage of stable regression. The relativistic program and quantum theoretical program were now at the frontiers of physics, and their systematic development led, on the one hand, to the relativistic rearrangement of all of physics and to the general theory of relativity and, on the other hand, to the systematic introduction of quanta into the physics of the microscopic world and, finally, to quantum mechanics.

UNIFIED FIELD THEORIES BASED ON THE ELECTROMAGNETIC FIELD PROGRAM

How did the electromagnetic program attempt to overcome the difficulties associated with special relativity and quanta? For a long time, even in the second decade of the century, many physicists continued to be skeptical about the reality of quanta, believing, like Planck, that only the processes of emission and absorption of light were discrete. Others believed in the reality of quanta but did not think they were incompatible with the electromagnetic program. Proof of the secondary nature of quantum phenomena, that they could be derived from the classical equations of electron theory, would be a serious confirmation of the electromagnetic program and would help stop its regression. The program in its widely accepted classical form based on the Maxwell–Lorentz equations did not lead to quanta, however, and therefore the idea arose of a modification of the hard core of the program, which, without changing it radically—preserving, for example, the idea of the primary nature of the electromagnetic field and reduction to it of all physical phenomena—would make it possible to obtain quantum properties in the behavior of radiation and matter. Such a problem was posed by Einstein in 1908–1910, when he attempted to construct a unified theory of fields, particles, and quanta, and by Mie in 1912–1913, who developed a nonlinear generalization of Maxwell's equations that appeared to offer hope of explaining the corpuscular and quantum aspects of matter on an electromagnetic field basis.

The electromagnetic program attempted to overcome the difficulties associated with special relativity differently. In essence, the special theory of relativity developed and deepened the classical field concept. The Maxwell–Lorentz equations were consistent with the requirements of Lorentz covariance, and the reality that they described appeared to fit entirely into the framework of the electromagnetic program, except for the restriction that special relativity imposed on the ether concept and hypotheses about the form and structure of electrons. Moreover, the electromagnetic program could now actually take special relativity as a tool, an important heuristic in searches for the necessary modification of the Maxwell–Lorentz equations capable of leading to quanta and, at the same time, realizing in a mathematically rigorous way the original idea of the electromagnetic program—the interpretation of charged particles possessing rest mass as certain field configurations. Thus, special relativity, which had originally arisen as a fundamental difficulty in the path of the electromagnetic program, was transformed into an important support of it. This was the role played by relativity theory in the attempts of Einstein and Mie to breathe new life into the electromagnetic program. Moreover, these theories of Einstein and Mie were the first field theories in which it was intended to obtain the corpuscular and quantum aspects of matter as consequences of generalized field equations; incidentally, the authors of these theories also hoped to obtain gravitation in their framework as a consequence of electromagnetism. Thus, these theories had much in common with the unified geometrized field theories of the 1920s and 1930s. Historically, they were also some of the most important preconditions of the geometrized field theories. Indeed, Hilbert, the author of the first unified field theory, which was based on general relativity, also made explicit use of Mie's theory, and Einstein, ten years after his first attempts to realize the field ideal of the unity of physics, returned to the same global ambition, but this time on a geometric and not electromagnetic field basis.

Before the attempts of Einstein and Mie, the reduction of material particles (in other words, particles of matter endowed with the property of mechanical mass) to the ether or field was considered basically in two aspects in the framework of the electromagnetic program. First, the concept of the electromagnetic origin of mass was developed very deeply, both mathematically and experimentally. Although the existence of particle-like solutions that could be interpreted as electrons did not follow from Maxwell's equations, the basic question of the field origin of these particles appeared to be settled. Second, several ether-mechanical attempts were made to find a mechanism of formation of stable localized entities possessing the properties of charged particles, these being based on a variety of assumptions about the ether. In this case, it was difficult to avoid mechanistic modeling, since the supporter of this approach did not proceed from Maxwell's equations but rather attempted to model both those equations (or consequences of them) and the phenomenon of charged particles by using definite ideas about the mechanical structure of the ether. This approach was most popular in Britain due to the influence of Maxwell himself, W. Thomson, who developed in detail the concept of a gyrostatic, or vortex, ether, J. J. Thomson, the pioneer of the theory of electromagnetic mass, and authorities like Larmor, Lodge, and others.[5]

Around the end of the first decade of this century, when the electromagnetic program encountered the difficulties described above, belief in the ether as a whole was strongly undermined.[6] The concepts of an electromagnetic mass and, *a fortiori*, the ether-mechanical model of charged particles receded into the background.[7] A less mechanistic position was adopted by Mie (in the period preceding his development of the previously mentioned nonlinear electrodynamics). He did not associate definite models with the ether and regarded electrons as "singular places [singular points] in the ether at which there meet lines of electric tension of the ether, in brief 'knots' of the electric fields in the ether" ([48], p. 171 of the Russian translation). His belief in the fundamental nature of the ether concept and the electromagnetic program still directed his investigations, however. He wrote in a book whose second edition was published in 1911:

> The entire diversity of the sensible world, at first glance only a brightly colored and disordered show, evidently reduces to processes that take place in a single world substance—the ether. And the processes themselves,

---

[5] For example, in 1902 Lodge wrote:

> Especially must the inner ethereal meaning both of positive and negative charges be explained: whether on the notion of a right and left-handed self-locked intrinsic wrench-strain in a Kelvin gyrostatically stable ether, at present being elaborated by Larmor, or on some hitherto unimagined plan. And this will entail a quantity of exploring mathematical work of the highest order. ([44], p. 115)

[6] The electromagnetic field was regarded more and more as a reality that did not require the concept of the ether for its justification. For example, in 1909, Planck wrote:

> Instead of the so-called free ether, there is absolute vacuum, in which electromagnetic energy propagates as independently as the ponderable atoms move. I regard the view that does not ascribe any physical properties to the absolute vacuum as the only consistent one. [45]

An extended criticism of the ether concept was published by Campbell in 1910 [46].

[7] Even in these years, however, the hopes of a return of the ether-mechanical view were not abandoned. For example, the atomistic model of the ether developed by Lenard enjoyed great popularity, and, in the view of some physicists, "indicated the way to a possible mechanical explanation in the future of the relativity principle itself in the form that it has been given by Einstein" ([47], p. 29).

for all their incredible complexity, satisfy a harmonious system of a few
simple and mathematically transparent laws. ([48], p. 174 of the Russian
translation)

The special theory of relativity, especially through Einstein, actually overthrew
the ether concept. The abandonment of the ether concept was also related to
Einstein's concept of light quanta as real structural elements of radiation. In
the previously quoted report of 1910, the Russian physicist Goldhammer ([47],
p. 31) said:

The question naturally arises of how the idea of the nonexistence of the
ether arose. However strange this may appear at first glance, this idea does
have to a certain degree experimental foundations.

He then spoke of the black body problem and quanta. Therefore, many associated the overthrow of the ether concept (or at least the loss of its strong
position) with the names of Einstein and Planck. It is curious—and this reflects
the extreme complexity of the actual process of the advance of knowledge,
which never completely agrees with even the successful model schemes and
methodological idealizations such as, for example, those of scientific research
programs—that it was precisely Einstein, soon after he had laid the foundation of the special theory of relativity and had advanced the idea of the reality
of light quanta, who began actively to develop a unified field theory of the
electromagnetic type, based, admittedly, not on Maxwell's equations but on a
certain generalization of them. In essence, the approach was new and was not
related either to the concept of electromagnetic mass or to an ether-mechanistic
modeling in the spirit of Kelvin, Larmor, or Lenard. Einstein worked with
Maxwell's equations and sought possible ways to generalize them, proceeding
from the requirements of Lorentz covariance, the idea of nonlinearity, certain
mathematical properties of partial differential equations of the second and the
fourth orders, etc. Thus Einstein worked for a time in the framework of the
electromagnetic program, though modified in a definite way. It is interesting
that during this period Einstein evidently occasionally considered the ether
idea. In a draft of a letter to Planck dated around 1910, he wrote:

Without an ether, energy continuously distributed in space is for me something impossible. One can easily show that Maxwell's equations are compatible with energy localization corresponding to the old theory of action at
a distance; I intended to publish this soon together with other things. ([49],
p. 25)

EINSTEIN'S ATTEMPTS

In 1907–1908, Einstein, attempting to extend the special theory of relativity to gravitation, established the equivalence principle, a new form of relativity, which required the extension of Lorentz covariance and the class of allowed frames of reference in the presence of homogeneous gravitational fields. On the basis of this principle, he calculated two previously unknown physical effects associated with the influence of the gravitational field on light: the deflection of light rays in a field (for example, by the sun) and a displacement of the spectral lines in a gravitational field toward the red end of the spectrum ("red shift"). The secure experimental basis and the theoretical depth of the equivalence principle, and also its heuristic strength, indicated the correctness and fruitfulness of the new approach to the solution of the problem of constructing a relativistic theory of gravitation. Einstein was to face great difficulties in this approach, however, associated with the attempt to go beyond the framework of special relativity and the relativistic program based on it (the fundamental principle of Lorentz covariance was shown to be invalid) and with the need to extend the equivalence principle to arbitrary gravitational fields (under conditions when there were no physical guides to a new class of admissible frames of reference and the space-time coordinates had lost their direct metrical significance). As a result, despite a very promising start, progress along the path opened up by the equivalence principle was much slower.

Actually, in those years (1906–1907) Einstein did not stop his work in the field of quantum theory. In 1906, he derived Planck's formula by replacing the integral in the expression for the entropy by a sum, and drew the conclusion that the "energy of an elementary resonator can take only integer values that are multiples of $(R/N)\beta\nu$" [50] (here $(R/N)\beta = h$). Maxwell's classical theory could not give a description of oscillators with discrete levels, however, and Einstein emphasized the sharp discrepancy between the classical formula and Planck's formula and, moreover, the agreement of the latter with the hypothesis of light quanta.

In 1907, Einstein applied Planck's formula to the "vibrations of atoms" in a solid and gave an at least qualitative explanation of the deviation of the specific heats of solids from the values predicted by the classical theory and their decrease with decreasing temperature. Thus, the heuristic value of the quantum concept was once again brilliantly confirmed, but the theoretical status of quanta remained obscure for Einstein. Therefore, in 1908 when Einstein encountered difficulties in the development of the theory of gravitation that seemed unsurmountable, he returned to the problem of quanta, which also agitated him.

The concept of the classical field, the deep analysis of which led Einstein to special relativity, was the most fundamental for him. Special relativity undermined the ether idea, especially in the mechanistic form, but not the field concept. Einstein's revolutionary recognition of the reality of light quanta and his pioneering studies in the application of the quantum idea to explain a rather large class of physical phenomena did not mean that he abandoned the fundamental nature of the classical field concept. Indeed, Einstein's papers during 1906–1909 and his correspondence in those years, mostly with Lorentz, show that the field concept was decisive for him, and it was precisely in those years, especially after he temporarily abandoned attempts to solve the problem of a relativistic theory of gravitation on the basis of the equivalence principle, that he seriously turned to a new, grandiose task—an explanation of the existence of quanta and electrons on the basis of some generalization of Maxwell's equations of the electromagnetic field [51].

This task was very close to the program formulation of the electromagnetic program, although, strictly speaking, it went beyond that framework, since it recognized the inadequacy of the classical Maxwell–Lorentz theory and aimed at a significant generalization of it. Nevertheless, if one judges by actual attempts made by Einstein, the fundamental unifying concept still remained the concept of a classical field described by partial differential equations that for large field amplitudes (and, possibly, long wavelengths) must go over into Maxwell's equations. Like the supporters of the electromagnetic program, he believed that in this way he could derive both electrons and their equations of motion from the field equations, reducing the corpuscular aspect of matter to the field aspect. Simultaneously, there should also be an explanation of the quantum structure of radiation, the mystery of which had long occupied Einstein. A certain hope of achieving this came from the proximity of the orders of magnitude of Planck's constant h and the constant e^2/c, which both have dimensions of an action: $h = 6 \times 10^{-27}$ erg sec, $e^2/c = 7 \times 10^{-30}$ erg sec. This fact was noted by Einstein in his paper "On the present status of the radiation problem," which was completed in January 1909. Emphasizing this, he felt it was possible to conclude that a theory capable of explaining the existence of the electron would also simultaneously answer the problem of the field nature of quanta. Einstein wrote:

> It is now necessary to recall that for Maxwell–Lorentz electrodynamics, the elementary quantum ϵ [i.e., the electron charge] is foreign. In order to construct the electron, it is necessary to use nonelectrodynamic forces. . . . But, in my view, it follows from the relation $h = \epsilon^2/c$ that a modification of the theory that will give as a consequence the elementary quantum ϵ will also contain within it the quantum structure of radiation. [52]

How did he intend to find this modification of the classical theory? First, attempts were made at the level of the field equations, in particular, their mathematical structure, which was an entirely new approach, especially against the background of the numerous ether-mechanical models of the type proposed by Kelvin, Larmor, Lenard, and others. Second, the idea was to generalize Maxwell's equations with effective allowance for a correspondence principle, i.e., in such a way that in a certain limiting situation the generalization would reduce to Maxwell's equations. Third, the required generalization must be Lorentz covariant—it was here that the basic idea of the relativistic program was used. Fourth, the modification of the linear second-order homogeneous equations (the form of Maxwell's equations for empty space) could take the form of a transition to nonlinear equations, or to inhomogeneous equations, or to equations of a higher order, or to a combination of these generalizations. Moreover, in the analysis of the different possibilities and the selection of suitable ones, one could use dimensional arguments, the fundamental methodological principles of physics (for example, the principle of energy and momentum conservation) and arguments of a physical nature associated with the corresponding physical properties of the solutions of the required equations.

At the end of 1908 and beginning of 1909, Einstein, as is clear from the January letter quoted above, was thinking of a nonlinear and inhomogeneous generalization of Maxwell's equations:

The fundamental equation

$$D(\varphi) \equiv \frac{1}{c^2}\frac{\partial^2\varphi}{\partial t^2} - \left(\frac{\partial^2\varphi}{\partial x^2} + \frac{\partial^2\varphi}{\partial y^2} + \frac{\partial^2\varphi}{\partial z^2}\right) = 0 \tag{1}$$

of optics must be replaced by an equation that also contains as a coefficient the universal constant ϵ (probably its square). The required equation (or system of equations) must be dimensionally homogeneous. Under a Lorentz transformation, it must go over into itself. It cannot be linear and homogeneous. It must—at least, if Jeans's law is indeed satisfied in the limit of small v/T—go over in the limit of large field amplitudes into the equation $D(\varphi) = 0$. [52]

Nonlinearity of the equations was required because linear equations would not be capable of ensuring stability of the electron. In addition, the mathematical properties of nonlinear equations, as was already known then, are much richer than those of linear equations, and it could be hoped that precisely nonlinear equations would lead to solutions possessing corpuscular and quantum properties.[8] The requirement of inhomogeneity of the equations was due to the fact that linear homogeneous equations always have solutions whose sum is also a solution, and in this case it would be difficult to describe the interaction between the particles. Persistent searches in this direction were not crowned with success, and Einstein concluded his considerations in the January 1909 paper with the following remark:

> I have not yet succeeded in finding a system of equations satisfying these conditions of which one could say that it is suitable for constructing elementary electric and light quanta. However, the range of possibilities is evidently not so great as to discourage one from this task. [52]

The final sentence indicates that Einstein had not yet exhausted the possibilities of the approach and in January 1909 continued to place his main hopes on a nonlinear and inhomogeneous generalization of Maxwell's equations.

In the spring of 1909, a correspondence between Einstein and Lorentz apparently began on questions of the quantum theory of radiation and the problem of a unified field theory [51]. In the January paper quoted above, Einstein derived an expression for the fluctuations of the light pressure on a mirror of area f placed in a cavity filled with thermal radiation. Einstein found that the mean square momentum $\overline{\Delta}^2$ transferred during time τ could be represented in the form

$$\overline{\Delta}^2 = \frac{1}{c}\left(h\rho\nu + \frac{c^3\rho^2}{8\pi\nu^2}\right)d\nu \cdot f \cdot \tau, \qquad (2)$$

where ρ is the energy density of the radiation at frequency ν, and c and h are the speed of light and Planck's constant. The second term of this expression

---

[8] It is appropriate here to recall the quasicorpuscular solutions of certain nonlinear wave equations of the Korteweg–de Vries or the nonlinear Schrödinger types. These are the so-called soliton solutions. The corresponding particle-like formulations, which are called solitons (since they can be regarded as solitary excitations), are very stable—they preserve their individuality even after collisions with one another. A soliton solution for waves on the surface of a liquid was probably first described by Boussinesq in 1872. The Korteweg–de Vries equation, which leads most directly to solitons, first appeared in a dissertation of de Vries written in 1894 under the supervision of Korteweg. These studies were practically unknown at the beginning of this century and even in the 1930s and 1940s. None of the obituaries that appeared after the death of Korteweg (the fate of de Vries, who was a teacher at a gymnasium, remained unknown) contained references to the studies (in collaboration with de Vries) in which the solitons and the associated differential equations were discovered. Knowledge of these studies by Einstein could have directed his thinking into a new path [53, 54].

corresponded to the light pressure of randomly interfering waves, and the first to the effect on the mirror of quanta with momentum $h\nu/c$. This formula was an important step on the road to the corpuscle–wave dualism. Einstein wrote about this in his first letter to Lorentz, but he also communicated his idea of relating Planck's constant to the constant e^2/c and his idea of a unified field theory capable of explaining the existence of both the electron and light quanta. By this time, however, Einstein's point of view with regard to the method of solving the problem of a unified field theory had undergone important changes, as is clear from his letter to Lorentz on May 23, 1909.

He returned to the use of linear homogeneous equations, not of the second but of the fourth order, and, moreover, equations that are written down separately for each component of the field (more precisely, the four-dimensional potential). It appeared to Einstein that these conditions would ensure propagation of concentrations of the field substance at the speed of light without dispersion. As a static model analog of such equations, he considered the fourth-order equation

$$\Delta \Phi - \lambda^2 \Delta \Delta \Phi = 0, \tag{3}$$

which has a solution of the type

$$\Phi = \epsilon \frac{1 - e^{-r/\lambda}}{r}, \tag{4}$$

where ϵ is a certain undetermined constant; λ is a constant having the dimension of length, regarded as a certain wave characteristic of the field. The constant ϵ, related to a characteristic of the field sources, should not occur in the field equations; it should be obtained by solution of these equations.

The model solution with the exponential term became identical at large r to the classical solution, i.e., led to a $1/r$ dependence, and at $r = 0$ did not give a singularity. This offered the hope of eliminating the difficulties associated with the infinite energy of the self-interaction of the electron and its stability (the latter was apparently ensured without the introduction of forces of nonelectromagnetic origin, in agreement with the aims of the electromagnetic program). In order to make the system of these equations Lorentz covariant, it was necessary to replace Δ by $(\Delta - 1/c^2 \, \partial^2/\partial t^2)$. Einstein used one further argument in support of his new approach. By substituting a plane-wave solution in the field equations, it was possible to obtain the relation

$$(-k^2 + \omega^2/c^2)(-k^2 + \omega^2/c^2 - 1/\lambda^2) = 0, \tag{5}$$

where k and ω are the wave vector and frequency of the wave. It was appealing to relate the first factor to the light quantum, and the second to the electron.

Einstein probably soon realized that even fourth order linear equations could not describe the source of the field and its motions, since the system consisting of the sources and field possesses essentially nonlinear properties. Therefore, it was also unclear how the system of equations would yield the source characteristic ϵ.

In his searches for a suitable system of equations, Einstein was, as we see, forced to use the method of mathematical hypothesis, or mathematical testing, but one senses that for him transparent physical images had, as before, the main significance. In his report "On the development of our views about the essence and structure of radiation" given at the 81st Congress of the Society of German Natural Scientists at Salzburg in October 1909, Einstein again referred to his attempts to give a field concept of the electron and light quanta. Significantly, he restricted himself to a lucid picture, which he unsuccessfully attempted to introduce into the framework of a rigorous mathematical scheme:

> As far as I know it has not yet proved possible to construct a mathematical theory of radiation describing both the wave structure and the structure that follows from the first term of our formula (quantum structure) [i.e., he has in mind the formula for the fluctuations of the light pressure]. . . . However, it still appears to me most natural that the occurrence of electromagnetic fields of light must be associated with singular points just as the occurrence of electrostatic fields is in accordance with electron theory. It is not impossible that in such a theory all energy of the electromagnetic field could be regarded as localized at these singular points, exactly as in the old theory of action at a distance. I imagine each such singular point surrounded by a force field, which basically has the nature of a plane wave with an amplitude that decreases with the distance from the singular point. If a large number of such singular points are present at distances small compared with the dimensions of the force field of one singular point, then the force fields will overlap and together will give a wave force field that might differ very little from the wave field in the sense of the modern electromagnetic theory of light. It need hardly be emphasized that until such a picture has led to a precise theory it should not be accorded a special significance. I merely wish to show by means of it that one must not regard as incompatible the two structures (wave and quantum) that radiation must possess simultaneously in accordance with Planck's formula. ([55], pp. 824–825)

A letter to Besso on December 31, 1909 shows that at the end of 1909 and the beginning of 1910 Einstein continued to think about the field conception of quanta and electrons. It is interesting that he hoped to use the gauge symmetry of Maxwell's equations, but this window was blind![9]

---

[9] In this letter to Besso, he concluded his exposition of the new idea with the fol-

Already by the end of the summer of 1910, and possibly even somewhat earlier, he began to think more and more about the problem of gravitation. At least, in a letter to Laub on August 10, 1910, he did not mention the subject that had agitated him for the last few years. Einstein returned to gravitation and the equivalence principle, which he reported to Laub ([57], p. 120 of the English translation). He still continued to work on particular problems of quantum theory, however. In May 1911, he wrote to Besso:

> I am currently attempting to deduce from the quantum hypothesis the law of heat conduction for solid insulators. I no longer question the real existence of quanta. I also no longer attempt to construct them, since I now know that I am not equal to this task. ([56], p. 17 of the Russian translation)

On June 21, 1911 he sent to the *Annalen der Physik* the paper "On the influence of the force of gravity on the the propagation of light," which opened a new cycle of investigations by Einstein on gravitation that was completed in November 1915 with the creation of the foundations of the general theory of relativity [38].

As McCormmach, whose fundamental study "Einstein, Lorentz, and the electron theory" we have widely used in this section, correctly notes, the experience acquired by Einstein in constructing a unified field theory in the period 1908–1910 proved to be a very valuable support in developing a relativistic theory of gravitation, in particular in solving the problem of the equations of the gravitational field. One can list here the idea, fundamental for general relativity, of nonlinearity of the field equations, and also the idea of deriving the equations of motion from the field equations, the use of very general principles of physics, quite often of methodological significance (for example, the principles of symmetry, correspondence, conservation, simplicity, etc.) for the construction of field equations, the ability to combine transparent physical ideas with a high level of abstract mathematical analysis, and so on [51].

It is possible (though direct evidence is absent) that the attack on the global problem of constructing a unified field theory during 1908–1910 was related to the attempts to solve the problem of gravitation to the extent that the equations of this unified field theory could give additional information

---

lowing words: "I hope that this is not a blind window." The commentator on this correspondence, P. Speziali, explains the meaning of these words with a reference to an aphorism of Pascal: "He who constructs an antithesis by distorting the meaning of words is like the person who to achieve symmetry draws blind windows: Their principle is not to speak correctly but to create regular figures" ([56] pp. 18–19 of the German original and p. 16 of the Russian translation).

about gravitation—ideally, these equations could also include gravitation. It could also be the case that the unified theory equations would go beyond the framework of Lorentz covariance, and the larger invariance group found as a result could be used in searches for the equations of the gravitational field. Einstein did not succeed, however, in constructing a unified theory of the electromagnetic field, the electron, and quanta on a field basis, and he returned to the investigation of more specialized problems, both in the field of quantum theory and in the field of gravitation.

At the same time, the return to more specific problems, above all the problem of constructing a relativistic theory of gravitation, did not at all mean that Einstein had become disenchanted with the program of a field synthesis of physics that, in its entirety, went beyond the aim of the electromagnetic program. It is possible that Einstein argued as follows: The equivalence principle indicated the need to extend the relativistic program based on special relativity. The attempts at the construction of a unified field theory during 1908–1910 were based on the requirement of Lorentz covariance alone. Thus, Einstein could have believed that this was precisely the reason why his global concept failed. It was necessary to return to the theory of gravitation and to the program of its construction based on the equivalence principle. The realization of this program should lead to a theory possessing a larger symmetry group than the Lorentz or Poincaré group. Perhaps on the basis of this extended group the attempts to construct a unified field theory would be more successful. Moreover, in this case the theory would be even more "unified," since it would also be possible to include the theory of gravitation within it.

Thus, the field-theoretical ideal of the unity of physics, intimately related to the electromagnetic program but, in its entirety, going beyond its framework, was very close to Einstein. The triumph of the classical fields concept, admittedly in the new geometrical guise, reawakened Einstein's hopes of realization of the old idea, to which he turned at the beginning of the 1920s.

MIE'S THEORY

In contrast to Einstein, Mie remained a consistent adherent of the electromagnetic program even after the creation of the special theory of relativity. In 1910 his *Lehrbuch der Elektrizität und des Magnetismus* (*Textbook of Electricity and Magnetism*) [58] was published, which contains references to special relativity but nevertheless very energetically insists on the need in physics for the ether concept and the electromagnetic field conception of the physical world. At the same time, he was an opponent of the explicit mechanistic interpretation of the ether. He regarded material particles, above all electrons, as certain ether configurations, certain "knots" in the ether, the nature of which must be clarified in the subsequent development of the theory. Mie did not believe that special relativity provided a reason to abandon the ether concept, which he valued highly and with which he associated a true understanding of physical reality.

Mie wrote in the cited textbook:

> According to electron theory, the inertia of a material particle is simply the inertia of the magnetic field and, perhaps, other as yet unknown phenomena in the ether that make up a significant proportion of the entire process of motion but are not restricted to the small volume of node points in the ether. According to this view, mechanics, which has always regarded the concept of inertia as the most elementary, simple concept, must lose its position as the foundation of science. It is not the ether that must be explained mechanically; rather, matter must be explained by electromagnetism. ([58], p. 415 of the Russian translation)

In another place he emphasized:

> To the extent that inertia is, at least in part, a particular manifestation of electromagnetic forces, we can regard the mechanical theorem of relativity only as a special case of a general electromagnetic principle. (Ibid., p. 428)

In 1911 Mie republished the public lectures *Moleküle, Atome, Weltäther* (*Molecules, Atoms, and the Universal Ether*) [48], which he had given in 1903 in Greifswald. Referring to the criticism of the ether concept made from the point of view of the special theory of relativity, he wrote:

> This concept [i.e., the ether] has been subject to some recent vilification because it has become customary to associate with the universal ether mechanical representations.... ([48], p. 99 of the Russian translation)

In the section "Modern view of the nature of matter" Mie describes the ether-field picture of particles and their motion in the following terms:

> Elementary material particles... are simply singular places in the ether at which lines of electric stress of the ether converge; briefly, they are "knots" of the electric field in the ether. It is very noteworthy that these knots are always confined within close limits, namely, at places filled with elementary particles.... The formation of the knots is probably associated with particular force manifestations of the ether, which counter the tendency to separation and hold the knots within tight boundaries. I shall call these as yet not entirely investigated force effects of the ether "concentrating pressures," and I assume that the universal attraction of masses, or universal gravitation, is intimately associated with them.... The only reason for the displacement of a particle can be the loss of the state of equilibrium in regions of the ether adjoining it.... (Ibid., p. 171)

As McCormmach correctly noted, "Mie emerged as a major figure in the later phase of the electromagnetic program" ([34], p. 491). At the same time, Mie regarded the special theory of relativity, in particular the requirement of Lorentz covariance and the four-dimensional field-invariant formulation of this theory, as effective theoretical tools of the electromagnetic field program. Thus, in contrast to many other supporters of that program, he did not oppose special relativity and the relativistic program; instead he tried to introduce them into the approach of the electromagnetic field program. In his *Lehrbuch der Elektrizität und des Magnetismus* he devoted an entire chapter to the theory of relativity, in which, in particular, he wrote:

> The principle of relativity is remarkable in that it asserts the existence of an inner connection between all properties of matter and all physical phenomena. ([58], p. 447 of the Russian translation)

In another place, he highly estimated Minkowski's concept:

> Einstein insisted in 1905 on the general principle of relativity as valid for the whole of physics. The mathematician Minkowski developed this principle in a very elegant and transparent mathematical form extremely convenient for further theoretical investigations. (Ibid., p. 439)

On October 31, 1912 Mie completed the first part of his famous paper "Die Grundlagen einer Theorie der Materie" ("Foundations of a theory of matter") [59], in which he succeeded to some extent in doing what Einstein had attempted during 1908–10. The qualification "to some extent" is needed because Mie's theory, for all the grandeur of its project and subtlety of its theoretical means, proved to be incorrect, although it did lead to a field model of particles. The idea of particles as special field configurations, knots in the ether, described above was given a precise mathematical expression in this paper based both on the basic ideas of the electromagnetic field program as well as the methods of the special theory of relativity and, thus, on elements of the relativistic program. At the start of the paper, Mie wrote:

> In accordance with this understanding [i.e., in accordance with the principles of the electromagnetic field program] electrons and, generally, the smallest particles of matter are not entities sharply different from the universal ether. They are not, as was still believed 20 years ago, bodies foreign to the ether; rather, particles are merely places at which the ether is in a particularly singular state, which we call electric charge. ([59], p. 511)

Slightly later, he described in more detail the manner in which particles exist in the ether:

> In my theory, the electron . . . is not a strictly bounded part of space in the ether; in contrast, it consists of a core that goes over continuously into an

atmosphere of electric charge which extends to infinity but even near the core is so rarefied that it cannot be detected experimentally in any way. ([59], p. 512)

In creating his unified theory, Mie, like Einstein in 1908–10, actually went beyond the framework of the electromagnetic program in its original form. This was not only in the use of elements of the relativistic program, but also in the fact that Mie tried to generalize Maxwell's field equations, which, as he believed, became invalid at very high strengths of the electric and magnetic fields. Nevertheless, he still regarded his equations as equations of the electromagnetic field—all the basic concepts remained those of Maxwell and Lorentz, and even the transparent representations of particles as field formations were entirely typical of supporters of the electromagnetic program. In fact, his theory can be regarded as nonlinear electrodynamics. Just as typical were the global aims Mie had of including in his theory gravitation and quanta, to which the third part of his study, published in 1913, was devoted [60].[10]

Mie's work had considerable resonance, and, as we shall see, directly influenced investigations in the field of unified geometrized field theories. It was therefore included in many classical monographs on the theory of relativity. It is sufficient to mention that Mie's theory was considered in detail by Weyl in his famous book *Raum-Zeit-Materie* (1918) [62] and by Pauli in his no less famous work on the theory of relativity—in an article written for the *Enzyklopädie der mathematischen Wissenschaften* (1921) [63]. Therefore, we shall not consider Mie's theory in all its details here but merely sketch the basic scheme, making use of the studies by Weyl and Pauli.

We list immediately some of the main features of Mie's theory that enabled it to achieve no mean success. Indeed, in this theory for the first time in the history of physics it became possible to give a mathematically rigorous field model of particles, which was based on a nonlinear generalization of Maxwell's equations derived from a Lorentz-covariant variational principle. We have actually already mentioned these features: (1) the postulates of the electromagnetic program are understood in an extended sense (the basis of everything that exists is assumed to be the electromagnetic field; (2) the basic electromagnetic concepts and quantities—the field strengths, potentials, etc.—retain their significance, but a certain modification of Maxwell's equations is allowed); (3) the universal equations can be derived from a variational principle, namely, from Hamilton's principle; (4) the requirements of Lorentz covariance can be used explicitly in the construction of the universal function,

---

[10] For the second part of Mie's study, see [61].

or Lagrangian, of the theory, i.e., the postulates of the relativistic program; and (5) the factual recognition of the need for a modification of the Maxwell Lagrangian such that at sufficiently large distances from the electron it leads to linear wave equations (namely, Maxwell's equations) but in the neighborhood of the electron gives nonlinear equations.

The main physical prerequisite of the theory was, as Pauli noted, the requirement

> that the Coulomb repulsive forces in the interior of the electrical elementary particles are held in equilibrium by other, *equally electrical*, forces, whereas the deviations from ordinary electrodynamics remain undetectable in regions outside the particle. ([63], p. 188)

In addition, Mie assumed that in a vacuum there exists a fundamental difference between the force (or intensive) and quantitative (or extensive) qualities, i.e., between the potential φ_i and the tensor $F_{ik} = \partial \varphi_k / \partial x_i - \partial \varphi_i / \partial x_k$ on the one hand, and the current s_k and the tensor H^{ik} on the other.[11] The Maxwell relations acquired an effectively new physical content if one assumed further, as Mie did, that the extensive quantities H^{ik} and s^k are universal functions of the intensive quantities F_{ik} and φ_i:

$$H^{ik} = u_{ik}(F, \varphi), \quad s^k = v_k(F, \varphi). \tag{6}$$

These equations appear to introduce into the theory not less than ten universal functions, but Mie was able to show that the Maxwell relation between the field strengths and the potentials and the law of energy conservation for the field are obtained only if there exists a Lorentz-covariant function $\mathcal{L}(F, \varphi)$ from which H^{ik} and s^k are obtained by differentiating with respect to F_{ik} and φ_i, i.e.,

$$H^{ik} = \frac{\partial \mathcal{L}}{\partial F_{ik}}, \quad s^i = -\frac{1}{2} \frac{\partial \mathcal{L}}{\partial \varphi_i}. \tag{7}$$

This means that

$$\delta \mathcal{L} = H^{ik} \delta F_{ik} - 2s^i \delta \varphi_i. \tag{8}$$

From this one can readily show that the field equations ultimately follow from the variational principle

$$\delta \int \mathcal{L} \, d\omega = 0. \tag{9}$$

---

[11] The tensor F_{ik} relates the vectors **E** and **B**, while H^{ik} relates **D** and **H**.

Thus, the problem reduces to finding a suitable Lorentz-covariant Lagrangian $\mathcal{L}$. The force, or intensive, quantities F_{ik} and φ_i were taken by Mie as the basic quantities that characterize the state of the field. An important difference of the new theory from the classical scheme of Maxwell in its widely adopted formulation was the inclusion, among the quantities that determine the state of the field, of the potentials alongside the field strengths. This meant that the Lagrangian $\mathcal{L}$ could contain, besides the classical electrodynamic invariant $\frac{1}{2} F_{ik} F^{ik}$, which leads to Maxwell's equations, additionally three types of quadratic invariants formed from F_{ik} and φ_i in accordance with the theoretical technique of four-dimensional invariants developed by Minkowski:[12]

$$\varphi_i \varphi^i, \quad F_{ir} \varphi_s F^{is} \varphi^r, \quad (F_{ik} \varphi_l + F_{kl} \varphi_i + F_{li} \varphi_k)^2. \tag{10}$$

As we have already noted, $\mathcal{L}$ can differ significantly from $\frac{1}{2} F_{ik} F^{ik}$ only in the neighborhood of charged (elementary) particles. Otherwise, the choice of the non-Maxwellian correction $\mathcal{L}_1$ to the Lagrangian $\frac{1}{2} F_{ik} F^{ik}$ remains rather arbitrary. Further particularization of the function $\mathcal{L}_1$ could be determined on the basis of the following considerations. First, it must be such that the field equations are consistent with particle-like solutions to which one can ascribe finite masses and a charge, which, ideally, should be equal to the corresponding quantities for the electron. Second, these solutions must be everywhere regular (in particular at $r = 0$ and $r = \infty$) and also (in the simplest case) static and spherically symmetric. Finally, there must be as many different solutions as types of charged particles exist. Searches for such a Lagrangian were not crowned with success, although it was clear to Mie that the non-Maxwellian correction must depend nonlinearly, even in the simplest case, on the square of the potential. In particular, he considered a function $\mathcal{L}_1$ with a cubic dependence on $\varphi_i \varphi^i$ (or containing the sixth power of φ). The adopted law was rather singular, since the nonlinear correction decreased with the distance as r^{-6}. If the nonlinear correction is written in the form

$$\mathcal{L}_1 = w\left(\sqrt{\varphi_i \varphi^i}\right), \tag{11}$$

---

[12] In note 20 to the English edition of his encyclopedia paper on the theory of relativity Pauli pointed out that the given list omitted the invariant

$$\left[\frac{1}{\sqrt{-g}} (F_{23} F_{14} + F_{31} F_{24} + F_{12} F_{34})\right]^2,$$

which subsequently appeared in the Born–Infeld theory and in quantum electrodynamics ([63], p. 223).

then, in the static case, since $\Delta\varphi = w'(\varphi)$, we arrive at the generalized Poisson equation

$$w'(\varphi) = -4\pi\rho, \tag{12}$$

which leads to a solution (and even to an infinite set of solutions) regular at $r = 0$ and $r = \infty$. From this infinite set of solutions one could select (how, admittedly, is not clear) a certain finite number of solutions corresponding to charged elementary particles. The theory made it possible to calculate the charge and mass of the particle for a solution φ of Eq. (12). The charge was calculated in spherical coordinates as follows:

$$
\begin{aligned}
q &= \iiint \rho r^2 \sin\theta \, dr \, d\theta \, d\psi \\
&= \frac{1}{4\pi} \int_0^\infty r^2 w'(\varphi) \sin\theta \, dr \, d\theta \, d\psi \\
&= -\int_0^\infty r^2 w'(\varphi) \, dr = \left[r^2 \frac{d\varphi}{dr} \right]_0^\infty .
\end{aligned}
\tag{13}
$$

The mass was calculated by an analogous integral of the energy density over the whole space:

$$m = \frac{1}{c^2} \iiint W r^2 \sin\theta \, dr \, d\theta \, d\psi = \frac{4\pi}{c^2} \int W r^2 \, dr,$$

where the energy density was taken to be

$$W = \frac{1}{2}(\operatorname{grad}\varphi)^2 + w(\varphi) + \varphi w'(\varphi),$$

which leads (after integration by parts) to the expression

$$m = \frac{4\pi}{c^2} \int_0^\infty r^2 \left[w(\varphi) + \frac{1}{2}\varphi w'(\varphi) \right] dr. \tag{14}$$

All this gave hope of realization of the field concept of particles, but the form of the function w was still unknown, and at this stage Mie's theory remained only a fairly fully developed and mathematically formulated program; it had not acquired the status of a fully valid physical theory.

    In those years, there were still many supporters of the electromagnetic program, but the supporters of the special theory of relativity and the relativistic program were also impressed by Mie's theory, at least in its inspiration. At the end of the 1910s and beginning of the 1920s Weyl and Pauli regarded this grandiose project highly. Weyl wrote:

These physical laws, then, enable us to calculate the mass and charge of the electrons, and the atomic weights and atomic charges of the individual existing elements whereas, hitherto, we have always accepted these ultimate constituents of matter as things given with their numerical properties.... The special hypothesis (67) [i.e., the expression for the "world function" of Mie's theory] from which we just now started was assumed only to show what a deep and thorough knowledge of matter and its constituents as based on laws would be exposed to our gaze if we could but discover the action-function. ([64], p. 214 of the English translation)

Despite the serious shortcomings of the theory, which Pauli noted in 1920 and to which we shall return, the 20-year-old student of Sommerfeld wrote:

This [the fact that even by 1920 it had not proved possible to find a suitable "world function"] in itself is not sufficient reason for abandoning Mie's electrodynamics, since it has not been proved that a world function *cannot* exist which is compatible with the existence of *certain* elementary particles. ([63], p. 192)

Mie's theory inspired great hopes in Hilbert, when in the middle of the 1910s he became seriously interested in the foundations of physics. We shall speak later about this. We now consider the question of the possibility of including gravitation and quanta in Mie's theory. Initially, it appears that Mie intended to include the theory of gravitation in the framework of his theory and thus reduce it to electromagnetism. The electromagnetic concept of mass also suggested the thought of an electromagnetic origin of gravitation. Mie's efforts in this direction, however, actually led him to the idea of a distinct, non-electromagnetic nature of the gravitational field despite a certain similarity of Newton's law of gravitation to the electrostatic law of interaction—Coulomb's law.

The laws of gravitation did not follow from the "basic equations of ether dynamics" (as Mie called the basic equations of his unified theory). To describe gravitation, he had to introduce special quantities characterizing the state of the gravitational field, namely, a scalar gravitational potential and a corresponding four-vector of the field strength, and also a four-vector of gravitational displacement (by analogy with the corresponding electric quantities). The equations of the gravitational field were found from the condition of Lorentz covariance, and, in essence, differed little from Nordström's first theory (scalar and Lorentz covariant). Speaking of his attempts to introduce gravitation into the electromagnetic approach, Mie referred to their affinity with the well-known electromagnetic theories of gravitation of Lorentz, Gans, and Heaviside. It is interesting that Mie rejected all these theories, including his own versions, since they did not agree with the two most fundamental principles of physics—the principles of relativity and conservation of energy and momentum. He wrote ([60], p. 26):

> I made many such efforts, which always lead to decidedly cumbersome equations [gravitational equations of the electromagnetic type] and am now convinced that on this [electromagnetic] basis it is quite impossible to develop a theory of gravitation that would satisfy both the principle of relativity and the principle of energy conservation.

Thus, it was necessary to give up the unification of gravitation with electromagnetism and defer this problem to better times. The situation was even worse with quanta. Since the quantum of action (or Planck's constant) had a rigorously fixed value, like the electron charge, and was related to the structure of electromagnetic radiation (whereas the electron charge was associated with the structure of an electrostatic field and was regarded as a certain concentration of the ether), Mie also thought to interpret the quantum of action as a certain configuration. It is possible that the closeness of the value of Planck's constant h to the ratio e^2/c (in order of magnitude), which was important for Einstein, also stimulated Mie's idea of an ether relationship between the quantum of action and the electric charge:

> The quantum of action h, or, more precisely, the square root of it, is a quantity that is entirely analogous to the elementary quantum of electric charge. Just as the latter gives the number of lines of force that converge at an electron, as a "knot," h gives the number of lines of force that, in the form of a tight configuration, make the dipole. ([60], p. 24)

Of course, Mie did not succeed in explaining the quantum nature of radiation on the basis of his equations of "ether dynamics," although there remained a certain hope that, once a "world function" that permitted accurate calculation of the electron charge and mass had been found, it would also make it possible to determine the value of Planck's constant and explain the quantum properties of radiation. As a result, gravitation and quanta remained outside the scope of Mie's theory.

Of course, both Mie himself and other physicists saw in this theory in the first place a promising program and hoped that searches for an adequate action function would in time be crowned with success. What was envisaged in the first place was an action function, that, for each species of charged particle, would lead to one quite definite solution permitting calculation of the charges and masses of these particles in agreement with the experimental values. Therefore, the fact that such a function had not yet been found could not, in the middle of the 1910s be regarded as a sufficiently serious defect of Mie's theory. The most serious shortcoming of the theory, which was seen by Mie himself and was emphasized especially clearly by Pauli, was its gauge (or gradient) noninvariance; for both the Lagrangian of the theory and the field equations contained the absolute values of the potentials, which, according to Mie, were interpreted as pressures in the ether. Therefore, if φ were a solution of the basic equations, the potential $\varphi +$ const could not be a solution. This could be interpreted as the impossibility of a material particle being able "*to exist in a constant external potential field*" ([63], p. 192). Moreover, this difficulty was organically inherent in Mie's theory, i.e., it was not a feature of a particular form of the theory but of its entire program. Mie himself, and all those who subsequently wrote about the theory, assumed that to find the necessary world function one needed either additional experimental investigations in very small volumes of space (near electrons or protons) or the appearance in the already existing material of new physical ideas or principles. Mie ended the quoted paper:

> The most immediate problem to which we are directed by the theory is the investigation of whether it is possible to find, in very strong electric or magnetic fields, or even in regions in which the field strengths are zero but have very large values of the potentials, deviations from Maxwell's laws that hold in the ideal vacuum.... ([60], p. 64 of the English translation)

Considering the basic proposal of Mie concerning the structure of the world function associated with the non-Maxwellian correction $\mathcal{L}_1$, Weyl wrote in 1918:

> The discussion of such arbitrarily chosen hypotheses cannot lead to any proper progress; new physical knowledge and principles will be required to show us the right way to determine the Hamiltonian Function. ([64], p. 214 of the English translation)

For all these difficulties, one cannot fail to recognize that Mie succeeded in advancing the field-theoretical ideal of the unity of physics further than anyone else in those years. Despite the use of the ether terminology, Mie essentially based his work on elegant invariant variational methods that entirely corresponded to the spirit of relativistic physics and the mathematical construction of its basic equations. Even after the creation of quantum mechanics, physicists returned more than once to Mie's ideas in their attempts to solve the problem of a field synthesis of physics. In particular, in the 1930s and 1940s an analogous direction was developed by Born (in collaboration with Infeld), and also by Bopp and Podolsky [65, 66]. Born and Infeld gave up the use of a Lagrangian and equations that depended explicitly on the potentials. The nonlinearity of the electromagnetic field that they introduced did not violate the gauge invariance of the theory. The Born–Infeld theory made it possible to calculate the electromagnetic mass and classical radius of the electron (of order 10^{-13} cm) and also to obtain the value of the field strength "at the edge of the electron."

Sommerfeld wrote about Mie's theory in 1948:

It would have been amazing if the fundamental problem of elementary particles could have been solved by inspired guesswork! Today we are convinced that much experimental work must still be done in this field. Nevertheless, it was a great service to have outlined for the first time the way to this problem, as is also evident from the fact that all following workers on the problem have followed Mie's lead. ([67], pp. 284–285)

What was Einstein's reaction to Mie's theory? Unfortunately, we know of only comparatively late reactions of Einstein to the investigations of the Greifswald physicist, which ought to have greatly interested the founder of the theory of relativity. It is probable that this delayed reaction can be explained by the fact that precisely in those years (i.e., 1913–1915) Einstein was working intensively on the general theory of relativity, and it was here, in his opinion, that "the principal line of the front" lay. In the fall of 1913, Einstein met Mie at the 85th conference of German natural scientists in Vienna. Mie entered into a discussion on Einstein's paper, "On the present status of the problem of gravitation," containing an outline of the Einstein–Grossmann theory, which provided the basis of the general theory of relativity. This discussion touched only questions related to the problem of gravitation; in particular, Mie spoke of his scalar Lorentz-covariant theory and criticized both the general principle of relativity and the equivalence principle [38]. For the next few years, Mie concentrated on the problem of gravitation, and essentially did not return to his unified theory.

Only in Einstein's letters from 1916, principally in connection with Hilbert's unified field theory, which relied essentially on Mie's theory, do we find Einstein's comments on Mie's theory. Thus, in a letter to Sommerfeld[13] on December 9, 1915 he wrote:

So far as I know about Hilbert's theory, it uses such an approach to electrodynamic phenomena, excluding consideration of the gravitational field, in which it is closely related to Mie's theory. From the point of view of the general theory of relativity, such a special approach cannot be justified. ([68], p. 194)

---

[13] Similar comments by Einstein on Mie's theory are contained in his letters to Ehrenfest on May 24, 1916 and to Weyl on November 23, 1916.

A curious estimate of Mie's theory is contained in a letter to Weyl on June 6, 1922, when Einstein himself was already working actively on the problem of a unified field theory. Discussing Eddington's efforts to construct a certain unified field theory, he remarked:

> I find the Eddington argument to have this in common with Mie's theory: it is a fine frame, but one cannot see how it can be filled. ([57], p. 177 of the English translation)

In other words, Einstein, at least at the beginning of the 1920s, was apparently attracted by the program of Mie's theory (field-theoretical ideal of unity, reliance on the requirements of Lorentz covariance, nonlinear modification of Maxwell's equations, etc.). They probably also impressed him in the middle of the 1910s. First, however, he did not see a clear way of constructing a suitable action function, and the particular forms appeared to him physically unjustified; second, he was very occupied with developing the general theory of relativity; finally, he initially believed that general relativity was exclusively a relativistic theory of gravitation and not directly related to the problem of the structure of matter. It could well be that during 1913–1915 Einstein, absorbed in the problem of gravitation, was quite generally skeptical with regard to the global problem of a field synthesis of matter, to which his own unsuccessful attempt to solve this problem during 1908–1910 could have played a significant part.

Mie's theory, or program, undoubtedly played a great role in the genesis of the program of unified geometrized field theories. First, it showed the possibility of reducing charged elementary particles to a field, at least in principle, and thus strengthened the field-theoretical ideal of the unity of physics. Second, it confirmed the heuristic power of the requirements of the relativistic program based on the special theory of relativity. After the confirmation and recognition of the general theory of relativity, the idea naturally arose of considering Mie's theory from the point of view of general relativity. The requirements of that theory might cast light on the structure of the required world function. At the same time, the consideration of general relativity led to the need to take into account space-time curvature, though this prompted a new question, that of the relationship between gravitation, understood geometrically, and the electromagnetic field. Further, Mie's theory demonstrated the heuristic possibilities of the variational approach based on the Hamiltonian principle. In general, it encouraged the method of construction of theories as a "game with equations," a method that had also been used by Einstein in his investigations of unified field theory in 1908–1910. Both the variational approach and the game with equations entered the arsenal of basic methods of the geometrized field program. Very important also was the idea of nonlinearity of the basic field equations that offered hope of a field description of particles and the possibility of obtaining the equations of motion of the particles from the field equations.[14]

THE THEORIES OF ISHIWARA AND NORDSTRÖM

Although the unified geometrized field theories of the 1920s, beginning with Weyl's theory, posed as the maximum problem that of a field description of the . elementary particles of matter, their main and direct problem, the minimum problem, was the attainment of a unified description of the electromagnetic and gravitational fields. In the theories of Einstein and Mie described above, a solution of the maximum problem was sought in the first place. In this section, we consider some less well known theories of 1912–1914, primarily the theories of Ishiwara and Nordström, in which the electromagnetic and gravitational fields were unified in a certain manner and posed the minimum problem. During 1912–1915, Einstein (together, in part, with Grossman) developed a tensor geometrical concept of gravitation. The theories of Ishiwara and Nordström were intimately related to scalar theories of gravitation: the former to a theory of Abraham, the latter to a theory of Nordström [38]. Whereas Ishiwara's theory was based explicitly on the electromagnetic field program, Nordström's appeared as a unification of the two fields on an equal footing in a single mathematical scheme and, thus, was not related to the electromagnetic program. In fact, as we shall see, nobody, including Nordström, believed that the achieved unification possessed any physical significance. Nevertheless, we shall consider Nordström's theory again in the chapter devoted to theories of the electromagnetic field type, doing so not only because it, like the theories of Mie and Ishiwara, was advanced in the crucial period 1912–1914 and was associated with scalar theories of gravitation but also because Nordström, like

---

[14] Hitherto, we have not dwelt especially on this aspect of Mie's theory. It follows, however, from the energy conservation law for the energy–momentum tensor of the field, expressed in the integral form

$$\frac{\mathrm{d}}{\mathrm{d}x^4} \int T_i^4 \, \mathrm{d}x^1 \, \mathrm{d}x^2 \, \mathrm{d}x^3 = -\int (T_i^k n_k) \, \mathrm{d}\sigma \qquad (k = 1, 2, 3),$$

where n_k is the unit vector of the normal to a surface sufficiently far from the particle (on which ordinary Maxwellian electrodynamics holds) that the surface integral on the right-hand side is equal to the Lorentz force. As a result, Mie's theory gives an electrodynamic explanation of the law of motion for an electron.

many other Göttingen scientists or students of the Göttingen tradition, largely shared the aims of the electromagnetic program.

The Japanese theoretician Ishiwara worked at the University of Tohoku at Sendai and, like two other physicists of that university, K. Tamaki and T. Mizuno, actively advocated and developed various aspects of the theory of relativity and the electrodynamics of moving bodies in the 1910s. It is probable that in those years close contacts existed between the universities of Tohoku and Göttingen. In any case, one senses the influence of Göttingen physicists in the work of Ishiwara—he adopted the position of the electromagnetic program and regarded the theory of relativity as a development of it; in addition, he published an appreciable proportion of his papers in German and, moreover, in the journal of Göttingen physicists—the *Physikalische Zeitschrift* [69].

It followed from the equivalence principle, restated by Einstein in 1911, that the speed of light can be regarded as a measure of the gravitational potential. On the basis of this idea Abraham, who had graduated from Göttingen and from the beginning of the century had been one of the leaders of the electromagnetic field program in Germany, developed a scalar theory of gravitation during 1911–12 in which the speed of light (in the second form of the theory, the square root of it) was actually regarded as the gravitational potential. As equations of the gravitational field, Abraham used a four-dimensional, though not Lorentz covariant, generalization of Poisson's equation. Einstein also developed, at least during the first half of 1912, a scalar approach, restricting himself, in fact, to consideration of the static case [38].

To Ishiwara, Abraham's approach, which was not associated with Einstein's "metaphysical" (as it appeared to Ishiwara) equivalence principle and had some points of contact with the electromagnetic program, appeared preferable. In the opinion of the Japanese theoretician, however, Abraham too did not fully exploit the possibilities of his theory, namely, the possibility of deriving the equations of the gravitational field and an expression for the energy–momentum tensor of this field from the equations of an electromagnetic field with a variable speed of light. Ishiwara wrote in 1912:

> We postulate... together with Einstein and Abraham that in a space with constant speed of light there exists only an electromagnetic field, but if the speed of light varies in space and in time, then these variations lead to the appearance precisely there of a gravitational field. Abraham now ascribes to the gravitational field a certain four-dimensional tensor, which depends solely on the speed of light and whose components are fictitious stresses, energy density, energy flux, and momentum density. However, he makes his propositions relating to the gravitational tensor the cornerstone of his theory without any further explanation, and as a result the relation-

ship of that tensor to the corresponding electromagnetic tensor remains completely obscure. But one must bear in mind that the electromagnetic 10-tensor [i.e., a second-rank tensor possessing 10 nonvanishing components, namely, the symmetric energy–momentum tensor] contains the speed of light in its components. Thus, variations of it must lead to the appearance of additional terms in the equations for the law of conservation of energy and momentum, which as Einstein correctly emphasizes, must together represent the gravitational field. ([70], p. 1190)

The task that Ishiwara posed to himself corresponded completely with the aims of the electromagnetic field program, of which the Japanese physicist made the following comments:

It still appears expedient to adopt the hypothesis of equality of these two speeds [the propagation speeds of light and the gravitational field, the presence of the latter following Abraham's gravitational equations] because of its simplicity and consonance with the conception of an electromagnetic basis of all physical theories. ([70], p. 1189)

"In what follows," wrote Ishiwara, explaining the aim of his paper,

I wish to adopt the position of Abraham to the extent that, in a physical theory, I do not wish to enter into a metaphysical space-time problem. I wish to attempt to derive the relations of Abraham's theory for the gravitational tensor, and also his thesis of proportionality between gravity and inertia from the fundamental equations of the electromagnetic field. [Ibid.]

Ishiwara wrote Maxwell's equations in the following form, having in mind variability of the speed of light and the special role of $\sqrt{c}$ in Abraham's second theory, which he took as his basis or, rather, attempted to derive:

$$\text{rot}(\mathbf{H}\sqrt{c}) = \frac{\partial}{\partial t}\left(\frac{\mathbf{E}}{\sqrt{c}}\right) + \rho\frac{\mathbf{v}}{c}, \quad \text{div}\left(\frac{\mathbf{H}}{\sqrt{c}}\right) = 0,$$
$$-\text{rot}(\mathbf{E}\sqrt{c}) = \frac{\partial}{\partial t}\left(\frac{\mathbf{H}}{\sqrt{c}}\right), \qquad \text{div}\left(\frac{\mathbf{E}}{\sqrt{c}}\right) = \frac{\rho}{\sqrt{c}}. \tag{15}$$

Naturally, for $c = \text{const}$, these equations go over into the ordinary equations of Maxwell. On the basis of these equations, Ishiwara obtained the following generalized expression for the energy–momentum conservation law:

$$\rho \mathbf{F}^e + \frac{\eta^e}{c}\,\text{grad}\,c = \text{div}\,T^e. \tag{16}$$

Here, T^e is the energy–momentum tensor of the electromagnetic field, η^e is the energy density of this field, and F^e is the electromagnetic four-force per unit volume. The second term on the left-hand side was interpreted by Ishiwara as a gravitational force. This meant that the electromagnetic field generated the gravitational field; moreover, Ishiwara, in contrast to Abraham, did not postulate the expressions for the components of the energy–momentum tensor of the gravitational field but derived them and made them consistent with Abraham's equations of the gravitational field. Of course, this was only a partial synthesis of electromagnetism and gravitation, since gravitation was generated not only by the electromagnetic field but also by matter. Complete reduction of gravitation to electromagnetism would have required solution of the maximum problem, i.e., the problem that was posed by Mie's theory. Ishiwara continued to develop this theory, both in the direction of making it more consistent with the theory of relativity, attempting to limit the validity of the special principle of relativity to infinitesimally small regions of space-time, as well as in connection with the maximum problem, returning in fact to the idea of an electromagnetic origin of the electron mass [71].

We now turn to Nordström's theory, which was a more purely formal, phenomenological unification of the electromagnetic and gravitational fields in a single mathematical scheme. Using Abraham's scalar approach as a basis, and taking into account the criticism of that theory by Einstein from the position of the relativistic program, the Finnish theoretician Nordström, who had been taught at Göttingen, developed during 1912–1913 two forms of a Lorentz-covariant scalar theory of the gravitational field. Referring the reader for details to our study in [38], in which we consider both of Nordström's theories and the part they played in the genesis of the general theory of relativity, we merely mention here that the second theory admits a generally covariant formulation, which was first given by Einstein and Fokker in 1914. Strictly speaking, this theory went beyond the framework of the special theory of relativity, since the line element in it was expressed in the form

$$ds^2 = \Phi \sum_i dx_i^2,$$

where Φ is the gravitational potential, and this meant that it possessed conformally invariant properties, like Maxwell's equations. From the gauge symmetry of Maxwell's equations, it therefore followed that gravitational fields had no influence on electromagnetic phenomena (as is well known, in Nordström's theory there is no bending of light rays). Similarly, because of the vanishing of the trace of the energy–momentum tensor of the electromagnetic field, this field did not influence gravitation.

The common invariance properties of Nordström's and Maxwell's equations, and also the structures of the equations themselves, enabled Nordström in a very elegant manner, namely by introducing a five-dimensional space, to give a unified five-dimensional formulation of the system of equations of gravitation and electromagnetism [72].

If the electromagnetic field strengths, which form a so-called six-vector (i.e., antisymmetric second-rank tensor), are denoted by $f_{xy}, f_{yz}, f_{xz}, f_{xu}, f_{yu}, f_{zu}$, where $u = ict$, and the four-vector of the gravitational field strengths (it is related to the scalar potential) are denoted by $f_{ux}, f_{wy}, f_{wz}, f_{wu}$, where w is a fifth, as yet completely formally introduced, variable, then the equations of the gravitational and electromagnetic fields can be written, as Nordström showed, in the following symmetric form:

$$\frac{\partial f_{xy}}{\partial y} + \frac{\partial f_{xz}}{\partial z} + \frac{\partial f_{xu}}{\partial u} + \frac{\partial f_{xw}}{\partial w} = \frac{1}{c} t_x,$$

$$\cdots$$
$$\cdots \tag{17}$$

$$\frac{\partial f_{ux}}{\partial x} + \frac{\partial f_{uy}}{\partial y} + \frac{\partial f_{uz}}{\partial z} + \frac{\partial f_{uw}}{\partial w} = \frac{1}{c} t_u,$$

$$\frac{\partial f_{wx}}{\partial x} + \frac{\partial f_{wy}}{\partial y} + \frac{\partial f_{wz}}{\partial z} + \frac{\partial f_{wu}}{\partial u} = \frac{1}{c} t_w.$$

$$\frac{\partial f_{yz}}{\partial x} + \frac{\partial f_{zx}}{\partial y} + \frac{\partial f_{xy}}{\partial z} = 0, \qquad \frac{\partial f_{zw}}{\partial y} + \frac{\partial f_{wy}}{\partial z} + \frac{\partial f_{yz}}{\partial w} = 0,$$

$$\frac{\partial f_{zu}}{\partial y} + \frac{\partial f_{uy}}{\partial z} + \frac{\partial f_{yz}}{\partial u} = 0, \qquad \cdots$$
$$\cdots \tag{18}$$

$$\cdots$$
$$\cdots$$

$$\frac{\partial f_{yu}}{\partial x} + \frac{\partial f_{ux}}{\partial y} + \frac{\partial f_{xy}}{\partial u} = 0, \qquad \frac{\partial f_{uw}}{\partial z} + \frac{\partial f_{wz}}{\partial u} + \frac{\partial f_{zu}}{\partial w} = 0.$$

Nordström commented as follows on the system of equations he had obtained:

Both systems of equations [i.e., systems (17) and (18)] are completely symmetric with respect to the variables x, y, z, u, w. Naturally, they do not as yet have any physical significance. If, however, we set all partial derivatives with respect to w equal to zero, then we find that these equations go over into the equations of the electromagnetic and gravitational fields [in the case of gravitation, into the equations of Nordström's theory] if the quantities t_x, t_y, t_z, t_u are identified with the components of the 4-current and $-t_w/c$

with the rest density of the gravitating mass. The first four equations in both systems are Maxwell's equations in Minkowski's form; the final equation of system (17) is the basic equation of the gravitational field, while the six remaining equations of system (18) express the irrotational nature of the gravitational field. This interpretation of Eqs. (17) and (18) shows that it is valid to regard the space-time world as a plane lying in a five-dimensional world. In this five-dimensional world, t_m are the components of a five-vector, and f_{mn} are the components of a 10-vector [i.e., an antisymmetric second-rank tensor]. The physical state of the ether is completely characterized by the f_{mn}. The five-dimensional world has a distinguished axis, the w axis; the four-dimensional space-time world is orthogonal to this axis, and at all points the total derivatives of the components of f with respect to w are equal to zero. ([72], p. 505)

If the concept of a five-dimensional potential is introduced, then it is simple to write down expressions for the energy–momentum conservation law in a unified form, and also wave equations of the fields. Bearing in mind the possibility of a generally covariant formulation of Nordström theory, and also its explicit generalization beyond the framework of Lorentz covariance, one can say that this five-dimensional unified theory of gravitation and electromagnetism was, essentially, the first prototype of unified geometrized field theories, although the idea of geometrization of the physical interactions was expressed in it in indirect rather than direct form. Nevertheless, an extended, five-dimensional space-time continuum was used in the theory, and it was on this geometrical basis that it proved possible to unify the electromagnetic and gravitational fields in a single structure described by a five-dimensional antisymmetric second-rank tensor. In addition, as Einstein and Fokker showed, Nordström's gravitational theory and, therefore, his unified five-dimensional theory admitted a generally covariant formulation, with, moreover, the line element in it differing from the line element of Minkowski space-time. This was in fact tantamount to the use of the geometry of curved space-time, in which the rate of clocks and the length of measuring rods depended on the gravitational potential (were inversely proportional to it).

Of course, the unification adopted by Nordström was purely formal in nature, the requirement that all the derivatives of the field strengths with respect to the fifth variable w should vanish did not have a physical justification, and the geometrical and, *a fortiori*, physical meaning of the fifth dimension remained completely obscure. Nevertheless, the very possibility of such a five-dimensional representation of gravitation and electromagnetism suggested the idea of a deep relationship between these interactions that could, perhaps, be rooted in the geometry of space-time, generalized in some way or another. Nordström wrote at the end of his paper,

> The theory developed above gives, as we have seen, a formal advantage because the electromagnetic and gravitational fields can be understood as a single (unified) field. Of course, a new physical content does not result from this. Nevertheless, I do not rule out the possibility that the form of symmetry that has been found has a deeper basis. ([72], p. 506)

Although Nordström, naturally, felt the influence of the electromagnetic conception of physics (in particular, he used the concept of the ether), his form of unification of the electromagnetic and gravitational fields actually went beyond the framework of the electromagnetic program and undoubtedly contained elements of the approach based on the program of unified geometrized field theories. It is true that it is not entirely clear whether this early attempt by Nordström to introduce a fifth dimension into physics in order to unify gravitation and electromagnetism influenced the first classical investigations of the geometrized field program (i.e., the studies of Hilbert, Weyl, and Kaluza). Hilbert and Weyl referred to Mie as a main precursor, though, in fact, Weyl does mention Nordström in his book *Raum-Zeit-Materie* in the exposition of his unified theory. Kaluza, whose name is firmly associated with the idea of a five-dimensional approach in the framework of the geometrized field program, did not mention Nordström in his classical paper in 1921 (see Chapter 4).

SUMMARY

In this chapter we have considered the development of the electromagnetic field program, which replaced the classical mechanical program and first gave a precise physical significance to the continuum ideal of scientific knowledge, an ideal that thematically (in the sense of Holton [73]) goes back to Aristotle and Descartes. The electromagnetic program was the first program based on a solid physical foundation (Maxwell's theory of the electromagnetic field and Lorentz's electron theory). Indeed, the formulation of the electromagnetic program was associated, as we have seen, with one of the most important stages in the unfolding of the scientific revolution in physics at the beginning of the 20th century.

On the basis of this program, projects for the synthesis of physics were developed in detail, for example, Einstein's unified field theories of 1908–1910, Mie's theory of matter of 1912, and the theories of Ishiwara and Nordström (1912–1914), which, as we have seen, played a certain part in the genesis of the program of unified geometrized field theories, in particular in the development of the theories of Hilbert and Weyl.

After a certain time (at about the beginning of the second decade of the century) the special theory of relativity and quanta cast doubt on the claims of the electromagnetic field program, but the "field ideal of unity" (Hilbert's expression), formulated in the framework of that program, attracted many theoreticians. It received significant support through the successful completion of a relativistic theory of the gravitational field based on an extended relativistic program.

CHAPTER 2

The General Theory of Relativity:
The Core of the Program of
Unified Field Theories

THE RELATIVISTIC PROGRAM AND THE
GENERAL THEORY OF RELATIVITY

We have already considered, in [38], the process of the development of the
general theory of relativity, which was simultaneously a consistent relativistic
theory of the gravitational field. Here, we shall merely consider some details
of this process associated with the genesis of the geometrized field program.
In essence, special relativity was a theory of program type. Its program nature was expressed in the fact that it required a restructuring of effectively
the whole of physics, the restructuring being such that the basic equations
of physical theories (mechanics, hydrodynamics and the theory of elasticity,
electrodynamics, the electron theory and optics, the kinetic theory of heat and
thermodynamics, etc.) become Lorentz covariant and, in the limit of low velocities of the particles of matter, reduce to the classical equations. Formally,
this was the new relativistic program, and compared with the electromagnetic
program it appeared to be significantly more phenomenological. It undoubtedly helped to synthesize physics but on some quite new basis, which to many
appeared to be formal and therefore inadequate. Therefore, the relativistic
program often coexisted with the concept of the ether and, to a certain extent,
with the electromagnetic field program itself, although consistent adherence to
the aims of the relativistic program appeared to preclude the use of the ether
concept and the universalistic pretensions of the electromagnetic program.
Many failed to grasp the important subtext of the relativistic program associated with the prominence now given to space-time concepts and invariance
principles, and also the fundamental status given to the concept of a classical
field (although not necessarily the electromagnetic field).

In essence, the relativistic program based on special relativity should be called a relativistic field program in view of the fundamental importance of the field concept in the structure of this theory. At the beginning of the 1910s the relativistic program had achieved great successes—the relativistic mechanics of a system and continuum had been created, and the foundations had been laid for a relativistic mechanics of the solid state, thermodynamics, and continuum electrodynamics. Special relativity itself finally obtained convincing experimental support and wide recognition.

The relativistic program did encounter certain difficulties, though not of a catastrophic nature, in the relativistic restructuring of the theory of gravitation. Poincaré and then Minkowski showed that Newton's law of gravitation could be generalized in a natural Lorentz-covariant manner, eliminating thereby the action-at-a-distance nature of gravitation. A corresponding elementary law of interaction could not, however, be obtained from field equations analogous to Maxwell's equations. In an attempt to extend the special theory of relativity to gravitation (1907), Einstein recognized the fact of the equality of the inertial and gravitational masses, in which he saw the basis for the construction of a field theory of gravitation that clearly had been underestimated both by earlier workers and contemporaries (for example, Poincaré and Minkowski). Pondering this fact from the point of view of the relativistic program, he arrived at the principle of equivalence (of a homogeneous gravitational field and a uniformly accelerated frame of reference), which, in its turn, led Einstein to conclude that the relativistic program must be extended in an attempt to apply it to the problem of gravitation. This actually undermined the principles of the relativistic program based on the special theory of relativity. Moreover, it was not clear precisely how the program should be extended (i.e., what class of allowed transformations should replace the Lorentz transformations).

Believing firmly in the correctness of his equivalence principle but not knowing how to overcome these difficulties, Einstein actually put aside the problem of gravitation for some time. As we have seen during 1908–1910 he worked actively on quantum theory and in fact even more intensively on the problem of a unified field theory, from which both the electron and the quantum aspects of the behavior of the electromagnetic field should be obtained naturally. These efforts of Einstein, which demonstrated with particular force his deep adherence to the "field-theoretical ideal of unity," were not, as we know, crowned with success, and he returned to the problem of gravitation by 1910.

The combination of the equivalence principle and the requirements of the relativistic program meant that in the development of a relativistic theory of gravitation Einstein was actually guided by an extended relativistic program, which in contrast to the original relativistic program did not have clear outlines, at least not before 1913, when Einstein, in collaboration with Grossmann, advanced a tensor-geometrical concept of the gravitational field that was indissolubly linked with the general principle of relativity and its mathematical equivalent—the principle of general covariance. Despite the explicit aim of general covariance, however, the Einstein–Grossmann tensor-geometrical theory of gravitation contained equations of the gravitational field that did not satisfy the requirement of general covariance. It was only in November 1915 that Einstein and Hilbert were able to obtain generally covariant equations of the gravitational field, thereby completing the construction of a relativistic theory of gravitation, which became known as the general theory of relativity. In essence, it was only after this that the contours of the extended relativistic program were clearly delineated.

In contrast to special relativity, which laid the foundation of the relativistic program, general relativity proved to be only a theory of gravitation, and the extended relativistic program in its initial form did not go beyond the problem of gravitation. The tensor-geometrical concept of gravitation, by which gravitation, a physical interaction having a field nature, is identified with a geometrical, space-time structure, signified a radical rethinking of the very concept of a classical field as it had been developed in electrodynamics. In the theories of Maxwell and Lorentz, the electromagnetic field had been regarded as certain perturbations in the ether, or, after the ether concept had been undermined in the eyes of many scientists by the special theory of relativity, it was ascribed the significance of an independent physical reality. In both cases, it was regarded as a certain substance embedded in space and propagating in this space and in time in a definite manner. (Alternatively, from the point of view of the four-dimensional formulation of special relativity developed by Minkowski, it was regarded as a certain substance embedded in the four-dimensional Minkowski "world.") The four-dimensional potential, or the antisymmetric second-rank tensor, the tensor of the electromagnetic field strengths, did not have any geometrical significance, although the transformation properties of these quantities were, of course, related to the space-time geometry. The geometry of the Minkowski world was external with respect to the field, and its geometrical characterization, for example, the line element (metric) or metric tensor, were given once and for all and depended neither on a field nor on matter.

The tensor-geometrical concept of gravitation radically changed the concept of a classical field, or rather it changed the meaning of the gravitational field, which after the discovery of the special theory of relativity had been understood in approximately the same way as the electromagnetic field. It was found that the requirements of relativism in the case of the gravitational field led to an interpretation of it in which it described not only the gravitational interaction *"but also the behavior of measuring rods and clocks, i.e., the metric of the four-dimensional world"* (Pauli, [63], p. 148). Indeed, the potential is identical to the metric tensor of a four-dimensional Riemannian space, and the field strengths are identical to an affine connection that determines the structure of the geodesics of this space. Thus, geometry loses its *a priori* nature and becomes dynamical, while the gravitational field is geometrized. "This fusion of two previously quite disconnected subjects—metric and gravitation—must be considered," wrote Pauli in 1920, "as the most beautiful achievement of the general theory of relativity" ([63], p. 148).

In those years, only two different fields were known to physics: the electromagnetic field, which was considered on an equal footing with matter in its relationship to space-time, and the gravitational field, which, thus, had a geometrical nature. Only a few years before the completion of general relativity, Einstein, Mie, and some other physicists, striving to realize the electromagnetic-field ideal of the unity of physical knowledge, sought ways to reduce both matter and gravitation to the electromagnetic field, but in light of the general theory of relativity the gravitational field appeared the richer and more fundamental structure because it was described by a much more complicated system of equations and had a deep space-time significance. Therefore, for those whom the problem of the field-theoretical synthesis of physics continued to agitate, general relativity opened up a new possibility for the solution of the problem and therefore acquired a program significance.

EINSTEIN'S ATTITUDE TOWARD THE UNIFYING POSSIBILITIES OF GENERAL RELATIVITY

Although Einstein regarded general relativity as a generalization of special relativity and frequently said that general relativity radically changed ideas about space and time and was a fundamental theory relating to physics as a whole, he also emphasized that in the physical respect general relativity was no more than a relativistic theory of gravitation. Essentially, this was how he thought from 1913, when (in collaboration with Grossmann) he developed the tensor-geometrical concept of gravitation, to 1919–1920, when his attitude toward the unified geometrized field program gradually began to change and, following Hilbert and Weyl, he adopted the position of the program.

We give some statements of Einstein, relating mainly to the period 1915–1916, that confirm this.

The general theory of relativity can tell us nothing more about the essence of the remaining phenomena of nature than was already known in the special theory of relativity. The opinion that I expressed recently in this connection was in this respect incorrect. Every physical theory that is compatible with the special theory of relativity can be included by means of the absolute differential calculus in the scheme of the general theory of relativity, and the latter does not give any criterion of the admissibility of a physical theory.

Einstein wrote this at the end of his famous paper in which for the first time he obtained the correct generally covariant equations of the gravitational field and which signified the completion of the foundations of general relativity ([74], p. 847).[1] Two weeks after this, he wrote in much the same vein

---

[1] In fact, during those dramatic November days there was a moment, even after Einstein had returned to the path of general covariance with respect to field equations, when he thought differently. This was what he had in mind when speaking (in the above quotation) of the error of his opinion expressed shortly before his final step to the correct gravitational equations. We are referring to the paper presented on November 11, 1915 [75]. In it, Einstein, attempting to eliminate the contradiction between the shortened form of the field equations

$$R_{ik} = -\kappa T_{ik},$$

advanced the hypothesis of an "electromagnetic-like" structure of matter, expressed in the assumption of a "traceless" energy–momentum tensor T_{ik} of matter. In essence, this brief note, which Einstein two weeks later regarded as completely incorrect, contained a proposal for an approach to the construction of a unified field theory of matter in accordance with which the gravitational and electromagnetic fields were considered on an equal footing and contributed equally to the formation of particles possessing rest mass. Essentially, Einstein's approach was that the tensor T_{ik} should be associated exclusively with the electromagnetic field, or at least with as yet unknown "massless" fields, which, like the electromagnetic field, do not contribute to the trace of the tensor T_{ik}; and the origin of matter (with rest mass, or nonvanishing trace T) was attributed to the interaction of the electromagnetic (or quasi-electromagnetic) and gravitational fields. In other words, the trace of the tensor $T + t$, where t_{ik} are the gravitational energy–momentum components, must be nonzero. Einstein wrote in this paper:

> We recall that in accordance with our understanding "matter" must not be understood as something given primordially and physically elementary. There are still not a few people who hope that it will be possible to reduce matter to purely electromagnetic processes, though these processes would at least take place in accordance with an improved theory compared with Maxwell's electrodynamics. ([75], p. 799)

Assuming that in this (not yet existing) theory, too, the trace of the energy–momentum tensor vanishes and "that an important constituent part of 'matter' . . . is represented by gravitational fields," Einstein concluded that "in reality only $\sum_\mu (T_\mu^\mu + t_\mu^\mu)$ is positive,

to Sommerfeld:

> This last [i.e., the general theory of relativity] essentially gives a law of
> the gravitational field, doing this, moreover, completely uniquely if the
> requirement of general covariance is satisfied.... On the other hand, any
> other theory that satisfies the special theory of relativity must go over by a
> transformation into the general theory of relativity, but the latter does not
> give any new criterion. Therefore, you see that I am not in any way able to
> help you. ([68], p. 194)

Sommerfeld was probably thinking of the application of general relativity to
the motion of an electron in an atom, especially since it was just at this time
that he obtained his remarkable result relating to the allowance for special
relativity in the electron motion (as a result of which the perihelion of the
elliptic electron orbit acquired an additional rotation, leading to an additional
splitting of the spectral lines of hydrogen-like atoms). Einstein, however,
believed that the transition to general relativity in this situation could not lead
to new physical results. In a letter to Besso on January 3, 1916 Einstein noted,
in particular, "Dimensional analysis, according to which electrons and quanta
require a special h hypothesis, independently of gravitation, thus remains
correct" ([56], p. 49 of the Russian translation). Here Einstein is evidently
referring to the relation $h \simeq e^2/c$, which connects Planck's constant, the
electron charge, and the speed of light, but not the gravitational constant. We
recall that he was guided by this relation when he developed his own unified
field theory of quanta and electrons during 1908–1910.

Thus, Einstein believed that

> the general principle of relativity does not indeed afford us a further limi-
> tation of possibilities; but it makes us acquainted with the influence of the
> gravitational field on all processes, without our having to introduce any new
> hypothesis whatever. ([76], pp. 151–152 of the English translation)

This phrase is taken from Einstein's famous March review "The foundation
of the general theory of relativity" (1916), which gave the first complete
systematic exposition of general relativity. In speaking of a "new hypothesis,"
he had in mind his hypothesis of an "electromagnetic-like" structure of matter.

---

while $\sum T_\mu^\mu$ vanishes everywhere" ([75], p. 799). Of course, this was not yet by
any means a unified theory, but it still contained without any doubt the idea that the
electromagnetic and gravitational fields were fundamental, primordial (it is also true
that Einstein believed that a certain modification of Maxwell's electrodynamics was
necessary) and that matter, or more precisely, particles possessing rest mass, could be
reduced to a certain combination of these fields.

Although Einstein did not apparently regard the question of the part played by gravitation in the structure of the microscopic world as definitively settled, during 1915–1917 (and possibly right up to 1919) he was very skeptical with regard to the possible significance of general relativity not only for the solution of the problem of the structure of matter but also for the development of a nontrivial unification of gravitation and electromagnetism. Einstein wrote in the March review:

> In particular it may remain an open question whether the theory of the electromagnetic field in conjunction with that of the gravitational field furnishes a sufficient basis for the theory of matter or not. The general postulate of relativity is unable on principle to tell us anything about this. It must remain to be seen, during the working out of the theory, whether electromagnetics and the doctrine of gravitation are able in collaboration to perform what the former by itself is unable to do. [Ibid.]

Nevertheless, the final sentence indicates that Einstein did not rule out the possibility of a unification of electromagnetism and gravitation that could furnish a "basis for the theory of matter."

In the following section, we shall consider the first such theory, developed by Hilbert, who in that remarkable November of 1915 when Einstein grappled with the generally covariant field equations, also solved the problem of these equations, actually as a side result, seeing his main task as the construction of a unified field theory of the physical world. Anticipating the events somewhat, we note that Einstein was very critical of Hilbert's form of field synthesis of physics. Moreover, in 1916, having sketched the theory of gravitational waves, Einstein noted:

> An atom must, because of the intra atomic motion of the electron, emit not only electromagnetic but also gravitational energy, albeit in a negligible amount. Since in nature nothing like this can occur in reality, it is evident that the quantum theory must modify not only Maxwellian electrodynamics but also the new theory of gravitation [i.e., general relativity]. ([77], p. 695)

Thus, in 1916 he was more inclined to believe that neither quantum theory nor the physics of the atom should be derived from general relativity or some unified field theory of the Hilbert type; on the contrary, recognizing the classical nature of general relativity, he considered that in a discussion of the problems of microscopic physics it should be augmented by quantum ideas.

Let us summarize what we have said. In 1915–1916 (and, apparently, up to 1918–1919) Einstein regarded the general theory of relativity as a theory of the gravitational field that was hardly capable of casting any new light on the problem of the structure of matter. Moreover, he believed that for the treatment of gravitational phenomena in the microscopic world it would be necessary to modify general relativity by means of quantum theory. It is possible that Einstein's clearly negative attitude to Hilbert's theory, and also his own past unsuccessful experience with the development of a unified field theory in 1908–1910, strengthened his negative estimation of the unifying possibilities of general relativity.

Hilbert's Theory: The First Unified Field Theory Based on the General Theory of Relativity

In earlier studies [38, 78] we considered Hilbert's fundamental paper "Die Grundlagen der Physik" ("The foundations of physics"), which he presented on November 20, 1915 at a meeting of the Göttingen Königliche Gesellschaft der Wissenschaften [79] and in which for the first time he obtained generally covariant equations of the gravitational field of the correct form and also clearly formulated the analog of Noether's second theorem, which two and a half years later was generalized and proved by his pupil Emmy Noether. It was, however, neither the generally covariant equations of gravitation by themselves nor the problem of the conservation laws in a tensor-geometrical theory of gravitation that interested Hilbert in the first place. He was concerned with the foundations of physics, and since he had in mind physics as a whole, he saw as his main task the development of a certain unified physical theory uniting not only electromagnetism and gravitation but also the theory of the electron and quanta. In the studies cited above, we in fact concentrated either on the problem of the gravitational equations or on the problem of the conservation laws in general relativity, i.e., on the positive results that Hilbert obtained in that remarkable paper. Its basic idea, associated with the construction of a unified field theory, remained unrealized.

We now return once more to an analysis of Hilbert's "Foundations of physics," concentrating above all on the way in which the paper solved the problem of creating a unified field theory, the first of that kind based on general relativity. One of the themes that runs through Hilbert's creative work was the axiomatization of scientific knowledge. Weyl divided his scientific biography into six periods: (1) the theory of algebraic invariants (1885–1893), (2) the theory of algebraic number fields (1893–1898), (3) the foundations of geometry (1898–1902), (4) integral equations (1902–1912), (5) physics (1910–1922), and (6) the foundations of mathematics as a whole (1922–1930). Three of these six divisions (the third and the last two) were entirely devoted to the investigation of foundations, i.e., axiomatics. Long before his studies in physics, in 1900, he advanced as the most important scientific problem that of the axiomatization of physics (in his paper "Mathematical problems" at the Second International Congress of Mathematicians at Paris). We are referring to the sixth problem, which according to Hilbert consisted of the "axiomatic construction after this example [i.e., the example of geometry] of the physical disciplines in which mathematics already now plays an eminent role" ([80], p. 306). He had in mind mechanics and the intimately related kinetic theory of gases and statistical mechanics (it is not by chance that he unified the axiomatization of physics and probability theory). It is worth noting that he foresaw the importance of the theory of continuous groups in the solution of this problem ([80], p. 307).

Moreover, in 1905, i.e., five years before the start of his "physical" period, Hilbert, together with Minkowski, led the famous Göttingen seminar on electron theory and the electrodynamics of moving bodies [81]. Although this activity did not leave a mark in Hilbert's publications (it ultimately led Minkowski to the four-dimensional invariant formulation of the special theory of relativity), one can hardly doubt the seriousness of his "physical" interest.

Hilbert's first physical studies were devoted to the application of the theory of integral equations to the kinetic theory of gases and the theory of radiation. In using the new mathematical methods he intended not only to improve calculations but also to establish on their basis the axiomatic structure of these physical theories. Gradually his interests shifted more and more to the most fundamental problems of physics, namely the problems of the structure of matter and the electromagnetic field synthesis of physics. In 1914 the seminar on the structure of matter organized by Debye at Hilbert's initiative began its work at Göttingen. Debye recalled subsequently that Hilbert was attracted by the problem of constructing a unified field theory based on a generalization of Maxwell's equations. It appears—and this is confirmed by Hilbert's "Foundations of physics"—that he was influenced by Mie's theory. We recall also that the electromagnetic field program was very popular among Göttingen physicists and mathematicians (it is sufficient to mention the names of Abraham, Wiechert, Kaufmann, Ritz, Schwarzschild, Minkowski, and others) [69].

Another theory that no less attracted Hilbert was the tensor-geometrical theory of gravitation, the first outline of which was given by Einstein in his joint paper with Grossmann in 1913. The invitation of Einstein to Göttingen in the summer of 1915 is evidence of this. Unfortunately, little is known about the encounter between Hilbert and Einstein that summer. We know of only one document—a letter from Einstein to Sommerfeld on July 15, 1915, in which we can read: "In Göttingen I have had great joy—everything has been understood down to the last details. Hilbert completely charmed me.

An outstanding man!" ([68], p. 192).[2] It may be recalled that, despite the successes of the tensor-geometrical theory, it was then at something of a dead end in connection with the problem of the equations of the gravitational field. Einstein incorrectly believed that in the generally covariant framework of the theory as a whole the field equations could not be generally covariant. Therefore the words to the effect that "everything has been understood down to the last details" could to some extent relate to the problem of the requirement of general covariance in connection with the field equations of gravitation and to Hilbert's having understood this theory.

In his "Die Grundlagen der Physik," Hilbert attempted to combine the ideas of Mie's theory of matter and Einstein's theory of gravitation, in which, as we know, fundamental roles were played by the "field-theoretical ideal of unity," invariance principles, and variational principles. In the combination of these theories, with a certain modification using, of course, powerful mathematical structures such as the theory of continuous transformation groups and their invariants, the calculus of variations, and Riemannian geometry, Hilbert saw the possibility of finally establishing the foundations of physics and constructing a unified physical theory.

We shall consider Hilbert's theory in somewhat more detail. The basic propositions and sources of the theory are already outlined by him in the first sentences of the paper:

> The grandiose problems posed by Einstein, and also the ingenious methods developed for their solution, his far reaching ideas, and the development of the concepts by means of which Mie constructed his electrodynamics, have opened up new ways for investigations into the foundations of physics. In this paper, following the axiomatic method and proceeding, in essence, from two axioms, I would like to establish a new system of basic equations of physics, which possess perfect beauty (*ideale Schönheit*) and contain simultaneously, as I believe, the solution of the problems of Einstein and Mie. ([79], p. 395)

Thus Hilbert is speaking of a theory that combines the relativistic tensor-geometrical theory of the gravitational field and Mie's theory, i.e., a nonlinear electrodynamics that would appear to promise the solution to the problem of matter on an electromagnetic field basis.

The theory is based on a space-time manifold with coordinates w_s, $s = 1$, 2, 3, 4. It is then postulated that a physical event at the point w_s is completely characterized by the ten gravitational potentials $g_{\mu\nu}$ and four electrodynamic

---

[2] The translation in [68] has been corrected ([38], p. 314).

potentials q_s, which are, respectively, a symmetric second-rank tensor and a four-vector with respect to arbitrary continuous transformations of the coordinates w_s. It is interesting that although a Riemannian structure of the manifold is not explicitly postulated, it is nevertheless assumed, since in what follows the Einstein–Grossmann theory is in essence used, and this subsequently provided the skeleton of general relativity; in particular, the curvature tensor of a Riemannian space and its two-index and scalar analogs appear in it. At the same time, the gravitational potentials have, in the light of general relativity or its original version, a transparent geometrical interpretation: they are the components of the metric tensor of a Riemannian space-time manifold.

In contrast, the electrodynamic potentials q_s, introduced by Hilbert on an equal footing with $g_{\mu\nu}$, do not have such a significance and appear in the theory as somewhat external elements. Further, it is assumed that the field equations can be derived from a generally covariant variational principle with a Lagrangian ("world function") that depends on the potentials $g_{\mu\nu}$, q_s and their derivatives $\partial g_{\mu\nu}/\partial w_l$, $\partial^2 g_{\mu\nu}/\partial w_l \partial w_k$, $\partial q_s/\partial w_l$. Hilbert divided this formulation of the dynamical law of the theory into axioms: "Mie's world-function axiom," in which the variational structure of the field equations is postulated with Lagrangian H:

$$\delta \int H \sqrt{g}\, d\omega = 0, \tag{1}$$

and the "axiom of general invariance," in which it is asserted that the function H must be "invariant with respect to any transformation of the world parameters w_s," and which he could truly call "Einstein's axiom." In a comment on this last axiom, he wrote that "only in the formulated Axiom 2 does Einstein's fundamental idea of general covariance find its simplest expression." He did not do this because, as he noted in the same comment, "for Einstein, Hamilton's principle plays only a secondary role, and his function H in no way consists of general invariants and does not contain the electric potentials" ([79], p. 396).

It was natural to identify the Euler–Lagrange equations of this variational problem associated with the gravitational potentials $g_{\mu\nu}$ with the ten equations of the gravitational field, and the four equations associated with variation of the electrodynamic potentials with the equations of the electromagnetic field. How it was proposed to relate or unify gravitation, electromagnetism (purely field phenomena by definition), and the "matter" associated with particles possessing rest mass, above all electrons? The scheme of unification was twofold. It was assumed that the "matter" could be reduced to the electromagnetic field in exactly the same way as in Mie's theory, and in this respect Hilbert did not introduce anything new. The gravitational and electromagnetic fields were unified in Hilbert's theory in an entirely new way, however, being based on a purely mathematical theorem in accordance with which the four equations of the electromagnetic field could be regarded as a consequence of the ten equations of the gravitational field in the same way that the equations of motion of particles in general relativity can be regarded as consequences of the gravitational field equations.

Let us say immediately that we are dealing here with Noether's second theorem, specialized for the case when the infinite continuous group depends on four arbitrary functions of space-time. "The guiding theme (*Leitmotiv*) for the construction of my theory, " wrote Hilbert, "is a mathematical proposition, the proof of which I defer to another place." There then followed the formulation of this theory:

> If the expression J is invariant with respect to arbitrary transformations of the four world parameters and contains n quantities and their derivatives, and if from the condition
>
> $$\delta \int J \sqrt{g}\, d\omega = 0$$
>
> the n Lagrangian variational equations for these n quantities are formed, then in this invariant system of n differential equations four are always a consequence of the remaining $n-4$ equations in the sense that four mutually independent linear combinations of these n differential equations and their total derivatives are always identically satisfied. ([79], p. 397)

Indeed, in this theory we recognize a special case of Noether's second theorem, proved two and a half years later by Emmy Noether, who had come to Göttingen.[3]

Thus Hilbert's "Theorem I," a special case of Noether's second theorem, made it possible to regard the equations of electrodynamics as consequences

---

[3] In Noether's own formulation, this theorem is stated as follows:

> If the integral I [i.e., the integral $I = \int f(x, u, \partial u/\partial x, \ldots)\, dx$, where u are the basic variables, in Hilbert's theory $g_{\mu\nu}$ and q_s] is invariant with respect to a group $G_{\infty\rho}$ [i.e., an infinite continuous group whose transformations depend on ρ arbitrary functions $p(x)$ and their derivatives], in which the arbitrary functions occur in derivatives up to the order σ, then there exist ρ identities between the Lagrangian expressions and their derivatives to the σ-th order." ([82], p. 239)

The genesis of Noether's theorems was considered in our book [78].

of the gravitational field equations. "On the basis of this mathematical proposition [i.e., 'Theorem I'']," wrote Hilbert, "*we can assert that in this sense electrodynamic phenomena are effects of gravitation*" ([79], p. 397). It is interesting that Hilbert believed that the problem of the unification of gravitation and electromagnetism had deep roots in Göttingen. After his words quoted above, there follows the comment: "In this insight, I see a simple and very surprising solution to the problem of Riemann, who was the first to attempt to establish theoretically the connection between gravitation and light" ([79], p. 398).

Hilbert then specialized the form of the Lagrangian, introducing thereby one further axiom. He postulated the relation

$$H = K + \mathcal{L} \tag{2}$$

where K is the scalar curvature (the gravitational part of the Lagrangian), and $\mathcal{L}$ is actually the electromagnetic part of the Lagrangian, which depends of course not only on q_s and q_{sk} but also on $g_l^{\mu\nu}$ (to simplify the calculations, Hilbert assumed that $\mathcal{L}$ does not contain g). Using the straightforward but somewhat cumbersome calculations characteristic of the calculus of variations, Hilbert also obtained an important result on the connection between the electromagnetic part of the Lagrangian and the action, on the one hand, and the energy–momentum tensor of an electromagnetic field, on the other. This connection was established by a variation of the metric tensor in the electromagnetic part of the action integral. In Hilbert's theory, matter was reduced to the electromagnetic field in accordance with Mie's theory. Therefore, this result was formulated as follows:

> Mie's electromagnetic energy tensor is thus none other than the invariant tensor obtained by differentiating the invariant $\mathcal{L}$ with respect to the gravitational potentials $g^{\mu\nu}$ on the indicated passage to the limit. ([79], p. 404)

In this result, Hilbert saw confirmation of his idea of a deep affinity between general relativity and Mie's theory. "This circumstance," he wrote, having in mind the result we have noted, "first indicated to me the necessary close connection between Einstein's general theory of relativity and Mie's electrodynamics and convinced me of the correctness of the theory developed here" [ibid.].

Representing then the Euler–Lagrange equations corresponding to the electrodynamic equations in the abbreviated form

$$[\sqrt{g}\mathcal{L}]_h = 0 \tag{3}$$

and using simple variational calculations based on two formal mathematical propositions (Theorems II and III),[4] Hilbert obtained the four identities between $[\sqrt{g}\mathcal{L}]_h = 0$ and their derivatives,

$$\sum_m \left\{ M_{m\nu}[\sqrt{g}\mathcal{L}]_m + q_\nu \frac{\partial}{\partial w_m}[\sqrt{g}\mathcal{L}]_m \right\} = 0, \tag{4}$$

guaranteed by his Theorem I (note that the interpretation of these identities as the equations of electrodynamics is based essentially on the equations of the gravitational field). *"This is the precise mathematical expression,"* he wrote at the end of the derivation of these identities, *"of the assertion made above concerning the nature of electrodynamics as a consequence* (Folgeerscheinung) *of gravitation"* ([79], p. 406).

As the simplest "material" Lagrangian, Hilbert considered the model Lagrangian of Mie's theory in the form

$$\mathcal{L} = \alpha Q + \beta q^3 \tag{5}$$

where α and β are constants, and Q and q are expressed by

$$Q = \sum_{k,l,m,n} M_{mn} M_{lk} g^{mk} g^{nl}, \qquad q = \sum_{k,l} q_k q_l g^{kl}, \tag{6}$$

and $M_{nm} = q_{mn} - q_{nm}$.

In Hilbert's theory it was assumed that the matter could be reduced to the electromagnetic field in exactly the same way as in Mie's theory. Thus Mie's nonlinear electrodynamics, understood as a unified field theory of matter, was an essential part of the new unified theory. The novelty of the new theory was the manner in which the problem of the relationship between gravitation and electromagnetism was solved. We may mention in passing that in Mie's theory this problem actually remained unresolved—gravitation was not included in the theory—but in Hilbert's theory gravitation was not only important, it acquired the nature of the most fundamental interaction, its equation—in Hilbert's opinion—including the equations of the electromagnetic field. The theory of the gravitational field was essentially identical to general relativity, but the geometry of space-time was not extended in any way; it remained

---

[4] These theorems establish the existence of certain fundamental identities that hold for generally covariant scalars that either are functions of $g^{\mu\nu}$, $g_l^{\mu\nu}$, $g_{kl}^{\mu\nu}$, q_s, and q_{sk}, or are functions of only $g^{\mu\nu}$ and their derivatives.

Riemannian. The theory was doubly reductionist—matter was reduced to the electromagnetic field, and the equations of the electromagnetic field were deduced from the equations of the gravitational field.

The method by which electromagnetism was reduced to gravitation in Hilbert's theory was very distinctive and had no analogs in the past. Reduction to mechanics signified either the construction of mechanical models or reduction to the equations of classical mechanics, with all the basic quantities receiving a mechanical interpretation. Reduction to electrodynamics and the inclusion of a theory in the framework of the electromagnetic field program was understood as the possibility of obtaining the basic equations of the theory from Maxwell's field equations in their standard or a generalized form. The basic elementary objects of the theory, such as particles possessing rest mass, like the very concept of mechanical mass, were to be obtained as certain solutions of the field equations or properties of these solutions.

The reduction of electromagnetism to gravitation in Hilbert's theory lacked many of these features. The main difference was, first, the postulation from the very beginning of the basic electrodynamic quantities q_s that could not be reduced to gravitation, for example, the gravitational potentials. Further, Noether's second theorem, in particular Hilbert's Theorem I, did not indicate precisely which four equations should be regarded as derived and which ten as fundamental. For example, one could take as fundamental six gravitational equations and four electrodynamic equations. Finally, the method of establishing the dependence of the equations of electromagnetism on the equations of gravitation was itself extremely formal. Physically, one could probably attempt to interpret it as a consequence of the principle of general relativity, somewhat like the manner in which the transition from the consideration of electrostatic phenomena in one frame of reference (at rest) to the use of a class of inertial systems leads to the need to introduce the magnetic field. It is true that gravitation could be accorded the more fundamental status, since it was gravitation that was identified with the geometry of space-time, while the electrodynamic potentials q_s did not have a direct relation to geometry.

The interpretation of electromagnetism as a consequence of gravitation adopted in Hilbert's theory (i.e., on the basis of Noether's second theorem) is based on the fact that the number of space-time dimensions is equal to the number of components of the electromagnetic potential and, accordingly, to the number of Maxwell equations obtained by variation from electrodynamic action. Indeed, the number of functions that determine the group of general covariance is equal to the dimension of space-time, and it is this number that determines the number of derived equations in accordance with Noether's second theorem.

In essence, Hilbert very clearly, and for the first time, used a property of general relativity that is now well known and is indeed associated with Noether's second theorem (although this is not always explicitly stated), namely, that by virtue of the requirement of general covariance "the field equations of gravitation contain four conditions which govern the course of material phenomena." "They give," continued Einstein (in the paper finished in March 1916), "the equations of material phenomena completely, if the latter is capable of being characterized by four differential equations independent of one another" ([76], p. 151 of the English translation). At this point, Einstein gave a reference to Hilbert's *Die Grundlagen der Physik,* in which Maxwell's electrodynamic equations, or rather their generalizations characteristic of Mie's theory, were used as these differential equations.

Following Einstein, such a point of view with regard to the four identities associated with Noether's second theorem gradually became established in the literature on general relativity. Since they are present in any generally covariant theory containing either the electromagnetic field or arbitrary matter, these identities should, by virtue of Einstein's gravitational equations, be regarded as four conditions on the matter-energy-momentum tensor, interpreted as a generalization of the differential energy–momentum conservation laws of matter in the presence of a gravitational field. It was in just this way that the identities associated with Noether's second theorem were interpreted by Pauli in his classic monograph on the theory of relativity ([63], p. 159). At the present time these identities are most often interpreted geometrically, as Bianchi identities (more precisely, as contracted Bianchi identities) ([83], p. 334).

Hilbert's theory did not appear to attack directly the problems associated with the structure of atoms and with quanta, but he hoped that further development of the theory would lead, in view of its fundamental nature (it was not by chance that he called his paper *Die Grundlagen der Physik*, "The foundations of physics"), ultimately to an explanation both of the electron and the quantum properties of radiation and microscopic particles. He wrote at the end of his paper:

> I am convinced that by means of the basic equations derived here the most intimate, hitherto hidden phenomena within the atom will be explained and in particular it must be possible quite generally to reduce all physical constants to mathematical constants. ([79], p. 407)

Comparatively recently, correspondence came to light between Einstein and Hilbert in November 1915 directly preceding Hilbert's lecture discussed here and the decisive papers of Einstein, in which the general theory of relativity was completed [84]. Without going into the details of this correspondence, which undoubtedly played no small part in both the process of completion of general relativity and the development by Hilbert of his unified field theory, we consider here only three questions associated with it.

First, it is clear from this correspondence, in particular from Hilbert's letter of November 14, that by that date, and possibly slightly earlier, Hilbert had already completed his work. The letter evidently gave a sketch of his theory, although basic equations and calculations were absent.

Second, after his paper given to the Berlin Academy of Sciences on November 11, in which he used the idea of a "traceless" energy–momentum tensor (in other words, the idea of an electromagnetic-like structure of matter) in order to achieve general covariance of the gravitational field equations, Einstein reported this to Hilbert in his letter of November 12, and on November 14 Hilbert told him, in turn, about his unified field theory. This gives some grounds for assuming that Einstein's paper (presented on November 11) could have been for Hilbert at least a certain additional stimulus.

Finally, on November 18, the day Einstein presented his third November academy paper (on the perihelion of Mercury), and after receiving a copy of the text of Hilbert's lecture containing his unified theory the day before, Einstein told Hilbert of the effective identity of his gravitational equations and the Hilbert equations (presumably he believed wrongly that the energy–momentum tensor of the "matter" in Hilbert's theory is traceless) and his successful calculation of the advance of Mercury's perihelion. For his part, in his reply of November 19 Hilbert wrote, in particular, that if he could calculate as rapidly as Einstein he would have explained why the electron in the hydrogen atom does not radiate.

Thus Hilbert, in contrast to Einstein—at least before the author of the general theory of relativity went over at the beginning of the 1920s to the position of the program of unified geometrized field theories—firmly adopted that position, although his field program, despite the fundamental significance of general relativity for it, was in a certain sense reductionist, since the electromagnetic field in Hilbert's theory did not have a geometrical status on an equal footing with the gravitational field.

How was Hilbert's theory received by the scientific community? In essence, the discussion was about the global theoretical design. Weyl, for example, recalled subsequently:

> Hopes in the Hilbert circle [which included F. Klein, Courant, Landé, Debye, Born, Ewald, and others] ran high at that time; the dream of a universal law accounting both for the structure of the cosmos as a whole, and of all the atomic nuclei, seemed near fulfillment. ([85], p. 283)

Hilbert's theory probably also made a great impression on Weyl himself, who frequently emphasized that Hilbert's theory could be regarded as a "forerunner of a unified field theory of gravitation and electromagnetism" [ibid.].

Einstein became acquainted with Hilbert's theory a few days before it was published. It is obvious—and this can be deduced from some of his statements in the final November publication of 1915 and the March paper "The foundations of the general theory of relativity"—that he did not accept Hilbert's theory, especially in the part that related to electrodynamics, its generalization in the spirit of Mie, and the way it was included in the theory of gravitation. In a letter to Sommerfeld on December 9, 1915, he expressed doubt with regard to the fruitfulness of applying general relativity to the problems of the electron and the structure of the atom, and he also expressed doubts about Hilbert's theory in the part in which it "relates intimately to Mie's theory" ([68], p. 194). In a letter to Ehrenfest on May 24, 1916, Einstein also characterized Hilbert's theory: "Hilbert's exposition does not please me. It is unduly specialized as regards contents [this is a reference to Mie's theory], unduly complicated, and dishonest... in construction (pretension of being a superman by concealing his methods)" ([57], pp. 165–166). Einstein probably saw undue complication in Hilbert's Theorems I and II, which appeared to him to lack physical content and to have a very complicated form, and he may have seen dishonesty of construction in the lack of justification for identifying precisely the four electrodynamic equations as Noether identities.

We give a very outspoken comment of Einstein about Hilbert's theory, addressed to Weyl (letter of November 23, 1916):

> The Hilbertian corollary for matter seems to me childish, in the sense of a child that knows none of the guile of the exterior world.... In any case it is inadmissible, if the solid reflections which originate from the relativity principle are to be combined with such audacious unfounded hypotheses on the construction of electrons or, alternatively, matter. I am prepared to admit that the search for the suitable hypothesis, or the Hamiltonian function for the construction of electrons, compromises one of the most important immediate tasks of the theory. But "axiomatic methods" can be of little help in this. ([57], p. 166)

Einstein again emphasizes the main point of Hilbert's theory that caused him to object—the "audacious unfounded hypotheses on the construction of electrons or ... matter" taken from Mie's theory. Einstein, for whom physical arguments had primary importance even in the creation of theory as abstract and mathematically sophisticated as general relativity, regarded as "childish" the self-sufficient attempts of the mathematician to solve, purely axiomatically, the fundamental difficulties of physics and construct a unified field theory.

At the same time, in emphasizing the importance of searches for a suitable Hamiltonian function, Einstein, we may suppose, had in mind the construction of some unified field theory of gravitation, electromagnetism, and matter. Moreover, he appears to have regarded searches for a suitable Lagrangian for the matter part of physical reality (excluding gravitation) as a problem fully worthy of attention. Thus, one can suppose that at the end of 1916 he did not regard the idea of a unified field theory that included a description of matter as absurd, but he regarded the path of Mie and Hilbert as having no promise. He probably also saw no physical grounds for geometrizing the electromagnetic field.

The diary entries of the Swiss writer R. Humm, who traveled in May 1917 to Berlin and met Einstein, to some extent contradict this assumption. Humm recounts in some detail his discussions with Einstein, in which in particular they talked of the attitude of the creator of general relativity to Hilbert's theory. In those years, the Swiss writer had heard lectures on physics and mathematics at the University of Göttingen and was apparently fully acquainted with the matters under consideration. Einstein spoke of his attitude to mathematics and of the preference that he gave to transparent physical notions. Humm, who at that time probably fully shared the high hopes that reigned in Hilbert's circle in connection with Hilbert's theory, wrote:

> He is very cautious and a physicist to the marrow; he does not plunge so recklessly into the general as we do in Göttingen.... I inferred that he had taken the quantum theory as an aid to modify the gravitation theory, whereas Hilbert, on the other hand, wanted to derive the quantum theory from the gravitation theory.

Later, Humm wrote about a detailed evaluation of Hilbert's theory by Einstein:

> That would not hold well [i.e., quantum theory cannot be deduced from the theory of gravitation], although the gravitation theory was the more general. The idea of relativity can give nothing more than the theory of gravitation....* The thought of constructing a picture of the world from his imagination is magnificent and could lead to something. But history teaches that such attempts usually come to grief..., the variety of tensor types [or corresponding Lagrangians] was far too great and one could not say which should be chosen for the foundation of electrodynamics [Einstein is evidently referring to the electrodynamic part of Hilbert's theory].

---

* *Translator's Note.* This sentence has been translated from the Russian. The corresponding sentence in the English translation of [57] is: "But he could no more abandon the idea of relativity than the idea of gravitation." I have no access to the German original.

Moreover, the experimental data were still too meager; in them one still
had no certain yardstick. ([57], p. 156 of the English translation)

Thus, the creator of the general theory of relativity and the geometrical
conception of gravitation, himself a supporter of the field-theoretical ideal
of the unity of physics, had a very negative attitude toward Hilbert's theory.
This was not only because the theory included Mie's nonlinear electrodynamics, which was rejected by the majority of physicists, including Einstein, but
also because Hilbert's unified theory was largely divorced from basic physical problems and because of its mathematical abstraction (the theory was
constructed axiomatically, the basic idea of the synthesis of gravitation and
electromagnetism had a formal mathematical nature, and so on).

Nevertheless, Hilbert's "Die Grundlagen der Physik" is distinguished by
its richness of content and is one of the most cited classical studies on the
general theory of relativity. We recall that in it there were obtained for the first
time generally covariant equations of the correct form for the gravitational field
(admittedly with a matter-energy-momentum tensor corresponding to Mie's
nonlinear electrodynamics occurring in the equations), the problems of the
energy–momentum conservation law in a generally covariant field theory were
considered for the first time and in great depth, Noether's second theorem for
the case of a group of general covariance was formulated, and the connection
was established between the matter-energy-momentum tensor and variation of
the gravitational potentials. Hilbert's derivation of the gravitational equations
from a variational principle with the scalar curvature as the gravitational part
of the Lagrangian has become classical. Thus, the main contribution of the
Göttingen mathematician to physics in this case was actually related to general
relativity, although it was obtained in the framework of a unified theory. We
note here a paradoxical circumstance: Hilbert's unified theory, in contrast to
all subsequent unified geometrized field theories, was in essence a narrower
theory than general relativity itself, differing from it only in the specialization
of the matter-energy-momentum tensor. In the absence of matter, the two
theories were identical.

Hilbert's contribution to the general theory of relativity was quite fully
reflected in the fundamental books of Pauli and Weyl on the general theory
of relativity, which were written in 1920 and 1918, respectively, and have
been cited above. Although he listed many of Hilbert's results obtained in the
framework of his unified theory, Pauli never even mentioned the theory itself,
whence one can conclude that he did not take seriously Hilbert's method of
unifying gravitation and electromagnetism. First, like Einstein, he also re-

garded Mie's theory as physically incorrect; second, he preferred to interpret the Noether identities, in view of their universal nature, as the expression for the differential energy–momentum conservation law. Weyl's evaluation was about the same in his classic paper on unified theory (1918) [86] and in his book *Raum-Zeit-Materie* [62]. In the 1918 paper, which contains the first exposition of Weyl's unified field theory, there is a reference to Hilbert's "Die Grundlagen der Physik," but only in connection with the energy–momentum conservation law, in particular to the fact that it corresponds to infinitesimally small transformations of the coordinates. As we have already noted, however, Weyl did subsequently call Hilbert's unified field theory a "forerunner of a unified field theory of gravitation and electromagnetism" ([106], p. 283), meaning by this both his own unified theory and subsequent unified geometrized field theories. Hilbert's idea was probably attractive to Weyl, but the specific way in which Hilbert realized the idea (on the basis of Noether's second theorem) did not appear promising to Weyl.

During 1916–1917, Hilbert published two further studies on general relativity: the second part of "Die Grundlagen der Physik" [87] and a fragment from correspondence with F. Klein on the problem of conservation laws in general relativity [88]. Questions of the unified field theory were not addressed in these papers. The main attention in them is concentrated on fundamental problems of general relativity such as those of causality and energy–momentum conservation.

Hilbert's interests gradually shifted more and more in the direction of an investigation of the foundations of mathematics. When, in 1918, Weyl, one of Hilbert's most eminent students, published his unified theory, which was truly geometrized in the accepted sense of the word, this could perhaps have stimulated Hilbert to return to his foundations of physics. As Born wrote to Einstein at the beginning of 1923, however, "Hilbert followed all this half-heartedly, as he is completely preoccupied with his new basic theory of logic and mathematics" ([89], p. 76). Despite this, in 1924 Hilbert somewhat unexpectedly republished, with slight changes, both parts of "Die Grundlagen der Physik" in one of the best known mathematical journals, *Mathematische Annalen* [90]. Thus, one may suppose that Hilbert wished to remind the scientific community that the direction, so popular at the beginning of the 1920s, associated with the development of unified theories of gravitation and electromagnetism and based on the general theory of relativity, went back to his paper of 1915 [91]. This suggests that at this time, too, he continued to regard his method of unification as fully correct and, in many respects, equivalent to the most recent versions of Weyl, Einstein, and others. Referring to studies on unified geometrized field theories, he wrote:

> Since the time of the publication of my first communication, there have been [some] important studies on this question. I mention only the brilliant and deep investigations of Weyl, and Einstein's works contained many new ideas. However, Weyl arrives ultimately at the equations that I established; and Einstein too, although he emphasizes the differences between the initial assumptions of our theories, returns in his latest publications directly to the equations of my theory. ([90], p. 259)

Hilbert believed that the unified field theories were developing along the lines that he had already indicated in 1915:

> I firmly believe that the theory that I have developed here contains a lasting core (*einen bleibenden Kern*) and provides a framework within which there is sufficient space for construction in the future of a physics that satisfies the field-theoretical "ideal of unity." [Ibid.]

This does not mean that Hilbert underestimated the most recent investigations of Weyl and Einstein. In 1926, the commission responsible for awarding the Lobachevski Prize (it included the Kazan scientists N.N. Parfent'ev, P.A. Shirokov, and others) approached Hilbert with a request to give his response to the studies of Weyl on differential geometry, group theory, and the general theory of relativity, and especially his book *Raum-Zeit-Materie*. The leader of the Göttingen mathematicians gave a very high estimation of the works of his student, noting here too the connection between Weyl's unified field theory, which he highly esteemed, and his own theory of 1915:

> Weyl's book *Raum-Zeit-Materie* (5th ed.) is a significant and outstanding investigation into Einstein's theory of gravitation, namely, Weyl's investigations directly join onto the direction of Einstein's theory of gravitation that I developed on the path to the unification of Einstein's theory of gravitation and Mie's electrodynamics. ([92], p. 66)

Noting further, and quite correctly, that Weyl, as Hilbert had done in 1915, used as basic variables of state the ten gravitational and four electrodynamic potentials—and this alone had already largely determined the structure of all future unified geometrized theories of gravitation and electromagnetism— Hilbert also emphasized the fundamental difference of Weyl's theory, namely, the essentially geometrical nature of electromagnetism. "However," he wrote, "Weyl succeeded in giving all 14 variables of state $g_{\mu\nu}$ and q_s a natural geometrical meaning and, thus, successfully fused this system of 14 potentials into a single organic whole" ([92], p. 66).

Einstein always associated Hilbert's approach, his unified field theory, with, on the one hand, Mie's theory and, on the other, Weyl's theory. Thus, in one of his first papers on unified field theories, published in 1919, Einstein wrote: "However elegant from the formal point of view this theory developed by Mie, Hilbert, and Weyl may be, the physical results still cannot satisfy us" ([93], p. 349).

SUMMARY

The general theory of relativity was the first example of a geometrized field theory, namely, a theory of the gravitational field. It arose as a result of development of the relativistic program, which itself underwent significant changes (from the relativistic program based on the special theory of relativity to a somewhat ill-defined extended relativistic program associated with recognition of the need for extension of the relativity principle to arbitrarily accelerated frames of reference, and then to the extended relativistic program based on the general theory of relativity). Einstein believed that general relativity was nothing more than a relativistic theory of gravitation and assumed, at least during 1915–1918, that it could not give anything new for the solution of the problem of the structure of matter. Rather, he believed that for the consideration of gravitational phenomena in the microscopic world it would be necessary to modify the general theory of relativity by means of the ideas of quantum theory.

Nevertheless, the idea of geometrization of physical interactions contained in the general theory of relativity appeared to open up to physics entirely new perspectives for realizing the "field-theoretical ideal of unity" of physical knowledge. The first attempt to unify gravitation and electromagnetism was not, however, based on the idea of geometrization itself but rather on one of the basic principles of the general theory of relativity, the principle of general covariance, and on the idea of an electromagnetic-like structure of matter such as had been developed in the framework of the electromagnetic field program, in particular in Mie's theory. This was done by the leader of the Göttingen mathematical school, Hilbert, who saw in the unification of general relativity and Mie's theory the possibility of constructing a unified field theory of physical phenomena and as a result an axiomatic scheme of physics as a whole. Thus Hilbert's theory was the first unified theory that combined gravitation, electromagnetism, and matter on the basis of the general theory of relativity.

Hilbert's theory, however, included as an important element the physically unsatisfactory theory of Mie. Moreover, the actual method of unification of gravitation and electromagnetism, or rather the reduction of the equations of the electromagnetic field to the equations of gravitation, was inadequately justified, arbitrary, and very formal. In fact, the theory did not attack the problem of matter, or at least in this respect it gave nothing new compared with Mie's theory. Except in Göttingen, where great hopes were placed on it, the theory essentially did not receive recognition. Nevertheless, it contained several important elements that were present in the later unified geometrized field theories, and it was subsequently correctly evaluated by Weyl, the author of the first truly unified geometrized field theory, as a "forerunner of a unified field theory of gravitation and electromagnetism" ([106], p. 283). At the same time, it was no less important that Weyl was a student of Hilbert, whose influence permeated Weyl's entire creative work.

CHAPTER 3

Weyl's Theory: The First Truly Geometrized Unified Field Theory

Although Hilbert's theory was a unified and to a certain extent geometrized theory, and also had a clearly formulated program for the reduction of particles to a field, it possessed two fundamental differences from the main direction in the development of the program of unified geometrized field theories. First, like unified electromagnetic field theories it had a reductionist nature. The electrodynamic equations, which were identical to those of Mie's theory, could be regarded as a consequence of the equations of the gravitational field, and the electron and other particles were reduced to the electromagnetic field (in accordance with Mie's theory). Second, Hilbert's theory was based on Mie's electrodynamics, which suffered from serious defects, as we mentioned above. As a result, the unification of gravitation and electromagnetism was not associated with an extension of Riemannian geometry and interpretation of the electromagnetic field as a geometrical phenomenon. The first unified field theory of that kind was Weyl's theory, which served as the model of the complete program of unified geometrized field theories (1918).

We start by mentioning some of the important circumstances that appear to have been associated with the creation of Weyl's theory. Weyl was a mathematician, a student of Hilbert educated at Göttingen, where he studied and taught for almost ten years (from 1904 to 1913 with a break of a year, when he was at Munich). From 1913 to 1930 he taught at Zurich. There he found Einstein, who at that time was working intensively with Grossmann on the creation of general relativity (in 1914, Einstein moved to Berlin). The deep and early manifestation of Weyl's interest in philosophy is also well known, in particular the philosophy of mathematics and the natural sciences. Concerning the foundations of mathematics, Weyl adopted the position of intuitionism

advanced by Brouwer, in opposition to Hilbert's formalism. We shall see that
these details of Weyl's biography and creative development played the part of
certain preconditions for his work on unified field theory, which thus was not
a chance episode in his activity.

.To this day, Weyl's theory astounds all in the depth of its ideas, its mathematical simplicity, and the elegance of its realization. The basic features of
the program of unified geometrized field theories are especially clearly manifested in it. Therefore, we shall dwell in some detail on the foundations of this
theory as they were formulated by Weyl in his first paper (1918). At the same
time, this first form of the theory encountered serious difficulties of a physical
nature, which was immediately noticed by Einstein, who shortly before had
criticized the unified theories of Mie and Hilbert. Weyl attempted to perfect
the conceptual structure of his theory in a subsequent series of papers (up to
1923), but Einstein and Pauli, giving due recognition to the depth and beauty
of Weyl's ideas, continued to regard the foundations of the theory as physically
unjustified and the consistent realization of the ideas (the construction of field
equations and their solution) as unrealizable in practice.

Despite this, Weyl's theory had a decisive influence on the subsequent
unified geometrized field theories that arose at the beginning of the 1920s,
above all, the theories of Eddington (1921), Kaluza (1921), Einstein (1923),
and others. In 1922 Schrödinger attempted to establish a more intimate connection between Weyl's theory and quantum theory, and as we shall see there
are good grounds for believing that this attempt played a part in the formation
of quantum mechanics.

SCIENTIFIC COMMUNICATION ASPECTS OF THE DEVELOPMENT OF WEYL'S THEORY

After completing his schooling at the gymnasium in Altona in 1903, Weyl
entered the University of Göttingen, which he had chosen because Hilbert,
one of the most famous professors of mathematics at Göttingen, was a cousin
of the director of Weyl's gymnasium. He appears to have begun with an
introductory course on the quadrature of the circle and the concept of number,
which was taught by Hilbert and was in fact very difficult. Weyl subsequently
recalled:

> Most of it went straight over my head, but the doors of a new world swung
> open for me, and I had not sat long at Hilbert's feet before the resolution
> formed itself in my young heart that I must by all means read and study
> whatever this man had written. ([106], p. 94)

Why did Hilbert attract him? Of course, he was an eminent mathematician, famous because of his studies on the algebraic theory of invariants, algebraic number fields, and the foundations of geometry, but Weyl was also attracted to him for purely human qualities:

> Optimism . . . his spiritual passion, his unshakable faith in the supreme value of science, and his firm confidence in the power of reason to find simple and clear answers to simple and clear questions. [Ibid.]

He spent his first vacation studying Hilbert's famous work on algebraic number fields, of which he subsequently wrote: "Indeed, his report is a jewel of mathematical literature. Even today, after almost fifty years, a study of this book is indispensable for anybody who wishes to master the theory of algebraic numbers" ([106], p. 254). The summer months given over to the study of Hilbert's work were, by Weyl's admission, the happiest of his life.

Weyl was struck just as forcibly by his study of Hilbert's *Die Grundlagen der Geometrie* (*Foundations of Geometry*) [94]. His "gymnasium Kantianism" turned to dust:

> Here [at Göttingen] taught David Hilbert, who shortly before this [i.e., before Weyl's entrance to the University of Göttingen] had published his epochal work *Die Grundlagen der Geometrie*. It infused me with the spirit of modern axiomatics.... The logical interdependences of axioms were investigated not only by means of the so-called non-Euclidean geometry, which at that time had existed for nearly a century, but, in addition, many other unusual geometries were also established, mainly on an arithmetical basis. Kant's tight connection to Euclidean geometry now appeared to me naive. Under this devastating blow, the edifice of Kantian philosophy, to which I was devoted with my entire heart, collapsed before my eyes. ([95], p. 633)[1]

These comparatively early events in the youthful Weyl's intellectual life largely determined his understanding of mathematical creativity, his interests, and the combination (characteristic for him) of mathematical investigations reflecting the physical roots of mathematical concepts. Hilbert became for him the paradigm of a mathematician and a true teacher. In 1907 Weyl graduated and in the same year defended, under Hilbert's supervision, a dissertation on singular integral equations. In 1910, he became a Privatdozent and taught at Göttingen until 1913.

---

[1] These fragments of Weyl's lecture are given in a Russian translation by A.V. Akhutin. *Translator's Note*: I have worked directly from the original.

The dissertation subject and his first scientific works were directly related to Hilbert's theory of integral equations. It so happened that from 1902 to 1912 (in accordance with the division into periods of Hilbert's creative work given subsequently by Weyl) the dominant theme in Hilbert's work was on integral equations ([85], p. 274ff.). As is well known, in this way there arose the famous "Hilbert space" of square-summable sequences, which, as was soon shown by E. Fischer and F. Riesz, was isomorphic to the space of functions that are Lebesgue square integrable. The formalism of the infinite dimensional Hilbert space found effective applications in the spectral theory of integral equations, and then in the theory of differential equations. "In the terrain of analysis a rich vein of gold had been struck, comparatively easy to exploit and not soon to be exhausted," wrote Weyl of these investigations ([106], p. 279).

> A large international school of young mathematicians gathered around
> Hilbert and integral equations became the fashion of the day, not only in
> Germany, but also in France. . . . The total effect was an appreciable change
> in the aspect of analysis. ([85], p. 280)

After one and a half decades "a sort of miracle happened: the spectrum theory in Hilbert space was discovered to be the adequate mathematical instrument of the new quantum physics inaugurated by Heisenberg and Schrödinger in 1925" ([85], p. 280).

The papers Weyl wrote during the first Göttingen period (1908–1913) related almost entirely to the field of this new branch of analysis. We mention that his investigations on the spectral theory of ordinary differential equations during 1909–1910, above all the investigation of singular cases, were subsequently used by Schrödinger in his classic papers on wave mechanics.

In accordance with Weyl's division of Hilbert's scientific work into periods that we have already mentioned, the period of integral equations (1902–1912) overlaps with the period of his physical investigations (1910–1922) ([85], pp. 282–283). Naturally, this attraction of Hilbert to physics did not leave Weyl unaffected. As in the case of Hilbert, Weyl's path to physics was associated with the applications of integral equations and spectral theory to physical problems. Weyl wrote several papers on the asymptotic behavior of the eigenfrequencies of vibrating continua (membranes, electromagnetic cavities, etc.) in the period 1911–1915.

Although Hilbert's influence on Weyl was decisive, one should also take into account the influence of the other leader of Göttingen mathematics—Felix Klein. Klein's outstanding geometrical investigations, his famous lectures on the theory of Riemann surfaces, and his emphasis on the constant contact between mathematics and physics undoubtedly struck a chord in Weyl. Lectures on the theory of Riemann surfaces in 1911–1912 formed the content of his first book, *Die Idee der Riemannschen Fläche* (1913), which was written more under the influences of Klein than Hilbert. Weyl was also open to Klein's ideas about the role of intuition in mathematics. In one of the first of his studies of a general nature, he emphasized that new ideas in mathematics arise not so much axiomatically as on the basis of deep but intuitive considerations.

> Despite this, I see the true value and significance of the present-day system of concepts of logicized mathematics in the fact that its concepts, without detriment to the truth of its assertions, remain transparent and open to reflection (*anschauungsmässig*). I believe that in no other way can the human mind raise itself from given reality to mathematical concepts. The validity of our science is then merely a symptom of its being rooted in the "ground," and not an independent feature of it. And mathematics, this proud tree with an extensive crown branching into the ether, draws up through its thousand roots true ideas and notions from the ground (*aus dem Erdboden wirklichen Anschauungen und Vorstellungen*). It would be a fateful error to cut these roots with the shears of a restricted utilitarianism, or to deprive this tree of the ground on which it has grown. ([69], p. 374)*

Weyl's lectures on the theory of Riemann surfaces were written in Klein's "geometrical" style and, in fact, indicated a certain turning by Weyl to geometry.

The spirit of Göttingen, most strikingly embodied in the works of Klein and Hilbert, entered the flesh and blood of Weyl's investigations, although Weyl, having left Göttingen in 1913, worked for a long time at Zurich (until 1930, when he returned to Göttingen to take over Hilbert's chair on his retirement). It is to the Zurich period that the "golden decade" in Weyl's creativity (1917–1927) belongs, associated with his distinguished investigations into differential geometry, the continuum problem and the foundations of mathematics, the general theory of relativity and the unified geometrized theory of the electromagnetic and gravitational fields, the group-theoretical analysis of the problem of space, the theory of representations of continuous groups, and quantum mechanics.

In 1913, there were two important events in Weyl's life and creative biography. He married the student Helene Joseph, a pupil of the famous Göttingen philosopher E. Husserl, and then (in the fall of 1913) transferred to Zurich,

---

* *Translator's Note*: Having no access to [69], I have had to translate this passage from Vizgin's Russian. *Added in proofs*: German original: H. Weyl. "Über die Definitionen der mathematischen Grundbegriffe." *Math.-Naturw. Bl.*, 1910, Bd. 7 (or: Weyl H. *Ges. Abh.*, Bd. 1, p. 304).

where he became a professor and occupied the chair of geometry in the famous Zurich Technische Hochschule, at which Minkowski and Hurwitz had taught earlier (and where the author of the theory of relativity had studied and who at that time also was teaching and working intensively with Grossmann on the relativistic theory of gravitation). The influence of the first event on the intellectual evolution of Weyl will be discussed below. The direct contact with Einstein and the study of the first version of the general theory of relativity, the Einstein–Grossmann theory, probably made an indelible impression on him. In the fall of 1914 Einstein transferred to Berlin. Soon after the start of the First World War, Weyl was called up to military service. In 1915, he returned to the Technische Hochschule and began to study the general theory of relativity; in particular, he gave a course on Einstein's theory (from 1916). In the same year, he began a correspondence with Einstein, fragments of which are quoted in Seelig's book [57]. Weyl's first published paper in this field was his *"Zur Gravitationstheorie"* ("On the theory of gravitation"), which was submitted to the editors of the *Annalen der Physik* in August 1917 [96].

It is probable that Weyl's interest in the general theory of relativity was not only due to his contact with Einstein. From the summer of 1914 Hilbert had manifested increased interest in fundamental problems of physics, and after the encounter with Einstein in July 1915 he had himself attempted to unify Mie's theory of matter, which had greatly impressed him, with the ideas of the general theory of relativity. As we know, this attempt was crowned in November 1915 by the discovery (practically simultaneously with Einstein) of the generally covariant equations of gravitation and the creation of the first unified theory of electromagnetic and gravitational fields based on the general theory of relativity (see Chapter 2). During 1916–1918, many mathematicians and physicists at Göttingen followed Hilbert and enthusiastically took up the ideas of Einstein's theory and joined in its development. Besides the leading lights at Göttingen, Hilbert and Klein, important studies were also made by Noether, Herglotz, and others [78]. Thus, the interest of the mathematician Weyl in the general theory of relativity was not anomalous. In light of what we have said, it would have been strange if he had passed by this theory.

A few words should be said about the paper *"Zur Gravitationstheorie."* Following Hilbert, Lorentz, and Einstein, Weyl established a simpler and more transparent form of a single variational principle from which the field equations of gravitation and electrodynamics, as well as the equations of motion of matter, could be obtained. There followed a deep discussion of the problem of energy–momentum conservation in the general theory of relativity. Weyl then considered a number of specific exact solutions of the gravitational equations, introducing in this connection Cartesian coordinates instead of polar coordinates; in particular, following Reissner, he calculated the field of a charged sphere and the static fields of charged and uncharged matter possessing axial symmetry. Many of these results became classics and have often been cited in subsequent studies of gravitation [63].

Weyl was probably also very interested in Hilbert's unified theory. By 1916 he appears to have written to Einstein about his delight in the general theory of relativity, and he asked Einstein about Hilbert's theory. This can be deduced from a letter from Einstein to Weyl on November 23, 1916, in which he expresses satisfaction in connection with Weyl's estimation of general relativity:

> I am delighted that you have accepted the general relativity theory with so much warmth and zeal. Although for the moment the theory has many opponents the following fact consoles me: the general thinking capacity of its supporters entirely eclipses that of its opponents. That is a kind of objective testimony for the naturalness and intelligence of the theory. ([57], p. 166 of the English translation)

After this, Einstein criticizes Hilbert, in particular his use of Mie's theory and the "axiomatic method" (this part of the letter was quoted earlier). Weyl obviously shared Einstein's point of view with regard to Hilbert's unified theory. At least, in his 1917 paper mentioned above, he did not follow Hilbert's idea of regarding electromagnetism as a consequence of the gravitational equations. Rather, according to Weyl, the law of conservation of energy–momentum of matter and its expression by an identity is the consequence of general covariance. Being a student of Hilbert and a true man of Göttingen, however, he nevertheless persistently pondered the possibility of constructing a unified field theory on a geometrical basis. As was correctly noted by Pyenson, the Göttingen tradition of mathematical physics, going back to Gauss, Riemann, and Weber, always had the aim of constructing mathematically refined theories that could encompass physical phenomena in a universal scheme [97, 98]. It was to such a synthesis of physics that Riemann, and to a certain extent Weber, strove; ideas about the electromagnetic field unification of physics were close to Minkowski, Abraham, Wiechert, and others. The geometrization of gravitation suggested a new possibility for realizing a field synthesis of physics.

THE IMPORTANCE OF WEYL'S PHILOSOPHICAL INTERESTS

If it is difficult to understand, without taking into account the influence of Hilbert, Klein, Einstein, and the "Göttingen spirit" as whole, how and why the mathematician Weyl came to work on the general theory of relativity and then advance his own remarkable unified field theory, the question of the stimulus played by the philosophical interests of the scientist is more problematic. Nevertheless, a brief analysis of Weyl's philosophical evolution would appear to be helpful.

Weyl is the author of many works on the philosophy and methodology of physics and mathematics, which, as it happens, were written after he had created the unified field theory. His book *Philosophie der Matematik und Naturwissenschaft* (*Philosophy of Mathematics and Natural Science*), first published in 1927 [99], became the best known. The author of one of the best accounts of the life and work of Weyl, M. Newman, wrote: "His life-long interest in philosophical problems, and his conviction that they cannot be separated from the problems of science and mathematics, has left its mark everywhere in his work" ([100], p. 501).

In 1954, Weyl's studies on the philosophy of science were recognized with the Arnold Reymond Prize awarded by the University of Lausanne. In this connection, he gave a lecture entitled *"Erkenntnis und Besinnung,"* which had the subtitle *"Ein Lebensrückblick"* [95]. In this lecture, Weyl spoke about his philosophical evolution and about the philosophical quests that accompanied his scientific work.

The first strong impression on this field was his acquaintance with Kant's teaching about space and time, "which immediately gripped me powerfully: a single jolt woke me from 'dogmatic slumber,' and for the youth's mind the world was put in question in a radical manner" ([95], p. 632). At that time Weyl was a pupil of the penultimate class at his gymnasium. Thus, the philosophical essence of space and time and the deep Kantian analysis of these categories were the subject of the future mathematician's youthful thoughts.

Having entered the university, Weyl, as we have already noted, came under the immense influence of Hilbert and in particular of his famous *Grundlagen der Geometrie*. The elucidation of the multiplicity and diversity of geometries revealed the vulnerability of the Kantian conception of space that, as it appeared to him, gave Euclidean geometry an *a priori* nature. It is difficult to imagine that the cardinal transformations in the teaching of space and time in physics associated with the names of Lorentz, Poincaré, Einstein, and the "Göttingenist" Minkowski completely passed Weyl by. One supposes that these events also played an important part in Weyl's abandoning Kant.

The disenchantment with Kant did not turn Weyl from philosophy. Of course, the study of mathematics and his first independent scientific investigations required great efforts, but his lively interest in philosophical, especially epistemological, problems did not disappear. In those years (1905–1913)

Weyl was captivated by the writings of Mach and Poincaré, whose philosophical aim he himself later classified as "constructive positivism" (among the works that he valued highly at that time, Weyl subsequently mentioned Poincaré's *Science and Hypothesis* [101] and Lange's *History of Materialism* [102]). He was probably impressed by the orientation of these thinkers toward the analysis of concrete methodological problems of natural science and mathematics, the antidogmatic direction of their work, and their deep understanding of the specifically scientific aspects of the philosophical problems discussed.

We have already said that 1913 was a turning point in Weyl's life—marriage to Helene Joseph, transfer from Göttingen to Zurich, and the meeting with Einstein. It was also a turning point in Weyl's philosophical evolution. "Sometime later, I married a student of philosophy, a pupil of Edmund Husserl, the founder of phenomenology, who worked at that time in Göttingen," he recalled 40 years after the described events. "Thus it was that Husserl was the person who drew me from positivism and led me back to a freer view of the world" ([95], p. 637).[2]

---

[2] At first glance, it would appear that fortuitous circumstances of his life played an important part in Weyl's philosophical evolution, but the turn to Husserl was in a way entirely natural for him. First, Husserl was an advocate of regarding philosophy as "rigorous science." Second, the main object of philosophical research, according to Husserl in his early period, was scientific knowledge, and the main problem was that of the "objectivity of knowledge." Weyl, who exhibited a genuine interest in philosophical problems of scientific knowledge, must have been impressed by such statements of Husserl as:

> For a philosopher it is not sufficient that we can orient ourselves in the world, that we have laws as formulae in accordance with which we can predict the future flow of things and recover the past. He wishes to bring into clarity what in essence is the nature of "thing," "event," "law of nature," etc., and if science constructs theories for the systematic realization of its problems, the philosopher asks what is the essence of a theory, what indeed makes a theory possible, etc. Only a philosophical investigation completes the scientific studies of the natural scientists and mathematician and makes theoretical knowledge pure and genuine. ([103], p. 6)

Finally, Husserl was initially a mathematician, having defended a dissertation with Weierstrass, whose assistant he was, in 1882.

Before his marriage to Helene Joseph, Weyl was for a time in love with a young singer, a very religious girl who was seriously occupied with philosophy. She visited a philosophical circle led by a "well-known Hegelian." Weyl recalled subsequently, "From this, as it happens, nothing came, partly because of my human immaturity, but partly also because of the opposition of our views of the world, between which it would have been difficult to build a bridge" ([95], p. 632).

It is possible that even before getting to know Helene Joseph, Weyl visited the lectures of Husserl, who enjoyed a certain popularity among the Göttingen physicists and mathematicians. From 1901 to 1906, Husserl was an extraordinary, and from 1906 to 1916 an ordinary, professor of philosophy. Among the Göttingen physicists who were influenced to some extent by Husserl we mention, for example, Voigt and Wiechert [69].

The turn from positivism and the interest in Husserl coincided in Weyl's life with the transfer to Zurich, where Weyl under the influence of the Zurich philosopher Medicus, became attracted by the philosophy of Fichte, in which he saw an anticipation of Husserl's phenomenology. Subsequently, he rated Fichte much higher than Husserl. Of course Weyl's philosophical occupations, in particular with Husserl, can hardly have had a strong influence on his investigations into general relativity and then unified field theory. But in the famous lectures on the general theory of relativity, which were then published in the form of *Raum-Zeit-Materie* (1918), a certain influence of Husserl can be detected.

We have already noted that the problem of space and time was of interest to Weyl before he entered the university and was related to his lively interest in philosophy. In his lectures on general relativity, as he wrote in the foreword, it was his "wish to present this great subject as an illustration of the intermingling of philosophical, mathematical, and physical thought" [64]. The very philosophical introduction contains several explicitly Husserlian pages. Discussing the philosophical problems of the knowledge of reality, Weyl uses terms of Brentano and Husserl: "intentional object," "essential analysis," "recognition of essence," etc. Omitting a discussion of Husserl's phenomenology and the way it was understood by Weyl, we quote some of the latter's statements, in which he considers the problem of knowledge of physical reality and space-time from the position of Husserl. He regarded the theory of relativity as a particularly striking example of the Husserlian "essential analysis."

> In the realm of physics, it is perhaps only the theory of relativity that has made it quite clear that the two essences, space and time, entering into our intuition have no place in the world constructed by mathematical physics. . . . Expressed as a general principle, this means that the real world, and every one of its constituents with their accompanying characteristics, are, and can only be given as, intentional objects of acts of consciousness. ([128], pp. 3–4)

Further, Weyl explains the difference between imminent and transcendental objects, and the discussion of the problem of reality ends with these words:

It is the nature of a real thing to be inexhaustible in content; we can get an ever deeper insight into this content by the continual addition of new experiences, partly in apparent contradiction, by bringing them into harmony with one another. In this interpretation, things of the real world are limiting ideas (*Grenzideen*). From this arises the empirical character of all our knowledge of reality. ([64], p. 4)

At this point Weyl gives a reference to Husserl's "*Ideen zu einer reinen Phänomenologie und phänomenologischen Philosophie*" (*Jahrb. für Philosophie und phänomenolog. Forschung.*, Bd. I, Halle, 1913).

In the recollections of his philosophical evolution, Weyl gives an extract from the same lectures that we have not been able to find in the book and which, perhaps, give a better-rounded description of Weyl's understanding of Husserl's essential analysis for the example of the space-time problem:

The investigations presented here about space seem to me a good example of the essential analysis (*Wesensanalyse*) to which phenomenological philosophy strives, an example typical for the cases when one is concerned with non-immanent being (*nicht-immanente Wesen*). The historical development of the problem of space shows us how difficult it is for us humans, caught up in reality, to identify what is essential. There had to be a long development of mathematics, the broad unfolding of geometrical investigations from Euclid to Riemann, the penetration (from the time of Galileo) of physics into nature and its laws, stimulated again and again by empirical discoveries, finally the genius of a few great minds—Newton, Gauss, Riemann, Einstein—to tear us from the fortuitous, unimportant features to which we were initially bound. ([95], p. 638)

In the Lausanne lecture in 1954, Weyl noted that the method by which Einstein constructed the general theory of relativity and derived in its framework the correct law of gravitation combined "experimentally supported knowledge, ontological analysis, and mathematical construction" [ibid.].

After the transfer to Zurich, Weyl, together with his wife, started to visit the seminar of the great Fichte specialist Medicus, and gradually Husserl receded into the background of Weyl's philosophical thoughts. In his opinion, a systematic development of Husserl's phenomenology must lead to the form of epistemological idealism that is most clearly expressed by Fichte. "In Fichte," wrote Weyl in 1954, "we find the most frank and strong expression of the metaphysical idealism to which at that time Husserlian phenomenology began timidly to feel its way" ([95], p. 637).

Weyl's interest in Husserl and especially Fichte was accompanied not only by his intensive work on general relativity, the unified field theory, and differential geometry, but also by the appearance of a new theme that took a deep hold on him. We are referring to the foundations of mathematics, with which Hilbert had long occupied himself, admittedly with interruptions, but here the pupil did not follow in the teacher's footsteps. On the contrary, having been attracted at the end of the first decade by the intuitionistic concept of mathematics advanced by the Dutch mathematician Brouwer, he came into conflict with Hilbert's formal axiomatic approach to the foundations of mathematics. In 1918–1919 there appeared Weyl's papers on the continuum, the point of departure of which was the recognition of a crisis in the foundations of mathematics, associated above all with the well-known antimonies of set theory. The approach to the foundation of analysis he put forward was in some respect related to the approach of Brouwer, the founder of intuitionism. Having gotten to know the works of Brouwer, Weyl became a decisive supporter of intuitionism. In 1921, he published a long paper *"Ueber die Grundlagenkrise der Mathematik"* [104], in which he popularized and developed Brouwer's ideas.[3]

In contrast to Hilbert, who believed that the intuitionist program was dangerous for mathematics, Weyl saw in this program the path to a resolution of the crisis in the foundations of mathematics and the future of metamathe-

---

[3] Here are some statements from this paper (translated into Russian in [105] by A.P. Yushkevich) characterizing the basic propositions of intuitionistic mathematics and Weyl's attitude to them.

> Brouwer's idea is simple but at the same time deep: here rises before us a 'continuum'..., which can, however, in no way be resolved into a set of already existing real numbers, but is rather a *medium of free becoming (Medium freien Werdens)*...([105], p. 153). What is valuable is not an existential theorem [Weyl had in mind the "existence theorems," which play a fundamental role in classical mathematics], but the construction made in the proof. Mathematics, as Brouwer says occasionally, is action more than theory (*mehr ein Tun denn eine Lehre*)...([104], p. 157). It is in this way that one must understand Brouwer's idea that *there are no grounds for believing in the logical principle of the excluded middle*...([105], p. 155). So I now abandon my previous attempt [Weyl is referring to the foundation of analysis in the book *Das Kontinuum*] and associate myself with Brouwer. ([105], p. 158)

Weyl saw the principal proposition of the intuitionist conception in the idea of the optional sequence that comes into being (*werdende Wahlfolge*) and the rejection of the principle of the excluded middle: "...the decisive stimuli [of this conception]—the optional sequence that comes into being and the disbelief in the axiom of the excluded middle—at least come from Brouwer" [ibid.].

matics. Most of his friends and colleagues regarded Weyl's new enthusiasm as a consequence of his brilliant, many-sided nature and his deep interest in philosophical problems. It seemed to them that he "could take an intoxicated pleasure in allowing himself to be carried away or merely tossed about by the opposing currents which disturbed the period" ([106], p. 148).

We can only speculate about the connections that there were between the following three spheres of Weyl's thoughts, studies, and interests: (1) the philosophy of Husserl and Fichte with his subsequent "ascent" (beginning in 1922) to the German mystical philosopher Meister Eckhart, who can be seen as a forerunner of German classical idealism (on the connection between the teaching of Eckhart and the philosophy of Fichte, see [107], pp. 209–210); (2) the general theory of relativity, unified geometrized field theories, investigations in differential geometry, and the problem of space; (3) the foundations of mathematics and the intuitionist conception of Brouwer.

At the very least, for Weyl, who so seriously and enthusiastically studied literature, music, and art and who always strove to connect the widely separated domains of art, scientific knowledge, and philosophical analysis,[4] these three areas of interest and investigation were most probably interrelated. In the general theory of relativity and the unified geometrized field theories, at least as understood by Weyl, one can to a certain extent recognize an unambiguous exchange with many ideas and specific statements of Husserl—we have already mentioned the understanding of real things as approximate (limiting) ideas (*Grenzideen*) and "ontological analysis."[5] The systematic mathematical and even axiomatic approach to physics, which was characteristic of the Göttingen theoreticians and which Hilbert and Weyl attempted to realize, is in harmony with the following statement of Husserl:

> The mathematical form of treatment ... is for all strictly developed theories ... the only scientific one, the only one that affords systematic com-

---

[4] Very characteristic for Weyl are the following statements: "My work has always tried to unite *the true with the beautiful* [our italics]" ([106], p. 161), or "I ... have always felt the need to reflect on the meaning and purpose of these (i.e., mathematical and physical] investigations" ([95], p. 631).

[5] An example of "ontological analysis" could also be the attempt made by Weyl to justify the need for the extension of Riemannian geometry to "pure infinitesimal geometry," which provided the foundation of his theory. The subsequent forced division of geometry into "phenomenal," which is discovered experimentally by means of clocks and rods, and "ontological," which cannot be revealed by measuring procedures but nevertheless grasps the deep essence of space-time, gravitation, and electromagnetism, is also in the spirit of Husserl's theory of knowledge [103, 108, 109].

pleteness and perfection and gives insight into all possible questions and
their possible forms of solution. ([105], p. 84)*

The general theory of relativity and Weyl's unified theory are close in their idea
and structure to Husserl's "*a priori* natural sciences," among which he included
"pure geometry" and "pure mechanics" and which he regarded as the "science
of the universal structure of the world," as a science "called to establish those
concepts and laws without which nature is altogether inconceivable" ([108],
p. 34).

Husserl's phenomenology with its characteristic declaration of intuition as
the basic method of knowledge ("every truly given intuition is a valid source
of knowledge," etc. ([108], p. 22) was a fertile ground for the reception of
Brouwer's intuitionism. In the intuitionistic studies of Weyl in the 1920s, this
connection can be readily recognized. Weyl wrote, for example, that "the
seeing of essence [in Husserl's sense], from which general theorems follow, is
always based on complete induction, on primordial mathematical induction"
([105], p. 26). Weyl related Brouwer's understanding of the continuous to
Husserl's conception of the relationship between the part and the whole ([105],
p. 78).

As we have seen, Weyl regarded Fichte's philosophy as a development of
Husserl's phenomenology. Reflection on Fichte's teaching could have been
an additional stimulus, both for development of the intuitionistic ideas and for
the philosophical underpinning of general relativity and unified geometrized
field theories. One of the most characteristic features of Fichte's teaching was
his dispassionate criticism, which could have attracted the attention of Weyl:

> The spirit of criticism, the desire to test everything in the court of the mind,
> the overthrow of all authorities if they do not pass this test—this is what
> relates Fichte to Kant and to the philosophy of the Enlightenment. ([107],
> p. 5)

Another important feature of Fichte's philosophy that must have impressed
Weyl was the recognition of the fundamental and primordial status of activity, the practical–ethical relationship to the world, although this recognition
was accompanied by an idealistic interpretation to practice. "We act," wrote
Fichte, "not because we know, but we know because we are predestined to
act; practical reason is the root of all reason" ([107], p. 6).

The first aspect of Fichte's teaching (his criticism) was probably in harmony with the critical direction of Weyl's thinking about the foundations of

---

* *Translator's note.* Quoted from p. 60 of the English translation of [99].

mathematics and the intuitionist conception and his clearly expressed critical attitude to the classical and formal axiomatic approaches to the foundations of mathematics. The second aspect (Fichte's emphasis on activity) finds direct expression in the constructivist principle of intuitionism: "Mathematics . . . is action more than theory" ([104], p. 157); another form of this intuitionistic proposition is: "to exist is to be constructed."[6] It was evidently not by chance that Weyl subsequently called Fichte "the constructivist of the purest water" ([95], p. 641). Deep interest in philosophy on the part of natural scientists is always accompanied by intense attention to the foundations of their science. The fact was that in those years Weyl studied the foundations of both mathematics and physics intensively.

One can identify a certain cross-fertilization between some of the ideas of intuitionism and Weyl's investigations in physics, above all the problem of unified geometrized field theories. The constructivist principle "to exist is to be constructed" is undoubtedly related to the methodological principle of observability, which played an important part in the creation of the theory of relativity and quantum mechanics. The elimination of "excessively arbitrary elements," i.e., elements that (in physics) cannot be given an operational (through measurement) meaning (for example, the concepts of "absolute space" and "ether") or which (in mathematics) cannot be obtained "constructively" (for example, the concept of least upper bound of each nonempty bounded set of real numbers) was a key point both in the history of the theory of relativity and in the development of the intuitionistic conception of mathematics.[7]

The intuitionistic understanding of the mathematical continuum as a "continuous medium of becoming" matches Weyl's adherence to the continualistic field program of synthesis of physics and with the desire to achieve this by

---

[6] In fact, the thesis "to exist is to be constructed" is the point of departure of the reforming activity of Brouwer; it made a very strong impression on his contemporaries and had a very important influence on the further development of objections to classical mathematics ([110], p. 110).

[7] In 1926, Weyl wrote:

It can be seen from the history of physics that intuition and theory must always go hand in hand. On the one hand, one cannot deny that Mach's phenomenalism was overthrown by the theory of the atom, but, on the other hand, Einstein's theory of relativity showed what an important part can be played by returning to the intuitive meaning of theoretical constructions (geometry) and the elimination of excessively arbitrary elements (absolute space)." ([105], p. 32; *Gesammelte Abhandlungen*, Vol. 2, p. 541)

introducing a more fluid four-dimensional continuum that generalizes the concept of Riemannian space and is described by Weyl's geometry. Characterizing Brouwer's theory of the continuum, Weyl wrote:

> The ice sheet . . . was smashed to pieces, and soon the fluid element became the complete master over invariability. Brouwer constructed . . . a rigorous mathematical theory of the continuum, regarding it, not as a frozen thing, but as a *medium of free becoming* (*Medium freien Werdens*). ([105], p. 22; *Gesammelte Abhandlungen*, Vol. 2, p. 528)

At the same time both physics, in the glow of the general theory of relativity and the field program, and intuitionistic mathematics could not (and to this day still cannot), despite the undoubted predominance of the aspect of continuity, completely exclude the effect of discreteness or reduce it to the continuum aspect. General relativity and the unified geometrized field theories, like unified electromagnetic field theories, could not give a field description of particles: "Against Brouwer's theory," wrote Weyl in 1926, "there may also arise the objection that it has not completely overcome the discrete" ([ibid.], p. 533; [105], p. 25).

What we have said is sufficient to establish a definite connection between Weyl's philosophical interest, his investigations into the foundations of mathematics, and his enthusiasm for intuitionism on the one hand and his work on the problem of the general theory of relativity and unified geometrized field theories on the other. Returning to Weyl's philosophical evolution, we note that from 1922 the deepest layer of his philosophical thought was associated with his study of the famous German mystical philosopher Meister Eckhart. In 1954 Weyl recalled:

> From all the events of my intellectual life, the happiest for me were two: when in 1905, as a young student, I read Hilbert's grandiose work "Report on the theory of algebraic numbers," and when in 1922 I read Eckhart. . . . ([95], p. 647)

But the new philosophical enthusiasm of Weyl, who by that time had become "perhaps the most generally known of the mathematicians of his generation" ([106], p. 160) —not only because of his contribution to mathematics but also because of the exceptional popularity of his brilliant book on the theory of relativity, which in the course of a few years went through five editions, and also on account of his active participation in discussions on the foundations of mathematics—belongs to a later period. It should be said that Weyl's occupation with Eckhart's mystical philosophy, far removed from the philosophical problems of scientific knowledge, did not lead to a loss of interest in those problems. Moreover, as we have already said, in 1926 Weyl completed a brilliant book on the philosophy of mathematics and physics that summarized his thoughts in this field and his own investigations into the foundations of mathematics and physics.

Thus, the remark of Newman quoted above, about the indissolubility of the philosophical and scientific preoccupations of Weyl and about the influence of philosophy on his work in the fields of mathematics and physics, appears to be entirely correct, at least if we speak of the period from 1914 or 1917 to 1922.

EARLIER STUDIES OF UNIFIED FIELD THEORIES

From the rise of the electromagnetic field picture of the world in the 1890s and the corresponding research program for the synthesis of physics, the attempts to unify the basic physical interactions on an electromagnetic field basis did not cease (see Chapter 1). Besides the radical ideas of Einstein (1908–1910) and Mie (1912–1913), who attempted to generalize Maxwell's equations in such a way that one could deduce from them the existence of electrons and the quantum structure of radiation, more limited ideas were also advanced for unifying the electromagnetic and gravitational fields or reducing one field to the other. In some cases, the authors of these ideas did not hide their own ultimate aim—to reduce particles to a field. These were very characteristic tendencies of the Göttingen scientists.

Strangely, in neither the first publications of his unified theory in 1918 nor in its subsequent expositions in the various additions of the book *Raum-Zeit-Materie* (from 1919 to 1923) did Weyl make any mention of Hilbert's unified theory, which was well known to him and was, in our opinion, one of the main sources of Weyl's theory. This is confirmed, in particular, by a statement of Weyl made in 1944:

> In his investigations on general relativity, Hilbert combined Einstein's theory of gravitation with G. Mie's program of pure field physics.... *Hilbert's endeavors must be looked upon as a forerunner of a unified field theory of gravitation and electromagnetism* [our italics]. However, there was still too much arbitrariness involved in Hilbert's Hamiltonian [or rather, Lagrangian] function; subsequent attempts (by Weyl, Eddington, Einstein himself, and others) aimed to reduce it. ([85], p. 283)

Weyl refers directly to several studies devoted to attempts to unify gravitation and electromagnetism (in the book *Raum-Zeit-Materie* and not in his 1918 publications):

> A similar tendency is displayed (although obscure to the present author in essential points) in E. Reichenbächer [Weyl here cites papers of the author

of 1917 [111] and 1920 [112]]. Concerning attempts to derive electricity and gravitation from a common root, cf. the articles of Abraham quoted in Note 4 [Weyl is referring to a review by Abraham of theories of gravitation written at the end of 1914 [113], in which there are references to the attempts to unify scalar theories of gravitation with Maxwell's electrodynamics made by Mie [114], Ishiwara [115], and Nordström [116] in 1912–1914], also G. Nordström [117], and E. Wiechert [118] [the indicated papers are cited]. ([64], p. 331)

The theories of Ishiwara, Nordström, and Mie were considered in Chapter 1. It is possible that Weyl knew of these unsuccessful attempts before the creation of his theory, but it is also possible that he did not and only learned of them subsequently. Whatever the truth, the logic of the construction of his theory is in no way related to the studies in that direction, although of course they are in the same line of development of synthetic field concepts as Weyl's theory.

Earlier Investigations in Differential Geometry

The general theory of relativity strongly stimulated purely mathematical investigations in differential geometry. One of the first significant achievements that resulted was the discovery of infinitesimal parallel transport of vectors in a Riemannian space. It was made during 1917–1918 by three mathematicians: the Italian T. Levi-Civita [119], a student of G. Ricci-Curbastro and a coauthor of his,[8] Hessenberg [121] in Germany (Weyl referred to Levi-Civita and Hessenberg in this connection), and the Dutch mathematician Schouten (1918) [122], who a few years later became the leader of one of the most important schools of differential geometry [123]. Schouten believed that this fruitful direction was initiated by a paper of Hessenberg published a year earlier (1916).[9]

---

[8] Their joint paper, "Méthodes du calcul différentiel absolu et leurs applications," published in 1901, is a classic and for a long time was the most cited work on tensor calculus [120]. It was especially on this paper that Einstein and Grossman based the development of the mathematical apparatus of the relativistic theory of gravitation.

[9] The new idea capable of generating the truly modern differential geometry arose only when it became clear that one could construct a geometry by establishing . . . a manifold transport independently of the fundamental tensor. This new idea was first put forward by Hessenberg in 1916 . . . in a completely different formulation . . . [i.e., in a formulation not associated with the concept of "infinitesimal parallel transport"]. ([124], p. 143)

Writing about the discovery of parallel transport, Schouten also mentioned Levi-Civita

Weyl relied first on the work of Levi-Civita and the paper of Hessenberg, both of which he regarded highly. Referring to the geometry that provided the basis of his unified field theory, Weyl wrote in his book (beginning with the third edition): "The development of this geometry was strongly stimulated by the following studies, which were written under the influence of Einstein's theory of gravitation: Levi-Civita [Weyl then cites the corresponding papers of Levi-Civita and Hessenberg]" ([125], p. 290). Nevertheless, Weyl, and subsequently other mathematicians, did give preference to the exposition of Levi-Civita, often associating the discovery of parallel transport in Riemannian geometry with his name. In the preface to the third edition (1919) of *Raum-Zeit-Materie*, Weyl wrote: "The discovery by Levi-Civita, in 1917, of the conception of infinitesimal parallel displacements suggested a renewed examination of the mathematical foundation of Riemann's geometry" ([125], p. vi). He then spoke of his own "pure infinitesimal geometry" based on an extension of the concept of infinitesimal transport and providing the space-time basis of his unified theory and that "every step ... [in its construction] follows quite naturally clearly, and necessarily" [ibid.].

Thus, Einstein and Grossmann used Riemannian geometry and the "absolute differential calculus" of Ricci and Levi-Civita to construct the general theory of relativity. The success of that theory stimulated interest in Riemannian geometry; Levi-Civita, Hessenberg, Schouten, and Weyl himself discovered and developed the concept of infinitesimal parallel transport. Weyl, as we shall see, guided not only by the mathematical possibilities of the extension of this concept but also by physical arguments (the concept of local interaction, a further extension of the relativity principle, and finally the idea of a geometrical synthesis of gravitation and electromagnetism), constructed his "pure infinitesimal geometry" and, on its basis, his unified field theory. This

---

and his own paper of 1918:

> Already Christoffel and Lipschitz in 1870 found that in V_n (and n-dimensional manifold with quadratic metric) there is an operation of differentiation leading to a covariant differential independent of the choice of the variables. Levi-Civita, in 1917, and, in 1918 independently of him, the author of the present paper, showed that this differentiation is associated with a certain "pseudoparallel," or "geodesic," transport characteristic of V_n. By this pseudoparallel transport, Levi-Civita ... understands transport under which the covariant differential vanishes." ([124], p. 142)

A similar definition of parallel transport was given by Schouten. In the fourth and fifth editions of *Raum-Zeit-Materie*, Weyl also refers to Schouten's paper of 1918 in this connection ([64], pp. 325–326).

success at the beginning of the 1920s gave an additional stimulus to purely geometrical investigations. Indeed, it was in this way that the pioneering studies of Schouten, Cartan, Veblen, and others arose, laying the foundations of the theory of spaces of first affine connections, and then conformal and projective connections (and, more generally, homogeneous connections) [126].[10] These achievements, in their turn, created the necessary treasury of geometric structures that were intensively used in the 1920s and 1930s for the development of unified geometrized field theories.

THE GEOMETRICAL FOUNDATION OF THE THEORY AND THE BASIC IDEA OF THE UNIFICATION OF GRAVITATION AND ELECTROMAGNETISM

In his first publication, completed in May 1918 and containing the foundations of his unified field theory [86, 127], Weyl argued as follows for the need for an extension of Riemannian geometry:

> If P and P^* are any two points connected by a curve, a given vector at P can be moved parallel to itself along this curve from P to P^*. But, generally speaking, this conveyance of a vector from P to P^* is not integrable, that is to say, the vector at P^* at which we arrive depends upon the path along which the displacement travels. It is only in Euclidean "gravitationless" geometry that integrability obtains. The Riemannian geometry referred to above still contains a residual element of nonlocal geometry—- without any substantial reason, so far as I can see. It seems to be due to the accidental origin of this geometry in the theory of surfaces. The quadratic form (2) [i.e., $ds^2 = \sum_{ik} g_{ik}\, dx_i\, dx_k$] enables us to compare, with respect to their length, not only two vectors at the same point, but also the vectors at any two points. *But a truly infinitesimal geometry must recognize only the principle of the transference of a length from one point to another point infinitely near to the first.* THIS FORBIDS US TO ASSUME THAT THE PROBLEM OF THE TRANSFERENCE OF LENGTH FROM ONE POINT TO ANOTHER AT A FINITE DISTANCE IS INTEGRABLE, MORE PARTICULARLY AS THE PROBLEM OF THE TRANSFERENCE OF DIRECTION HAS PROVED TO BE NON-INTEGRABLE [our capitalized emphasis]. ([86], p. 203 of the English translation)

---

[10] The general concepts of linear connection, geometrical object, and differentiable manifold were introduced in those years by König, Schouten, Cartan, Veblen, and others [124].

Thus, the new approach to Riemannian geometry based on the concept of infinitesimal parallel transport exhibits, according to Weyl, a certain inadequacy from the point of view of a generalized concept of local action. The field, or local, nature of gravitational interaction leads, in light of the general theory of relativity, to a local geometry (*Nahegeometrie*) of space-time, which is realized by Riemannian geometry. As Weyl showed, however, the approach to this geometry from the point of view of the conception of infinitesimal parallel transport of a vector reveals the presence in it of a certain "nonlocal" element associated with the absence of change in the length of a vector under this transport. The desire to implement most fully and systematically in geometry the idea of local interaction (locality) leads to the requirement that this "nonlocal" (*ferngeometrisches*) element be eliminated, and thereby the geometrical structure of space-time extended. As we shall see, this extension leads not only to a further generalization of the principle of relativity and thus reflects the spirit of the extended relativistic program [38], but also to a theory that "explains in a surprising manner *not only the phenomena of gravitation, but also those of the electromagnetic field*" [86], p. 203 of the English translation). A synthesis of gravitation and electromagnetism is achieved on a unified geometrical basis (*"In this theory all physical quantities have a meaning in world geometry"*) [ibid.]. Further, Weyl constructs this extended geometry "without any thought of its physical interpretation" (he notes at the same time that "its application to physics will then follow of its own accord"). The giving up of the possibility of comparing the lengths of vectors at different points means that in the new geometry "only the ratios of the components g_{ik} have direct physical meaning," and not these quantities themselves. As a consequence,

> the formulae which emerge must possess a double property of invariance: (1) they must be *invariant with respect to any continuous transformations of coordinates* [as in general relativity]; (2) they must *remain unaltered if* λg_{ik}, where λ is an arbitrary continuous function of position, *is substituted for the* g_{ik}. ([Ibid.], p. 204)

The requirement of locality leads to an extension of the relativity principle. The vector spaces at different points are now related, not by a congruence mapping (as in Riemannian geometry), but by a similarity mapping. The concept of infinitesimal parallel transport is modified accordingly. First, it realizes a similarity mapping of the vector spaces at neighboring points: the change $d\xi^i$ of the vector ξ^i when transported from point P to point P' is given by

$$d\xi^i = -\sum_r d\gamma_r^i \xi^r. \tag{1}$$

Second, the $d\gamma_r^i$ must be linear differential forms:

$$d\gamma_r^i = \sum_s \Gamma_{rs}^i \, dx_s, \qquad \Gamma_{rs}^i = \Gamma_{sr}^i. \tag{2}$$

If in accordance with the first requirement it is now assumed that in the case of parallel transport of two vectors ξ^i and η^i from point P to point P', their scalar product at point P',

$$\sum_{ih} (g_{ik} + dg_{ik})(\xi^i + d\xi^i)(\eta^k + d\eta^k),$$

must be proportional to their scalar product at point P, $\sum_{ik} g_{ik}\xi^i\eta^k$, and the coefficient of proportionality is taken to differ infinitesimally from 1 and be equal to $(1 + d\varphi)$, then the following equation is obtained:

$$dg_{ik} - (d\gamma_{ki} + d\gamma_{ik}) = g_{ik} \, d\varphi. \tag{3}$$

From this, in particular, taking into account Eqs. (1) and (2), we obtain the important conclusion that $d\varphi$ is a linear differential form:

$$d\varphi = \sum_i \varphi_i \, dx_i. \tag{4}$$

If the function φ is assumed given, then Eq. (3) also uniquely determines the Christoffel symbols in the new geometry. They can be found from the equations

$$\Gamma_{i,kr} + \Gamma_{k,ir} = \frac{\partial g_{ik}}{\partial x_r} - g_{ik}\varphi_r, \tag{5}$$

which follow from Eq. (3). We see that if $\varphi = 0$, then the relations (5) become the Riemannian relations. Denoting by $\Gamma_{i,rs}^*$ the Christoffel symbols (or the coefficients of the affine connection) of the Riemannian geometry, we obtain for the coefficients of the affine connection of Weyl's "pure infinitesimal geometry" the expressions

$$\Gamma_{i,rs} = \Gamma_{i,rs}^* + \tfrac{1}{2}(g_{ir}\varphi_s + g_{is}\varphi_r - g_{rs}\varphi_i). \tag{6}$$

In other words,

the internal metrical connection (*Masszusammenhang*) of space thus depends not only on the quadratic form (2) (which is determined up to an arbitrary coefficient of proportionality), but also on the linear form (7) [(4)] [i.e, on the differential forms ds^2 and $d\varphi$].

Freedom in the choice of g_{ik} is determined by the factor λ, which is a positive function of position. Replacing g_{ik} by λg_{ik} in Eq. (3), we find that when this is done it is necessary to add to form (4) the total differential of $\ln \lambda$. As a result, the invariance properties in Weyl's geometry are established: Besides the arbitrary smooth transformations of the coordinates that underlie Riemannian geometry, one must also allow gauge, or scale, transformations:

$$g'_{ik} = \lambda g_{ik}, \qquad \varphi'_i = \varphi_i - \frac{\partial \ln \lambda}{\partial x_i}. \tag{7}$$

Therefore it is not the φ_i that have an unambiguous invariant meaning, but

$$F_{ik} = \frac{\partial \varphi_i}{\partial x_k} - \frac{\partial \varphi_k}{\partial x_i},$$

which is an antisymmetric tensor, or the corresponding bilinear form

$$F_{ik} \, dx_i \, \delta x_k = \tfrac{1}{2} F_{ik} \, \Delta x_{ik},$$

where Δx_{ik} is the element of surface spanned by the two arbitrary displacements dx and δx. Therefore, the invariant condition starting when Weyl's geometry reduces to Riemannian geometry is the vanishing of the tensor F_{ik}.

In light of Einstein's fundamental idea of the geometrization of physical interactions, the transformation properties of the vector φ_i and the invariant nature of the antisymmetric tensor F_{ik} lead naturally to the thought of the possibility of identifying the vector φ_i with the potential of the electromagnetic field, and the tensor F_{ik} with the field-strength tensor of the electromagnetic field.[11] The antisymmetry of the tensor F_{ik} allows us to write down the first pair of Maxwell equations:

$$\frac{\partial F_{kl}}{\partial x_i} + \frac{\partial F_{li}}{\partial x_k} + \frac{\partial F_{ik}}{\partial x_l} = 0. \tag{8}$$

Weyl then shows how the geometry discovered by him and the associated tensor calculus can be developed in all details.[12] As Weyl notes, it is not

---

[11] Weyl writes in this connection: "This naturally suggests *interpreting φ_i in world-geometry as the four-potential, and the tensor F consequently as electromagnetic field*" ([86], p. 208 of the English translation).

[12] In particular, he introduced the concept of the weight l of a tensor a_{ik} (here, the rank of the tensor is not important) as the exponent of the factor λ that appears as a factor of the tensor a_{ik} under the transformation (7), i.e., on the transition from a_{ik} to $\lambda^l a_{ik}$.

difficult to show that in this geometry the curvature tensor decomposes in an invariant way into two terms:

$$R^i_{jkl} = P^i_{jkl} - \tfrac{1}{2}\delta^i_j F_{kl}, \tag{9}$$

where P^i_{jkl} is the geometric characteristic of the gravitational field (P^i_{jkl} is antisymmetric both with respect to indices k, l and with respect to indices i, j), and F_{ik} is the analogous characteristic of the electromagnetic field (the field-strength tensor of the electromagnetic field). The vanishing of F_{ik} leads to Riemannian geometry, in which the problem of the transport of length is integrable; the equation $P^i_{jkl} = 0$, which signifies the absence of gravitation, leads to Euclidean geometry, in which the problem of the transport of direction is integrable. From this point of view, Euclidean space is absolutely empty, since the presence of particles possessing rest mass and an electromagnetic field would result in its having curvature.

The Problem of the Field Equations and Conservation Laws in Weyl's Theory

In deriving field equations, Weyl proceeded from a variational principle for the action that must be an "absolute invariant," i.e., a scalar of weight zero.[13] Since the element of four-dimensional volume has weight 2, the Lagrangian W must be an invariant of weight -2. This condition is satisfied by the expressions[14]

$$\tfrac{1}{2}F_{ik}F^{ik}, \qquad R^i_{jkl}R^{jkl}_i, \qquad R_{ik}R^{ik}, \qquad R^2.$$

Whereas the first invariant leads to Maxwell's equations, none of the other three invariants can lead to Einstein's equations of the gravitational field. In

---

[13]   Passing now from geometry to physics, we have to assume, following the precedent of Mie's theory, that all the laws of nature rest upon a definite integral invariant—the action $\int W \, d\omega \ldots$—in such a way that the real world is distinguished from all other possible four-dimensional metric spaces by the characteristic that for it the action-quantity contained in any part of its domain assumes a stationary value.... ([86], p. 211 of the English translation)

[14] In the paper under discussion, Weyl considered only the first two invariant. The two other invariants appeared for the first time in subsequent studies, in particular of R. Weitzenböck, who showed that the listed invariants are the only expressions of the required form ([63], p. 198).

place of them we must, in view of the quadratic nature of the gravitational Lagrangians, obtain fourth-order differential equations in the g_{ik}. Indeed, if as W one takes the expression $W = R^i_{jkl} R_i^{jkl}$, then in accordance with Eq. (9) it can be reduced to the form

$$W = |P|^2 + 4L, \tag{10}$$

where $L = \frac{1}{4} F_{ik} F^{ik}$, and the gravitational part of the Lagrangian is quadratic in its curvature. Subsequently, this circumstance greatly hindered development of the theory. Initially, it is true, Weyl did not regard it as a serious difficulty, believing that the indicated generalization of the gravitational equation could be better suited to the problem of elementary particles than the field equations of general relativity:

> Indeed, it is very improbable that Einstein's equations of gravitation are strictly correct, because, above all things, the gravitation constant occurring in them is not at all in the picture with the other constants of nature, the gravitation radius of the charge and mass of an electron, for example, being of an entirely different order of magnitude (10^{20} or 10^{40} times as small) from that of the radius of the electron itself. ([86], p. 215 of the English translation)[15]

Subsequently, however, it became clear that even in the simplest cases integration of the corresponding fourth-order equations was extremely difficult ([63], p. 202). In addition, fourth-order equations must have many more solutions than second-order equations and "it becomes very difficult to explain why the solutions of these hypothetical equations of the fourth order are so closely approximated in nature by solutions of equations of the second order" ([12], p. 253). One further difficulty is the absence of natural arguments for a unique choice of the quadratic Lagrangian. Finally, we note that the problem of integrating fourth-order equations is greatly complicated by the fact that in this case the question of the proper posing of initial conditions is much more complicated.[16]

Weyl regarded as an important advantage of the theory the establishment of the connection between the law of conservation of electric charge and

---

[15] On the gravitational radius of the electron charge, see [128], pp. 261, 268.

[16] In the case of fourth-order field equations, the following aphorism from the book by Misner, Thorne, and Wheeler becomes even sharper: "Children of light are differential equations that predict the future on the basis of the past. Children of darkness are the factors that specify the initial conditions" ([129], p. 207). [*Translator's note*. I have not been able to locate the original of this quotation and have had to translate the Russian.]

the invariance of the action with respect to gauge transformations. By direct calculation he showed that the continuity equation for the density of the four-current, interpreted as a differential form of the charge conservation law, follows in exactly the same way from the invariance of the action with respect to the infinitesimally small gauge transformation

$$\delta g_{ik} = g_{ik}\delta\varphi, \qquad \delta\varphi_i = \partial(\delta\varphi)/\partial x_i,$$

as the analogous form of the energy–momentum conservation law is obtained from invariance of the action with respect to infinitesimally small coordinate transformations.

> The manner in which the latter [i.e., the conservation law of electric charge] associates itself with the principles of energy and momentum seems to me one of the strongest general arguments in favor of the theory set out here . . .

emphasized Weyl ([86], p. 212 of the English translation).[17]

We mention two further circumstances associated with Weyl's understanding of the advantages of his theory. Studying the possibility of a variational derivation of Maxwell's equations in the framework of his theory, he notes that the electromagnetic action $\int L\, d\omega$ is an invariant of weight zero (i.e., an "absolute invariant," as this quantity must be) only in four-dimensional space. This is explained by the fact that the volume element $d\omega$ in n-dimensional geometry has weight $n/2$, and L is an invariant of weight -2; therefore it is only in four-dimensional geometry that the action is an "absolute invariant." "Thus," he summarized, "on our interpretation the possibility of the Maxwell theory is restricted to the case of four dimensions" ([86], p. 211 of the English translation).

---

[17] In speaking of the energy–momentum conservation law, Weyl had in mind the four identities for the energy–momentum tensor of matter:

$$\frac{\partial(\sqrt{g}\,T_k^i)}{\partial x_i} - \Gamma_{kr}^s(\sqrt{g}\,T_s^r) = 0 \qquad (k = 1, 2, 3, 4).$$

Weyl interpreted the analogous identity (having, incidentally, divergence form) for the density of the four-current as the law of conservation of charge. In view of the fact that in accordance with Noether's first theorem conservation laws must be associated with finite-parameter continuous transformations, however, it must be recognized that, strictly speaking, neither the energy–momentum conservation law follows from the invariance of the action with respect to arbitrary smooth transformations nor the charge conservation law from gauge invariance. The true symmetry of the charge conservation law was found to be gauge symmetry of the first kind (see Chapter 6).

The second remark relates to quantum theory. Noting the "absolute invariance" of the action in his theory, and pointing out that it can be regarded as a "pure number" (*reine Zahl*), Weyl continued: "thus our theory at once accounts for that atomistic structure of the world to which current views attach the most fundamental importance—the action–quantum" ([ibid.], p. 212). Therefore, in Weyl's opinion, the theory should not on further development conflict with quantum theory.[18]

It appeared premature to discuss new physical consequences that could be verified experimentally, since the question of the field equations of the theory had not been resolved. For the choice (10) of the Lagrangian, these equations would be exceptionally complicated (fourth-order nonlinear differential equations). Weyl wrote at the end of the paper,

> The problem naturally presents itself of deducing the physical consequences of the theory of the basis of the special form for the action-quantity given in (14) [i.e., the Lagrangian (10)], and of comparing these with experience, examining in particular whether the existence of the electron and the peculiarities of the hitherto unexplained processes in the atom can be deduced from the theory. ([Ibid.], pp. 215–216)

Thus, Weyl assumed that his theory was capable not only of unifying in a natural geometrical manner electromagnetism and gravitation but also, in a further development, of explaining the existence of the electron and quantum phenomena.

Weyl soon published a paper devoted to his theory in the *Mathematische Zeitschrift* [130]. It contained a more detailed exposition of the geometrical aspects of the theory. The main aim of the program of unified geometrized field theories is expressed there more clearly than in the first paper. Having his theory in mind, he wrote:

> According to this theory, all reality, i.e., everything that exists in the world, is a manifestation of the world metric; physical concepts are nothing other than geometrical concepts (. . . *alles Wirkliche, das in der Welt vorhanden ist, Manifestation der Weltmetrik: die physikalische Begriffe sind keine andern als die geometrischen*). ([130], p. 2)

There are some new aspects in this paper. First, Weyl uses the concept of affine connection (*affine Zusammenhang*), and also very clearly emphasizes

---

[18] This remark of Weyl's is to be understood in the sense that only the Weyl action (with quadratic Lagrangian) is an absolute invariant in "pure infinitesimal geometry"; the action of general relativity does not, for example, possess this property. This gave hope that Weyl's theory could lead in a natural way to the quantum of action as well.

the connection nature of the gravitational field: "A space with an affine connection is, from the physical point of view, a world in which a gravitational field has been introduced" ([130], p. 10). For a unified geometrical description of gravitation and electromagnetism, it is necessary to make a further step: to introduce a metric. The obtained "metric manifold, in the language of physics, is none other than the ether that fills the world" ([130], pp. 13–16). At the same time, he considered "as a basic proposition of infinitesimal geometry [i.e., Weyl's geometry]" the assertion that

> the metric is also an affine connection, that the principle of transport of length leads directly to the principle of the transport of direction, or, in the language of physics, the state of the ether determines the gravitational field. ([130], p. 17)

Second, Weyl introduced the conformal curvature tensor (Weyl's tensor), which subsequently played am important part in the development of mathematical aspects of general relativity. Finally, there appeared in this paper for the first time Weyl's division of physical quantities into "intensive quantities" (*Intensitäts-Grössen*) and "extensive quantities" (*Quantitäts-Grössen*), associated respectively with tensors and tensor densities. This classification of quantities, related to the corresponding classification of Mie [131], was subsequently used by Weyl to eliminate the difficulties of the theory related to the discrepancy, first noted by Einstein, between experiment and the basic proposition of the theory concerning transport of length.

THE PROBLEM OF EXPERIMENTAL VERIFICATION OF THE FOUNDATIONS OF THE THEORY (EINSTEIN'S CRITICISM)

Weyl's first paper contains a supplement written by Einstein, who obviously had the possibility of learning about the paper before its publication [132]. He was the first to draw attention to a certain discrepancy between the foundations of the theory and experience. He noted first of all that the arbitrary factor in the expression for the interval ds in Weyl's theory would be justified if light were the "unique means of transmitting empirical data about the metrical relations in the neighborhood of a world point" ([132]). This arbitrariness vanishes, however, if measurements are made with infinitesimally small rods and clocks. Einstein further wrote:

> Such a definition of the elementary interval ds would then be illusory only if the concepts of the "standard rod" and "standard clock" were based on essentially false suppositions, and this would be the case if the length of the standard rod (or the rate of the standard clock) depended on their prehistory.

If nature were indeed such, then there could not exist chemical elements
whose spectral lines have a definite frequency, and the relative frequency of
two atoms (neighbors in space) of the same type would not, in general, be
the same. Since this is not the case, it seems to me that the basic hypothesis
of this theory is, unfortunately, unacceptable, although its depth and daring
must delight any reader. [132]

Einstein's remark can be readily understood if one recalls that in accordance with relations (1) and (2) the square of the element of length changes
when it is transported from point P to a point P' in an infinitesimally small
neighborhood in accordance with the formula

$$\frac{dl}{dt} = -l\frac{d\varphi}{dt}, \tag{11}$$

where $l = ds^2$. Then if the path over which a rod is transported is finite,
integration of (11) gives for the rod at point P'

$$l_{P'} = l_P \exp\left(\int_P^{P'} \varphi_i \, dx^i\right). \tag{12}$$

Now suppose there exists an electrostatic field described by the fourth com-
ponent of the potential φ_i, $\varphi_4 = \varphi$, and associated with it is a certain static
gravitational field $(\partial g_{ik}/\partial x^4 = 0)$. Applying formula (12) in this case to a
clock at rest with period τ, we obtain

$$\tau = \tau_0 \exp(\alpha \varphi t), \tag{13}$$

where τ_0 is the period of the clock at the initial time and α is a certain constant
that depends on the nature of the clock ([63], p. 196). The obtained relation
means that on transport of the clock U from the point P_1 with the potential φ_1
to the point P_2 with the potential φ_2 during the time t; during its subsequent
return to P_1 the period of the clock changes (increases or decreases depending
on the sign of α and the difference $\varphi_2 - \varphi_1$) by $\exp(-\alpha(\varphi_2 - \varphi_1)t)$ times
compared with the rate of an identical clock U_0 that remain at the point P_1.
In other words, the period of a clock would depend on its prehistory. Such
clocks could be atoms emitting light of a definite frequency and therefore
characterized by definite spectral lines. Even in the case of very small α,
one would not require too much time to note the smearing of spectral lines
([63], p. 196) (there is a very nominal quantitative estimate of the effect due
to Eddington [11], pp. 206 ff.).

In his answer to Einstein, concluding the publication, Weyl attempted to eliminate the difficulty. He pointed out first of all that when a clock is displaced it could be subject to strong accelerations, and therefore would no longer measure the proper time $\int ds$. In his opinion, this also applied to an atom that entered a region with a strong variable electromagnetic field. One could speak of measurement of proper time by a clock with confidence only in the case of a clock at rest in a static gravitational field and in the absence of an electromagnetic field. "How a clock will behave in the case of arbitrary motion and under the simultaneous influence of arbitrary gravitational and electromagnetic fields we shall only learn," wrote Weyl "when a dynamical theory based on physical laws has been developed." Until such dynamics exist, a restriction should be made, in his opinion, "to observation of the arrival of light signals as the fundamental basis for measurement of the components g_{ik}" [133].

Weyl was in fact forced to adopt such a point of view:

> It must be borne in mind that the mathematically ideal process of vector transport, which must form the basis of the mathematical construction of geometry, does not bear any relation to the real process of the motion of a clock, the rate of which is determined by the laws of nature. [Ibid.]

The means that the metric field (the components g_{ik} and φ_i) in Weyl's theory cannot be directly identified with the reading of rods and clocks. Thus, the geometrical foundations of the theory lost physical content. It is true that Weyl could defend the theory from the criticism of discrepancy with experiments, but in contrast to general relativity, in which the interval ds^2 could be identified with direct measurements, Weyl's geometry was to a large degree a purely formal mathematical structure.

At the same time, Einstein highly valued Weyl's theory for its "depth and daring" [132]. Weyl himself saw as the main argument in support of his theory the fact that, developing consistently the concept of local interaction in geometry, it described physical fields in a unified geometrical manner:

> The geometry that I have presented ... is the true "local geometry" (*wahre Nahegeometrie*). It would be curious if instead of this true local geometry there were realized in nature some half-way and inconsistent local geometry with an electromagnetic field tacked onto it (*mit einem angeklebten elektromagnetischen Feld*). [133]

Weyl nevertheless admitted the possibility of incorrectness of his theory and saw the need for comprehensive comparison of it with experimentation. He believed that it was the consequences of the theory, however, above all the solutions of its fundamental equations, that should be experimentally tested.

Of course, I may be on the wrong path in my approach. Here, indeed, we have pure speculation (*reine Spekulation*) and, of course, comparison with experiment is necessary. For this, it is necessary to deduce consequences from the theory, and in this difficult matter I hope for the assistance of colleagues. [Ibid.]

With these words, Weyl concluded his answer to Einstein. As we have seen, the question of the field equations was not yet unambiguously resolved.

Einstein's objection was a serious argument of a physical nature against Weyl's theory. It is interesting that even at the beginning of the 1950s, when discussing his theory of 1918, Weyl almost literally reproduced his dialogue with Einstein. His arguments in defense of the theory remained essentially the same[19] with, of course, allowance for the reconsideration of gauge symmetry on its basis following the discovery of quantum mechanics and then the abandonment by Weyl of the development of the program of unified geometrized field theories (see Chapter 6).

During the period 1918–1920, Einstein's attitude to Weyl's theory was largely negative. As we have seen, he saw its main shortcoming in the discrepancy between the geometrical foundations of the theory and experiment. He very clearly formulated his objection in the Supplement to Weyl's first paper. It is also interesting in this connection to quote some comments of Ein-

---

[19]   Describing the essence of Einstein's objection, Weyl noted:

The definition of the metric field in the ether by means of real rods and clocks can, of course, be regarded as having only a preliminary, indirect relation to experience. Only when the physical laws of action have been established will it be possible to find on its basis what relationship there is between the results of measurement of certain quantities of the theory.... I have no desire to defend this theory, in which I have long ceased to believe. But I could at that time say with justification that the theory—so to speak, after the event—gave in the form of the radius of curvature of the world an absolute local scale (gauge) measure to which spectral frequencies and other quantities having the dimensions of length [or duration] could adjust themselves.... ([18], p. 429)

In a supplement to the republication of this paper, written in June 1955, Weyl, having emphasized the pioneering nature of the paper ("This theory was one of the first attempts to construct unified theories..."), mentioned that "the strongest argument in support... of the theory appeared to be that gauge invariance was related to the conservation law of electric charge in the same way as coordinate invariance was associated with the energy–momentum conservation law" ([134], p. 192; also in Vol. 2 of *Gesammelte Abhandlungen*, p. 40).

stein from his correspondence with Weyl himself, Besso, Sommerfeld, Born, and others and from some of his papers during 1918–1920.

On June 30, 1918, he wrote to Weyl:

> Could one really accuse the Good Lord of being inconsequential if he rejected the opportunity discovered by you for harmonizing the physical world? [We recall Weyl's argument for going over to pure infinitesimal geometry as the genuine local geometry.] I do not think so. If he had made the world to your specification, Weyl II would have come to Him and said reproachfully: "Good Lord, if it did not lie within Thy power to give an objective meaning to the congruence of infinitely small rigid bodies, so that when they are removed from each other one cannot say if they are congruent or not, why hast Thou, oh Inconceivable, not disdained to bequeath to the angle this property or that of similarity?" When two infinitely small bodies K, K', which can originally be brought to coincidence can no longer be brought to coincidence because K' has made a round trip through space, why should a similarity between K and K' remain true during this round trip? It seems far more natural that the transformation of K' relative to K is a *general affine* one [our italics]. ([57], p. 168 of the English translation with slight alteration at the end)

We have given such a long extract because it contains, in a detailed and clear form, one further strong argument, which has a logical nature and was not given before, against Weyl's theory.

Nevertheless, some months earlier, having been sent the proofs of the first edition of Weyl's book *Raum-Zeit-Materie*, Einstein wrote to him with great enthusiasm:

> It is like a classical symphony. Each word is related to the whole, and the design of the work is grandiose. The magnificent use of the infinitesimal parallel displacement from vectors to the deduction of the Riemann tensor! How naturally it all falls into position! ([57], p. 166 of the English translation)

Weyl's theory, which fits directly into the geometrical scheme of the construction of the book, delighted Einstein, as we recall, in its mathematical perfection. "Weyl is a brilliant and splendid fellow, but his views on electricity are no good at all," wrote Einstein to Besso in a letter of June 23, 1918.

In a further letter to Besso (on August 20, 1918), Einstein discussed Weyl's theory in great detail, criticizing it above all from the point of view of the correspondence between its geometrical foundations and experimentation (in the spirit of the "Einstein objection" already considered), and he pointed out another weak point:

So far as I know, there are no physical grounds indicating that it is suitable for the gravitational field. Against the theory is the fact that the gravitational field equations acquire the fourth order, for which there is no warrant from the currently available data; moreover, there is no satisfactory formulation of the energy principle if the Hamilton function for the gravitational field contains derivatives of an order higher than the first. ([56], p. 85)

Indeed, the fourth order of the gravitational equations did not have physical justification and led not only to computational difficulties but also difficulties with the energy conservation law. As we can see, most of Einstein's arguments against Weyl's theory were physical in nature and were associated either with experiments or with fundamental physical principles of methodological significance (the need for experimental justification of the basic propositions of the theory, consistency with the energy–momentum conservation principle, correspondence, etc.) [135]. This is completely in accord with a famous letter that Einstein wrote to Besso on August 28, 1918, in which he spoke of the fundamental importance of experiment in the creation of physical theory. It is possible that Einstein's words, "in reality, no truly fruitful and deep theory has ever been found by anyone by pure thought" ([56], p. 88), also had a polemical subtext addressed to Weyl's theory.

In September 1918 Einstein wrote to Sommerfeld ([68], p. 202) saying that the physical assumptions of Weyl's theory did not stand up to testing by experiment, despite the mathematical perfection of the theory. At approximately the same time, apparently after becoming acquainted with Weyl's second paper [130], Einstein wrote in a letter to him:

It gives me indescribable pleasure to read your beautifully thought out work.... You know my ideas on the relationship [of the theory] to reality; they have never changed. I know how much easier it is to convince men than to find the truth, particularly for one who is such a master of demonstration as yourself. ([57], p. 168 of the English translation)

Although in the same letter Einstein wrote diplomatically to Weyl that he was "finally ... far from being arrogant" and that here too he could "be mistaken as I have already been mistaken countless times," in a letter to Besso of December 4, 1918 he wrote of his "firm conviction" of the nonexistence of the "invariant Weyl measure in nature" ([56], p. 92).[20]

In April 1919 Einstein himself, despite his negative attitude to Weyl's theory, made an attempt to formulate a version of a unified field theory that was based on general relativity and attempted to reduce particles to a field [93] (see Chapter 5). Einstein did not take up the specific proposal made by Weyl

---

[20] Evidence for the patience and good will of Einstein toward his colleagues is the

for the synthesis of gravitation and electromagnetism, but the idea itself of a geometrical unification of these fields with its aim of deducing from the unified equations the "elementary particles of matter" and their quantum properties undoubtedly made a great impression on him and took an ever greater hold on his thoughts.

In 1919, while still a student of Sommerfeld at Munich, the youthful Pauli began to develop Weyl's theory (see the following section). In October 1919 Sommerfeld reported to Einstein: "Pauli is calculating the orbit of the perihelion of Mercury and the bending of light according to Weyl's theory. Perhaps he will refute Weyl ([68], p. 207). On September 4, 1919 Einstein wrote to Ehrenfest about one of Pauli's results:

> A study by Pauli on the Weylian theory already shows the consequence of the initial fallacy of this theory: in general, static solutions do not exist for nonvanishing potentials. It is quite incomprehensible to me that Weyl himself and all the others cannot immediately sense that the theory is contrary to experience. ([57], p. 168 of the English translation)[21]

Einstein wrote in a letter to Besso about the artificiality, indeed the fanciful method, used by Weyl to eliminate the difficulty associated with the Einstein objection: "First data (changes of scales by the potential) are introduced and then, at the price of tremendous computational efforts, everything ends as was foreseen" ([56], p. 93).

PAULI AND WEYL'S THEORY: FROM ACTIVE SUPPORT TO CRITICISM (1919–1920)

Although the course on general relativity at the University of Munich was taught only in the second semester of 1918–1919, Pauli, a student of the

---

letter Einstein wrote to Weyl on December 16, 1918:

> I can only say to you that all my remarks, from a mathematical standpoint, are made with the greatest respect for your theory and *I also admire it as an intellectual construction* [our italics]. You do not have to battle, at least with me. There is no question of anger on my side; true admiration, but disbelief, is my feeling towards the whole subject. When we are together in Zurich we shall meet with or without gauge invariance. ([57], p. 169 of the English translation)

[21] He also reported this to Besso (in a letter of December 12, 1919): "Concerning Weyl's theory, it gradually becomes clear that there do not exist static solutions with nonvanishing different electrostatic potentials (see Pauli's paper in *Phys. Z.*)" ([56], p. 93).

philosophy faculty, gave a talk in December 1918 at Sommerfeld's seminar on the general theory of relativity ([136], p. 10). He had already studied this theory in Vienna, before entering the university, and then he wrote a paper on the energy–momentum conservation law in general relativity that was published at the beginning of 1919 in the *Physikalische Zeitschrift* (Pauli's first publication) [137]. In this paper, following Schrödinger, H. Bauer, and E. Kretschmann, he noted the difficulties of a correct definition of energy and momentum in the general theory of relativity associated with the nontensor nature of the energy–momentum components of the gravitational field.

Weyl's theory immediately stimulated a lively interest in Pauli. He was attracted by the grandeur of Weyl's idea and the mathematical perfection of the theory. At the same time, the physical aspects of the theory were undeveloped. To this Pauli devoted his first paper on Weyl's theory, which was received by the editorial board of the *Physikalische Zeitschrift* in June 1919 [138], and a second paper on this subject, which was completed in November 1919 [139]. The main subject of the first paper was on analysis of the cosmological consequences of the theory. Pauli was able to show that under the natural cosmological assumptions characteristic of Einstein's cosmology the equations of Weyl's theory gave a space of constant positive curvature, i.e., led to a closed world without the introduction of a cosmological term. In this feature of the theory Pauli saw a "particular merit" ([63], p. 201). Another important physical result was the establishment of charge symmetry of Weyl's equations, which Pauli saw as a certain physical defect of the theory:

> Until an entirely new principle [that could in some way break this symmetry] appears in the theory, hopes of explaining the large difference in the masses of the atomic nucleus and the electron must be abandoned. ([130], p. 11)[22]

In this connection, we may recall that it was precisely in 1919 that Rutherford, in experiments on the artificial transformation of nuclei, obtained protons, which he called "positive electrons" ([140], p. 111).

Pauli sent his paper to Weyl, who in an answering letter (of May 10, 1919) welcomed with great joy the gifted supporter of his theory and expressed surprise that in such youth it was possible to penetrate so deeply into the epistemological side of the problem and possess such free scientific thought ([136], p. 12; see also [141], p. 3).

---

[22] He also wrote about this in his classic encyclopedia article published in 1921: "The differential equations [of Weyl's theory] are the same for positive and negative electricity, so that the completely asymmetric conditions which obtain in reality are certainly not represented correctly" ([63], p. 202).

In the second (November) paper Pauli investigated two experimentally verifiable effects of Weyl's theory that were well known in general relativity: the anomalous advance of the perihelion of Mercury and the deflection of light in a gravitional field. The explanation of the anomalous precession of Mercury was the first experimental confirmation of general relativity. The second effect was discovered in observations of the solar eclipse of May 29, 1919 by two British groups of astronomers led by Eddington, Dyson, and Davidson. If Weyl's theory (with fourth-order gravitational field equations) gave results in disagreement with those of general relativity for these effects, then it would be physically unacceptable.[23] Pauli showed, however, that the Schwarzschild solution for a static, spherically symmetric field of a material point was also a solution to the field equations of Weyl's theory (in the absence of an electromagnetic field). This meant that both effects remained correct in this theory. Thus, direct contradictions between the consequences of the theory and experiment were not found, although in the presence of an electromagnetic field static solutions appeared to be absent.[24]

The "Einstein objection" continued to be a serious argument against the experimental foundations of the theory (it was analyzed in detail by Pauli in his encyclopedia article ([63], pp. 195–196). In the article we are considering, however, Pauli advanced one further argument against Weyl's theory, which does, in fact apply to all classical unified field theories:

> In Weyl's theory, we constantly work with field strengths within the electron. However, from the physical point of view a field strength is defined as a force that acts on a test charge. Since test bodies smaller than the electron do not exist, the concept of the strength of the electric field at a certain point [within the electron] appears as an empty fiction void of content. However, one would wish to insist that in physics only essentially observable quantities are introduced. ([136], p. 12)

He included this argument against classical field theories of elementary particles almost word for word in his encyclopedia article.

Pauli's deep thought, which to some extent marked the switching of his interests to the quantum, atomistic conception, was not met with understanding

---

[23] One could suppose that in undertaking this work Pauli was already more skeptically inclined towards Weyl's theory (cf. Sommerfeld's comment from the letter to Einstein in October 1919 quoted above).

[24] Einstein wrote about this result of Pauli in his letter, mentioned above, to Ehrenfest at the beginning of December 1919.

by Weyl ([136], p. 12). Einstein, also, did not support Pauli's conclusion. In a letter to Born dated January 27, 1920 he wrote:

> Pauli's objection is directed not only against Weyl's, but also against anyone else's continuum [or field] theory. Even against one which treated the electron as a singularity. ([89], p. 21 of the English translation)

"I do not think the theory can work without the continuum" [ibid., p. 26], continued Einstein in his letter to Born on March 3, 1920, maintaining the aim of his program, which at that time Weyl also shared. Born, however, was sympathetic to Pauli's idea, since it had something in common with his own ideas about the impossibility of the common simple transfer of macroscopic space-time notions to the physics of the microscopic world. He told this to Pauli in a letter of December 23, 1919, in which he also invited him to his Institute of Theoretical Physics at the University of Frankfurt ([136], p. 13). Two years later, Pauli became an assistant of Born, but at Göttingen, where they both worked on the problems of quantum theory.

Pauli became personally acquainted with Weyl and Einstein at the 86th Congress of German Natural Scientists at Nauheim in September 1920, at which, incidentally, the dispute between the supporters of the theory of relativity (Einstein, von Laue, Born, Weyl, and others) and its opponents (Lenard, Mie, M. Palágyi, and others) reached its peak [142]. In the discussion on Weyl's paper "Electricity and gravitation," in which Weyl attempted to weaken the consequences of the Einstein objection, which appeared to lead to a contradiction between his theory and experiment, Pauli made an objection, which he addressed, however, to Einstein, not to Weyl [143]. Giving his argument, which showed the physical inconsistency of the concept of field strength within the electron, and complementing it with an analogous argument about the impossibility of measuring space at distances less than the electron diameter ("... since there do not exist arbitrarily small rods" [144]), the 20-year-old Pauli posed a cardinal dilemma to those present: adherence to a continuum, purely field position in the solution of the problem of matter or the development and deepening of the quantum conception, which might require a modification of ideas about space, time, and fields that recognized the fundamental importance of discreteness.[25]

---

[25] I should like to ask Professor Einstein if he agrees that one can expect a solution to the problem of matter only by a modification of our ideas about space (and, perhaps, time) and the electric field in the sense of atomism, or whether he regards such doubts as unfounded and assumes that one must adhere to continuum theories. [144]

The answer was rather flexible—Einstein recognized the possibility of a realization of both alternatives, although, as we know from his correspondence of that time and from his papers on the problem of unified geometrized field theories, he preferred to maintain the continuum approach. Weyl kept to the same position. The disagreements between the leaders of the geometrical field program of the synthesis of physics (Weyl and Einstein, who by this time had actually adopted the position of this program (see Chapter 4), concerned only the form in which the program should be realized.[26]

Pauli, who had begun his investigation as an active supporter of this program, very rapidly (probably already by the end of 1919) discovered the seriousness of its limitations, which became obvious only after the creation of quantum mechanics.

Pauli's encyclopedia article, which was completed at the end of 1920 and published in 1921, together with the detailed exposition of Weyl's theory, summarized the main critical comments about it (including the Einstein objection) ([63], pp. 195–196). Pauli regarded Weyl's abandonment, associated with this objection, of the interpretation of the "ideal process of congruent transference of world lengths" by means of rods and clocks and, thus, the abandonment of the establishment of a direct connection between the metric field and the readings of these measuring devices, as having "very serious consequences" ([63], p. 196). Pauli continued:

> While there now no longer exists a direct contradiction with experiment, the theory appears nevertheless to have been robbed of its inherent convincing power, from a physical point of view. For instance, the connection between electromagnetism and world metric is not now essentially physical, but purely formal. For there is no longer an immediate connection between the electromagnetic phenomena and the behavior of measuring rods and clocks. There is only an interrelation between the former and the ideal process which is mathematically defined as congruent transference of vectors. [Ibid.]

The reason for such a pronounced difference between gravitation and electromagnetism resided, in Pauli's opinion, in the fact that

---

[26] We mention that, at Nauheim, Einstein modified his objection against Weyl's theory. Taking into account Weyl's counterargument mentioned above, he did not speak of a discrepancy between the theory and experiment, but said that the abandonment "of this empirically justified comparison . . . takes from the theory one of the most reliable empirical supports and possibility of verification" [144].

there exists only formal, and not physical, evidence for a connection between world metric and electricity. This is quite in contrast to the connection between world metric and gravitation, for which strong empirical support can be found in the equality of gravitational and inertial mass, and which is a rigorous consequence of the principle of equivalence and of special relativity. [Ibid.]

Weyl's theory also lost the important advantage that general relativity had in identifying the world lines of material points and light rays with geodesics. An even more serious difficulty was the problem of the gravitational field equations. It was necessary to give up Einstein's equations and use a Lagrangian of the quadratic type, which led to fourth-order equations. It is true that, as we have seen, Pauli succeeded in showing (for the Lagrangian $L = \frac{1}{2}F_{ik}F^{ik} + cR_{hijk}R^{hijk}$) that Weyl's theory gave the correct result for the motion of the perihelion of Mercury and the deflection of light in a gravitational field.[27]

Several years after the creation of Weyl's theory, it had also "*not succeeded in getting any nearer to solving the problem of the structure of matter*" ([63], p. 202). In this, too, Pauli saw a further important defect of the theory. Moreover, both Pauli's intervention at Nauheim in September 1920 and the final sentence of the section of Weyl's theory in the encyclopedia article together with the last section (§67)[28] indicated a complete reorientation of Pauli and a transfer of his adherence from the position of the unified geometrized field theories program to the position of the quantum theoretical program.

In this (final) section, he summarized the main difficulties of purely classical field approaches to the solution of the problem of matter, including the approach based on unified geometrized field theories (Pauli's article specifically discussed the theories of Mie and Weyl and Einstein's theory of 1919, which will be considered in the next chapter). "Their joint failure prompts us, however, to summarize specifically those shortcomings and difficulties which are common to them all" ([63], p. 205). Having clearly formulated

---

[27] Pauli found that the corresponding solutions of Einstein's field equations were also (in the absence of an electromagnetic field) solutions to the gravitational equations of Weyl's theory.

[28] "There is," he wrote, concluding the analysis of Weyl's theory, "... something to be said for the view that a solution of this problem cannot at all be found in this way" ([63], p. 202).

the maximum problem of the field program,[29] Pauli made a remarkably deep comment:

> It is clear that differential equations which have this property [i.e., meet the requirements listed in the footnote] must be of a particularly complicated structure. It seems to us that this complexity of the physical laws in itself already speaks against the continuum theories. For it should be required, from a physical point of view, that the existence of atomicity, in itself so simple and basic, should also be interpreted in a simple and elementary manner by theory and should not, so to speak, appear as a trick in analysis. [Ibid.]

In this very sharp comment addressed to field theories of matter (and the program of unified geometrized field theories) one senses both a certain irritation in the former adherent of Weyl's theory against its excessive mathematical complexity (with, at the same time, serious loss of physicality even compared with general relativity) and the great critical power Pauli was to develop in the future (as Bohr wrote on the eve of Pauli's 60th birthday, "he more and more became the very conscience of the community of theoretical physicists.... Everyone was eager to learn about Pauli's ... reactions to new discoveries and ideas..." ([145], p. 3).

Against field theories that attempted to explain the stability of charged elementary particles by the action of gravitational forces (and the theories based on the program of unified geometrized field theories actually belonged to such theories), Pauli also noted the following empirical argument:

> For one would expect, in such a case, that a simple numerical relation would exist between the gravitational mass of the electron and its charge. Actually, the relevant dimensionless number $e/(m\sqrt{k})$ (k = ordinary gravitational constant) is of the order 10^{20}! ([63], p. 205)

Pauli also assumed that generally covariant field equations could not reproduce the "asymmetry (difference in mass) of the two kinds of electricity."

Finally, Pauli's last argument against field theories in connection with their attempts to solve the problem of matter was the absence, in his opinion, of an operational, physical meaning of the concept of a classical field within the

---

29 It is the aim of all continuum theories to derive the atomic nature of electricity from the property that the differential equations expressing the physical laws have only a discrete number of solutions which are everywhere regular, static, and spherically symmetric. In particular, one such solution should exist for each of the positive and negative kinds of electricity. ([63], p. 205)

particles themselves: "The field strength in the interior of … a particle would seem to be unobservable, by definition, and thus be fictitious and without physical meaning.…" Pauli concluded:

> This much seems fairly certain, new elements which are foreign to the continuum concept of the field will have to be added to the basic structure of the theories developed so far, before one can arrive at a satisfactory solution of the problem of matter.

SUMMARY

Two and a half years after the completion of the foundations of general relativity and the unified (but only partly geometrized) field theory of Hilbert, Weyl, a student of Hilbert and successor of the Göttingen tradition of mathematical physics, first realized the programmatic idea contained in general relativity. He succeeded in generalizing Riemannian geometry in such a way that the two fields known at the time, the gravitational and electromagnetic, could be regarded as geometrical phenomena. The unification of the fields was not reductionist in nature—neither was reduced to the other. Both were regarded as manifestations of the world metric on an equal footing. At the same time it was assumed, in the spirit of the theories of Mie and Hilbert, that the problem of matter must also find its solution on the basis of the equations of single metric field. It was merely necessary to find the corresponding particle-like solutions of these equations.

Einstein might have been expected to have adopted Weyl's idea with enthusiasm, but in fact he was the first to advance objections of a physical nature against Weyl's theory. The geometrization of the electromagnetic interaction did not have such secure empirical roots as in the case of gravitation (in electrodynamics, there was no relation similar to the fact of the equality of the inertial and gravitational masses). Ultimately, this led to the need to postulate an ideal nature of the connection between electromagnetism and geometry and the impossibility of verifying it by physical measurements.

The youthful Pauli attempted to strengthen the physical component of Weyl's theory but soon (not later than the middle of 1920) concluded that there were insuperable difficulties for it in the solution to the problem of matter; this conclusion applied not only to Weyl's theory but, quite generally, to all classical field theories, in particular to those based on the program of unified geometrized field theories. Pauli believed that it was not possible to avoid the introduction of an aspect of discreteness from the very beginning. Effectively he had in mind some form of a quantum theoretical program.

In 1920, however, the position of the unification program, of which Weyl's theory became the first paradigm, was rather strong. In 1921 a whole series of theories created after the example and manner of Weyl's theories came into existence.

It is interesting that the first unified geometrized field theory, which actually gave rise to the unification program as a whole, was developed neither by Einstein, the creator of the geometrical concept of physical interactions and an undoubted adherent of the field-theoretical ideal of the unity of physics, nor by Hilbert, the author of the first unified field theory based on general relativity. Discussing the preceding scientific work of Weyl, the decisive influence of the Göttingen tradition and especially Hilbert and Klein on him, the direct contacts with Einstein, the philosophical enthusiasms of Weyl, and his interest in intuitionism, we have attempted to show that Weyl was not a figure that appeared by chance in this history. In many of his mathematical studies (especially in the field of geometry, foundational questions, and the epistemology of mathematics) and in his physical investigations and philosophical ideas at the end of the second decade of this century, there were certain thematic resonances that to some degree explain why it was Weyl who became the author of the first unified geometrized field theory.

CHAPTER 4

1921: The Pivotal Year of Unified Geometrized Field Theories

Until 1921, Weyl's theory was essentially the only unified geometrized field theory. Therefore, it would clearly be premature to speak of the research program of unified geometrized field theories as a scientific-historical reality. A certain conception acquires the character of a research program when a whole series of different theoretical schemes is developed on its basis.

Just such a situation arose around 1921. Weyl's theory continued to remain at the center of interest of investigators, despite the criticism of its physical foundations, above all by Einstein and Pauli. In 1921, Weyl himself sketched a new form of his theory that appeared less vulnerable. Parallel to this, Eddington in Cambridge advanced his "affine" unified field theory, in which the coefficients of an affine connection, rather than metrical quantities, played the fundamental role. Einstein also attempted to modify Weyl's theory and proposed a generalization of it associated with the recognition of the fundamental significance of ratios of the components $g_{\mu\nu}/g_{\lambda\sigma}$ (and not the components $g_{\mu\nu}$ themselves) and the abandonment not only of the assumption that the length of a rod is independent of the path along which it is transported but also the assumption that such rods exist at all. Eddington completed his fundamental paper in February 1921, Einstein his in March. Finally, in December there appeared a further form of unified geometrized field theory based on five-dimensional Riemannian geometry. This was the famous theory of Kaluza, which opened up the direction of five-dimensional geometrical constructions. All these theories in fact followed Einstein's idea of geometrization of a field and Weyl's concept of a geometrical synthesis of fields. This set of theories, developed in the spirit of Weyl's theory, does enable us to speak of the appearance of the program of unified geometrized field theories, or the transformation of the project of this program into a real phenomenon of the historical process in the development of scientific knowledge.

Anticipating somewhat, let us briefly describe the main features of the program. Weyl's theory, the first form of which was considered in detail in the previous chapter, gives a clear picture of it. In its "maximalistic" expression, the program (understood as a "maximum program") was undoubtedly global. In other words, it had the aim of solving the most fundamental problems of physics on the basis of the construction of a unified physical theory. In this respect it followed the tracks of the classical mechanical and electromagnetic field programs. Beginning with the unification of gravitation and electromagnetism on a geometrical foundation ("minimum program"), it then strove to explain the corpuscular forms of matter and the quantum properties of radiation and matter. Thus, the geometry of the space-time continuum, also manifested in the form of classical fields, was advanced to the foreground. The discreteness inherent in many physical phenomena (atomistics, electron physics, quantum theory) was assumed to be secondary and reducible to continuum physics (classical fields are regarded as a manifestation of a single space-time continuum). On the one hand, the method of this reduction went back to the tradition of the electromagnetic field program; the space-time continuum was likened to an electromagnetic ether, and particles had to be interpreted as certain "bunches," "vortices," "knots," and other particle-like formulations of a continuum. On the other hand, the basic equations of the unified field, clearly nonlinear and having a large number of field variables (from 14 to 40) and also, possibly, an order higher than two, had to be sufficiently complicated to contain within them certain particle-like solutions together with quantum-type behavior of them. These equations must be determined by the invariance properties of the geometry of the space-time continuum. Thus, the search for a suitable mathematical structure (different types of affine-connection geometries, five-dimensional Riemannian geometries, etc.) would make it possible, it appeared, subsequently to derive, purely deductively, the entire manifold of physical phenomena.

Besides geometricality and the continuum assumption, the program of unified geometrized field theories also contained the idea of a correspondence between the structure of physical reality and some ideal mathematical structure (one can say there was the idea of a "pre-established harmony" between these two structures) and the concept of classical determinism. This last feature was associated with the fact that the state of a physical system in the general theory of relativity and in the unified theories was described by a finite set of field variables (for example, gravitational and electromagnetic potentials) at a certain space-time point, while the evolution of the state was described by a system of partial differential equations.

From the point of view of this program, the discrete and probabilistic nature of many physical phenomena, especially in the field of atomic physics, was reducible to continuous and deterministic structures. In the searches for an adequate geometry of the world and the associated differential equations of the unified field, the experimental aspects receded into the background. In the overwhelming majority of cases, they had only to fulfill the function of a touchstone of the theoretical constructions. The manifold of solutions of the basic equations must ultimately be identical to the manifold of observable phenomena. The principal, creative role in the realization of the program was ascribed to mathematics, specifically geometry, and certain fundamental principles of physics bordering on methodological principles (causality, symmetry, simplicity, conservation, etc.). The characteristic feature of the program were: the aim of the unity of physical knowledge (in its maximalistic expression), i.e., the construction of a unified physical theory; the idea of a "pre-established harmony" between the mathematical structure and physical reality and the assigning of the main creative role to mathematics; the continuum (a classical field as primary reálity); geometricality (the classical fields should be geometrized, like the gravitational field in general relativity, and the unified field was identified essentially with the space-time continuum); and classical determinism (naturally associated with the classical nature of the unified field).

We note, in addition, that both general relativity and the unification program were inseparably associated with an extension of the relativity principle, i.e., the idea of relativism and invariance (symmetry) were undoubtedly among the pivotal ideas of the program. Since the theories based on this program had as their aim the identification of the primordial, fundamental physical essence and the associated mathematical structures, one can also say that fundamentalism was inherent in the program. Of course, any global research program, for example, the classical mechanical or electromagnetic field programs, expresses fully the tendency to unity and synthesis and possesses the property of fundamentalism. As a rule, these features are not present in program that do not have global aims and are concerned with more special problems.

Despite the theoretical attraction of the unified geometrized field theories program and the successes of general relativity, which laid the foundation of the program (its observational confirmations and cosmology), the overwhelming majority of physicists at the beginning of the 1920s preferred to work on more local problems. One of these, however, acquired a more and more global nature—the quantum theoretical program, which aimed at the construction of a consistent quantum theory of atoms and radiation.

THE QUANTUM THEORETICAL PROGRAM

We shall briefly describe the state of quantum theory in 1921, and we shall then characterize the quantum theoretical program in general terms.

There is no universally accepted view as to when the old quantum theory entered a state of crisis. Heisenberg, for example, in 1927, soon after the crisis had been resolved, assumed that the break occurred in 1923. Until then, in his opinion, the theory had developed very successfully, although

> Bohr himself drew attention to the existence of these frontiers (i.e., the frontiers of the "new territory—where logical difficulties prevented new advance") already in his first paper devoted to the hydrogen atom. ([146], p. 113)

Another participant in these events, Hund, believed that the turning point came during 1921–1922; indeed, 1921 was from his point of view a "sort of high point in quantum theory," while in 1922 "they began to talk of 'crisis' " ([147], pp. 87–88 of the English translation). In October 1921 Sommerfeld, one of the founders of quantum theory, wrote to Einstein that the introduction of internal quantum numbers made it possible to elucidate "details of the quantum magic in spectra" and that thus "in spectroscopy the day, or rather, probably the dawn, comes" ([68], pp. 228–229). At the same time, he recognized that although "everything works out in the details" the "deep foundations [of quantum theory] remain obscure" (from the letter to Einstein on January 11, 1922 ([68], pp. 228–229). In the third edition of Sommerfeld's famous book *Atombau und Spektrallinien* (*The Structure of Atoms and Spectra*), which was completed at the beginning of 1922, optimistic notes are sounded but the empirical nature of the quantum-theoretical systematization of spectra is emphasized [148].

The successes of spectroscopy based on quantum theory were indeed impressive. "By 1921," wrote Hund subsequently, "the n, l model of simple spectra was understood in order-of-magnitude terms" ([147], p. 95 of the English translation). In 1921, Sommerfeld introduced the internal quantum number j and discovered a formal way to explain the "fine structure" of terms; his ideas were used by Landé and Bohr, who related this quantum number to the total angular momentum of the atom ("By 1921, the structure of the terms of simple spectra was understood by modeling them as the combination of the angular momentum of the path and of the atomic core" ([ibid.], p. 99). In the same year Wentzel completed the systematics of x-ray spectra on the basis of the same quantum numbers n, l, j.

The greatest triumph of the old quantum theory, however, was justly assumed to be the explanation, developed by Bohr in 1921, for the properties of the elements of the periodic table. Hund wrote that it was received "as a sort of 'high point of quantum theory.' " Bohr based his work on the systematics of the simple spectra and the correspondence principle, which he used with great virtuosity.

The Third Solvay Congress, which took place in April 1921, Bohr's lecture in Copenhagen in October 1921, and especially the "Bohr festival" in Göttingen, which was planned for spring 1921 but actually took place in June 1922, appeared as the drawing up of a balance sheet of the development of quantum theory, which at this time reached an "apogee" but soon exhibited clear and increasingly alarming signs of crisis. The younger generation, for example Pauli and Heisenberg, did not receive Bohr's theory with such optimism. Heisenberg, recalling the Göttingen festival, wrote subsequently that Bohr's theory appeared to a large degree a qualitative, nonrigorous, and incomplete construction.

> One could sense directly that Bohr had not obtained his results by calculation and proofs but by feel and guesses, and that it was now difficult for him to defend them in front of the school of higher mathematics at Göttingen. ([149], p. 59; [150], p. 61)

In the discussion between Heisenberg and Bohr that took place during their walk in the neighborhood of Göttingen and that so strongly influenced Heisenberg, Bohr clearly understood, according to Heisenberg's recollections, the preliminary nature of the old quantum theory and saw many of its difficulties. Referring to his reconstruction of the elements of the periodic table, Bohr said:

> These images are derived, or, if you wish, guessed, on the basis of available data and are not obtained from the position of any theoretical calculations. I hope that these pictures describe the structure of the atom as accurately, but at the same time only as accurately, as is possible using the transparent language of classical physics. ([149], p. 63; [150], p. 65)

In answer to these words, Heisenberg noted that physics must nevertheless be an exact science and that the system that had taken shape could not be regarded as satisfactory. Bohr agreed and said:

> It must be expected that the paradoxes of quantum theory, the features that we do not understand associated with the stability of matter, will be brought into an ever clearer light with each new discovery. If this happens, then there is the hope that with the passage of time new concepts will arise by means

of which we can in some way or other also understand these processes in the atom in which we cannot achieve clarity. However, we still have a long way to go to this. ([149], p. 63; [150], p. 66)

Even in 1924 the majority of physicists thought in exactly the same way, believing that one would still have to work for a long time with this eclectic theory, which subsequently became known as the "old quantum theory." In fact on the very eve of the appearance of quantum mechanics, Born, having completed in November 1924 the preparatory work for the publication of his *Vorlesungen über Atommechanik* (*Lectures on Quantum Mechanics*), which contained an exposition of the theoretical foundations of quantum theory, called the book a first volume and noted that the second volume, in which he intended to give a rigorous deductive exposition of quantum mechanics, will "probably remain unwritten for many years yet" ([151], p. 3 of the Russian translation). As is well known, however, not more than a year had passed before Born (together with Jordan) developed, on the basis of Heisenberg's brilliant paper, the first form of such a "deductive" theory, subsequently called the matrix formulation of quantum mechanics.

Let us return to the events of 1921: the accumulation of empirical successes and achievements of a formal and qualitative nature continued. Some of them were very significant, although they received a complete clarification only after the creation of quantum mechanics. Stern proposed an experiment to test the principle of "spatial quantization" directly in a study of the motion of atoms in an inhomogeneous magnetic field. He soon successfully completed the corresponding experiments, in collaboration with Gerlach. These experiments also established the quantization of the magnetic moment of atoms.

In 1920–1921, doublet and triplet terms and their splitting in a magnetic field (the anomalous Zeeman effect) were discovered. Back and Landé developed a systematics of the effect. Pondering the anomalous Zeeman effect at the end of 1921, Heisenberg discovered half-integer quantum numbers for the first time; in light of the old quantum theory, these appeared very mysterious. He subsequently recalled:

> The results I obtained were to the highest degree unexpected. Instead of integer numbers, we had also to allow their halves as quantum numbers, in complete contradiction to the spirit of quantum theory and the Sommerfeld mystique of integer numbers. ([149], p. 55; [150], p. 57)

About one and a half years later, there arose, on the basis of the work of Back, Landé, and Heisenberg, the Landé vector model, which explained the phenomenological systematics of the multiplets and their Zeeman splittings. In this way physics had come right to the threshold of the discovery of spin.

The successes of the theory were permanently linked to its difficulties; crisis phenomena were also noted by many in the year of the high point of the old quantum theory. "Pride and doubts," wrote Hund subsequently, "were very closely intertwined" ([147], p. 106 of the English translation). For all that, the crisis phenomena did remain in the background of the uninterrupted progress of the quantum theoretical program, especially in the interpretation of atomic spectra and the periodic table of the elements; this was true at least until 1922–1923, when the further advancement slowed down considerably.

The main contradiction to the old quantum theory, which was recognized by many before the beginning of the 1920s (above all by Bohr himself), was the use of the concepts and formalism of classical mechanics and electrodynamics in a domain in which they were invalid in principle. As Born wrote in 1924,

> The application of classical theory to atomic processes leads to a contradiction with stability [of the atom], as a result of which there arose the problem of creating a single "atomic mechanics" that did not contain these contradictions. ([151], p. 3 of the Russian translation)

Despite this, the theory continued to achieve new successes, especially in the hands of Bohr himself. Heisenberg explained this subsequently (in 1927) as follows:

> Characteristic of Bohr's investigations was the way in which he applied the concepts of quantum mechanics qualitatively and only to the extent to which their applicability could be justified by the correspondence principle. Only this freedom in operating with the concepts of classical mechanics— or, it would be better to say, the consequences of classical physics that flowed from them—made it possible to establish a connection between the chemical and spectroscopic properties of atoms. ([146], p. 114)

Thus, the quantum theory of periodic and multiple periodic motion was based on the classical concept of the phase integral $\oint p\, dq$. As Ehrenfest showed in his paper at the Third Solvay Congress in 1921, quantization by means of the phase integral led in a number of cases to clear contradictions, whereas in the same situations the correspondence principle gave the correct result. In this connection, Hund recalled:

> Gradually Bohr came to feel that the phase-integral should not be taken anything like so literally, and in Göttingen this feeling was expressed in the form of a slogan: "up with the correspondence principle, down with $\oint p\, dq$!" ([147], p. 84 of the English translation)

Concerning the beginning of the twenties, Heisenberg wrote in 1927:

> If we study papers on theoretical physics written at that time, then we clearly see how the classical concepts began to totter and how there was a gradual

liberation from the prejudices inherited from the old physics, which, as became obvious, were the very cause of the contradictions. ([146], p. 118)

This fundamental contradiction was manifested with particular sharpness, for example, in the theory of dispersion. In 1921, Epstein noted that the calculation of frequencies characteristic for dispersion led to frequencies of revolution in stationary orbits, in disagreement with the quantum-theoretical absorption and emission frequencies observed in reality. A promising direction appeared to be associated with the application of perturbation theory to systems more complicated than the hydrogen atom, for example, to the helium atom or the ion of the hydrogen molecule. Heisenberg, recalling his discussions with Pauli, wrote:

> Wolfgang set himself a very difficult task. He wanted to test whether Bohr's theory and the Bohr–Sommerfeld quantum conditions led to experimentally correct results when one considered a more complicated system that could be calculated using astronomical methods. The point was that at the time of our Munich discussions we wondered whether the previous successes of the theory were restricted to extremely simple systems and whether the theory would fail in the study of more complicated systems, which Wolfgang intended to examine. ([149], p. 56; [150], p. 58)

Such work was already being done by Bohr and his collaborators, who were then joined by Pauli, and also Epstein, who had applied perturbation theory to problems of a periodic external potential and the dispersion of light. The fact that this path also led to a blind alley became clear in 1922–1923 when, first, Kramers calculated the ground state of the helium atom and obtained an incorrect value for the ionization energy, and then Born and Heisenberg obtained a discrepancy with experimentation for excited states of the helium atom calculated on the basis of perturbation theory. The calculation of the states of the molecular hydrogen ion H_2^+ made by Pauli in 1922 also did not agree with the spectroscopic terms.

There was also no satisfactory explanation found for the anomalous Zeeman effect and the multiple structure of spectra. As Heisenberg emphasized in 1927, although "Landé's formulae were an extremely fruitful means for bringing order into complicated spectra—their model interpretation was still unknown," and from their mathematical structure "it was seen that classical mechanics could not lead to these formulae even in any new model" ([146], p. 114).

We have already mentioned the qualitative nature of Bohr's construction of the periodic table of elements. Many of its details remained unclear, for example, the electron numbers 2, 8, 18, and 32 associated with the closing of shells, the details of the multiplet structure of spectra, etc. A significantly deeper understanding of the problem was achieved only through the discovery of the Pauli principle and the electron spin (1925).

No less complicated was the situation regarding the light quanta, which in Bohr's opinion suggested a possible violation of the energy–momentum conservation law in microscopic processes. For a long time the efforts of Bohr and the majority of adherents of the old quantum theory were directed toward the problems of the structure of the atom and atomic spectra. The attitude toward the hypothesis of light quanta was generally skeptical, especially on the part of Bohr, although in his Solvay lecture in 1921, which was read by Ehrenfest, Bohr noted:

> Such a concept [i.e., of light quanta] seems on the one hand, to offer the only possibility of accounting for the photoelectric effect if we stick to the unrestricted applicability of the idea of energy and momentum conservation. On the other hand, however, it presents apparently insuperable difficulties from the point of view of the phenomena of optical interference.... ([152], p. 19)

In other words, Bohr accepted the thought that one could retain classical notions of light by supposing the possibility of the validity of the laws of energy–momentum conservation only on the average, i.e., statistically. These ideas acquired particular popularity in 1922–1923. In his "Translator's supplements" to Bohr's *Three Papers on Spectra and the Structure of Atoms*, translated into Russian in 1923, Vavilov wrote:

> Further, one must not forget that the actual initial idea of Bohr's theory, of stationary states and the resulting discontinuity of the absorption of light, contradicts the laws of mechanics and electrodynamics in the most striking manner. The phenomenon of the dispersion of light in a material medium, for example, can be reconciled with the existence of stationary states only by the abandonment of the law of conservation of energy in an elementary system, as has been recognized recently by several investigators. ([153], p. 152)

(Vavilov cited papers of Webster, 1920, Darwin, 1922, Bohr, 1923, and Born and Heisenberg, 1923.)[1]

It is interesting to note in this connection that the corpuscle–wave dualism of light was unacceptable to many theoreticians. Einstein, who made the most important contribution to this concept, in December 1921, proposed an experiment relating to the elementary process of the emission of light and intended to establish precisely which concept of light was correct, the wave

---

[1] Vavilov characterized the crisis situation that had arisen in quantum theory as

or the corpuscle concept [154]. Sometime later, however, under the influence of criticism by Ehrenfest and von Laue, Einstein recognized that his idea was wrong [152].

In a book written with Holst, Kramers, one of the most eminent scientists among Bohr's collaborators, summarized the difficulties of the old quantum theory two or three years before the discovery of quantum mechanics, i.e., precisely at the very beginning of the crisis we are considering:

> Bohr has not succeeded in finding basic laws whose mathematical formulation can completely replace the laws of electrodynamics and which can be used to derive and explain atomic processes and other phenomena of nature. The motion of an electron in a given stationary state can be calculated to a large degree using the laws of mechanics. However, these laws are not capable of explaining why certain orbits are preferred to others, why electrons jump from outer to inner orbits, why electrons sometimes go over from one stationary orbit to a neighboring orbit and sometimes jump over several orbits; it is also unclear why the electron cannot approach closer to the nucleus than a distance equal to the diameter of the first orbit, and, finally, why transitions are accompanied by radiation whose frequency is determined by the above rule.

The fundamental problems were not exhausted by this list. The authors continued:

> We are immeasurably far from being able to describe an atomic mechanism enabling us to follow, for example, the entire motion of the electron in the atom or understand the role of the stationary states and their connection with the whole, and not as a *deus ex machina*. We know nothing about the

follows:

> The unexpected exacerbation of the crisis that has unfolded during the last decades in theoretical physics arises in this case from the collision between empirical data, the principles of the theory of quanta, and the remnants of classical notions. The abandonment of these classical notions is all the more difficult in that we have nothing definite to put in their place. If, for example, the model of helium calculated on the basis of the laws of mechanics and Coulomb's law gives an incorrect value of the ionization potential and is unstable, then we do not yet know on the basis of what new laws the same model will give correct results for the value of the ionization potential and be stable. At the very least, the failure with the helium model deprives Bohr's theory of a powerful tool for investigation, the methods of classical mechanics, and the entire theory is reduced almost to intuitive guessing of true relationships. ([153], p. 152)

behavior of the electron between transitions from one stationary state to another, etc. ([155], pp. 98–99 of the Russian translation)

Planck, summarizing a decade of development of the old quantum theory and giving its successes their due, nevertheless noted: "At the least, there can be no talk at the present time of any satisfactory solution of the problems that have arisen through the introduction of the theory of quanta in atomistics" ([156], p. 43). Also interesting is Bohr's point of view, formulated, for example, in the foreword to *Three Papers* at the beginning of 1922:

> The considerations developed here [essentially the theoretical constructions contained in the Copenhagen lecture of October 8, 1921] have a manifestly incomplete nature, not only with respect to the development of the details but also in the sense of the employed theoretical concepts. . . . It appears that in every success achieved in the field of the structure of the atom the well-known "mysteries" of the theory of quanta become even more pronounced. It is desirable that . . . the reader should receive an impression of the particular attraction of the study of atomic physics precisely on account of the existence of these "mysteries." ([157], pp. 9–10 of the Russian translation)

Thus, according to Bohr, progress in quantum theory did not consist of the successive solution of the fundamental mysteries, listed, for example, in the above statement of Kramers and Holst, but in an ever greater exacerbation of these mysteries and an ever clearer formulation of them. Thus, the "indissolubility" of the "pride and doubt" and of the "apogee" and "crisis" was a very characteristic feature of the development of quantum theory of 1920–1922. In fact, somewhat later, when the feeling of a crisis of the old quantum theory was overwhelming, namely in 1924, Sommerfeld nevertheless highly estimated the importance of this theory, which, in the words of Planck, attracted "an ever increasing number of fresh courageous forces":

> Whatever may be the solution of this burning problem [i.e., the problem of the corpuscular or wave nature of light], whatever the changes there may be in our notions in the future, there is no doubt that quantum theory and Bohr's model of the atom will always remain in one form or another in the treasure-house of the achievements of physics. ([158], p. 15)

We now briefly discuss the question of the sense in which one can speak of the quantum theoretical program as a global research program competing with the program of unified geometrized field theories. At first glance, the quantum theoretical program, which is based (in the "hard core"), on the one hand, on the quantum theory of radiation of Planck and Einstein and, on the other, on the Rutherford–Bohr atomic model and the old quantum theory, was oriented toward the solution of the problem of atomic structure and the theoretical interpretation of atomic spectra, and also questions of the interaction of light with matter (atoms).

For all their importance, these problems did not appear as global as that of the construction of a unified field theory capable not only of unifying the electromagnetic and gravitational fields on a single geometrical basis but also, on the same basis, of obtaining the basic elementary particles (above all the electron) and the quantum features of their behavior. The adherents of quantum theory recognized more and more strongly the fundamental nature of the quantum ideas and the fact that they could not be reduced to classical form, be it classical mechanics or classical electrodynamics. This recognition raised the quantum theoretical program to a global level, giving it the status of an alternative to the unified geometrized field theories program. The well-known physicist Haas was one of the first to apply the quantum conception to the problem of the connection between the structure of the atom and the laws of spectra (the Haas atomic model, 1910). He clearly emphasized the program nature of the old quantum theory (in the global respect) in the lectures that he gave in 1920 on the physical picture of the world, first in Vienna and then in Leipzig:

> Both relativity theory and the theory of quanta, which arose almost simulta-
> neously, are not simply a part of physics but physics as a whole, considered
> from a quite new point of view. The theory of quanta does not have as
> its subject of investigation a certain group of physical phenomena but the
> entire field of physics; it applies to it a new fundamental principle, which
> has proved to be hardly less fruitful than relativity theory. ([159], p. 68 of
> the Russian translation)

Having begun with the theory of thermal radiation, the quantum concept embraced several new fields in the following two decades, relating mainly to the microscopic structure of radiation (the electromagnetic field) and matter (solids, molecules, atoms), and to the interaction of radiation with matter. The energy distribution in the black body spectrum, the phenomena of fluorescence and photochemical reactions, the photoelectric effect (in both the optical and x-ray regions of the spectrum), the behavior of solids at low temperatures, the anomalous behavior of the specific heats of solids, a wide range of phenomena in atomic and molecular spectroscopy, the structure of atoms and molecules, and the explanation of the periodic table of the elements—this is a list by no means complete of the phenomena and branches of physics to which quantum theory had made a very weighty contribution during the fifteen to twenty years of its existence.

Because of this, quantum theory attracted both theoreticians and experimentalists, chemists, people working in optics, physicists concerned with the problems of solids, and those interested in the problem of the structure of the atom in the first place. After the creation in 1913 of Bohr's theory of the atom, which of course made essential use of the ideas of Planck and Einstein about the quantum of action and the quantum nature of radiation, quantum theory appeared to develop in two directions: (1) the quantum theory of light and quantum statistics and (2) the structure of the atom and spectra. Moreover, the leader of the second direction, Bohr, like many of his students and collaborators, ignored studies relating to the first direction, or even could be said to have had a negative attitude to the very idea of light quanta. Einstein, however, who had contributed so much to the development of the first direction, still continued at the beginning of the 1910s to regard light quanta as a key concept of quantum theory corresponding to physical reality.

It should be said that by no means all of those who worked in the first direction shared Einstein's view about light quanta, for example, Lorentz, Planck, and others (they were apparently the majority). There were also not a few physicists who contributed to development in both directions, for example Planck, Ehrenfest, and others. Among those who worked in one way or another on the stream of quantum theory there was no single understanding of its fundamental propositions or of its position in the system of physical theories. With the passage of time, however, an increasing number of physicists came to the firm conviction that the quantum of action represented a primary, fundamental concept, an "elementary physical quantity that in a certain sense is measured by a physical process" ([159], p. 69 of the Russian translation), and that the specific quantum properties of radiation and atoms could not be reduced to the principles of classical physics. This applied equally to the notion of a quantum, atomistic structure of radiation and the extremely mysterious behavior of electrons in atoms associated with nonemission during motion in stationary orbits and emission of quantum portions in jumps from one stationary orbit to another.

Therefore, it is difficult to speak of the quantum theoretical program as an integral entity, at least during the decade and a half before the appearance of quantum mechanics. A strong idealization would be to consider this program as having a two-component core (one component being the Planck–Einstein quantum theory of radiation, with allowance, however, for the fact that the majority did not recognize the real existence of quanta of radiation itself, and the second component being the Bohr theory of the atom with its underlying basis of the "Bohr postulates," the correspondence principle, adiabatic principle, etc.). In reality, there was not yet any clear picture of the program in those years, especially since the old quantum theory, the Bohr–Sommerfeld theory, was partly based on classical mechanics and electrodynamics and was itself modified in the process of its development. Therefore, before the creation of quantum mechanics, i.e., before 1926–1927, one can only speak of a certain prototype of a quantum theoretical program.

Nevertheless, as quantum theory developed, there gradually appeared specific features of the program that were in sharp contrast with programs based on theories of a classical type. Above all, the concept of atomism, discreteness, was developed further and very significantly. This feature was common to both the directions mentioned above. Independent of the question of the reality of Einstein's light quanta, all agreed on one thing—that the Planck–Einstein theory of quanta "decisively breaks with previous views; in it, an assumption of the existence of discontinuity is introduced for the first time in the formulation of general laws of nature" ([160], p. 319 of the Russian translation). This statement is taken from Bohr's lecture at Copenhagen on October 1921. Both postulates of Bohr's theory are, in the words of its creator, "of the form in which the theory of quanta is applied to problems of the structure of the atom" and are a direct succinct expression of the discrete structure of the atom and atomic processes (for example, the emission and absorption of light).

The fundamental status of the concept of action was also recognized in classical physics, but there it was a quantity that varied continuously in space and time and depended continuously on other physical variables. Quantum theory emphasized for the first time that the "action of physical processes is made up of elementary quanta of action" ([159], p. 69 of the Russian translation).

The nonclassical nature of quantum theory had been established with considerable clarity by the beginning of the 1920s, and it was expressed not only in discreteness but also to a certain extent in the special role of probability in the formulation of the laws of microscopic processes, for example, the laws of interaction between radiation and matter. It is interesting that none other than Einstein, who from the beginning of the 1920s until the end of his life firmly believed in classical causality, regarding probabilistic laws as secondary, was one of the pioneers who introduced probabilistic methods into quantum theory. In two papers on the quantum theory of radiation published in 1916, in which he introduced the coefficients A and B characterizing the probabilities of spontaneous and stimulated transitions, one senses a clear understanding of the probabilistic nature of the connection between the wave and corpuscular theories of radiation. In fact, Einstein regarded this circumstance as a shortcoming even then: "A weakness of the theory is, on the one hand, that it does not lead us to a closer unification with the wave theory and, on the other, that the time and direction of the elementary process are given over to chance" ([161]). In one of his studies, Forman collected extensive material demonstrating that in Germany at the end of the 1910s a disenchantment with the ideal of classical causality arose in the community of natural scientists, and the preconditions were created for the adoption of the idea of a fundamentally probabilistic causality, which was completely developed only in the framework of quantum mechanics (the probabilistic interpretation of quantum mechanics by Bohr, 1926) [162]. Indeed, at the beginning of the 1920s the idea became more and more popular that only the assumption of a statistical nature of the energy–momentum law could reconcile optical phenomena, such as the dispersion of light in a material medium, with the Bohr theory of the atom. Although this idea was soon shown to be incorrect, its popularity indicates that in those years physicists were prepared to abandon classical causality in favor of its statistical equivalent.

From the point of view of rigor and logical consistency, the old quantum theory was, as we have already noted, unsatisfactory to a large degree. It used fragments of classical mechanics and electrodynamics, and too much was postulated; extremely paradoxical notions that clearly did not fit in the framework of classical physics were combined with rather simple, transparent models of mechanical type. Essentially, the mathematical formalism was taken from classical physics. Many rules had a mainly empirical origin. Vavilov wrote at the beginning of 1923, in the foreword to the Russian edition of *Three Papers on Spectra and the Structure of the Atom,*

> New unexpected laws and rules were needed, the complicated and loosely related conglomeration of which is called the theory of quanta. . . . To make successful advances in one field of phenomena, it is necessary to close one's eyes to a neighboring field and forget its existence for a time. In this sense, the theory of quanta is opposite to the theory of relativity, with its elegance, consistency, and breadth.
>
> The *raison d'être* of the theory of quanta is its exceptional fruitfulness; it not only explains but also indicates new facts. One can assert with a high probability that the present form of the theory will be short lived. In its present form, it is a collection of empirical rules and facts awaiting a general law ([163], p. 5).[2]

---

[2] Vavilov also compared the position in quantum theory at the beginning of the 1920s with the position in optics in the 18th century, when the dominant theory, Newton's corpuscular theory of light, explained the empirically clearly formulated laws of interference by means of the notion of strange periodic "fits," "which lost all their strangeness in the transparent language of the wave theory of Young and Fresnel." "The atom of modern physics," continued Vavilov, "with its discrete stationary states, quantum absorption, etc., also suffers from similar 'fits.' " ([163], p. 6)

Thus, we have a predominance of the experimental–empirical aspect over the theoretical; this was characteristic for the prototype of the quantum theoretical program (in what follows, we shall call this prototype simply a program, bearing in mind, however, that between the quantum theoretical programs of 1921 and, say 1926–1927, there was in reality a huge difference).

The empiricism of quantum theory—and, more broadly, the quantum theoretical program as a whole—was especially striking against the background of the well-developed theoretical framework of the general theory of relativity and the program of unified geometrized field theories. In a comparison of the quantum theoretical program with the unified field program, we can, besides the already noted contrasts, also mention the following, on which we shall not dwell. First, as we have already noted, the primordial "quanticity," the discreteness of the structure of fields (above all the electromagnetic) and of the structure of microscopic processes, was associated with a prominence being given to the impossibility of their reduction to a continuum concept, be it the electromagnetic field or the space-time continuum. This aspect was in sharp contrast with the geometrical of the unification program.

Second, the insufficiently developed and unsystematic theorization of the old quantum theory and its conceptual eclecticism were also manifested in the very small part played in it by the ideas of symmetry and relativism. It was only after the foundations of quantum mechanics had been created that Dirac, von Neumann, and others developed a systematic formulation of quantum theory, after which it achieved a level of theorization comparable with that of the general theory of relativity. The absence of invariance concepts in quantum theory contrasted with the geometrical, invariant formalism of unified geometrized field theories.

Finally, the problems that faced the quantum theory, both solved and unsolved by it, were quite specific and special—explanation of spectral laws, dispersion, absorption and scattering of light, etc. A global problem such as the construction of a unified physical theory was not directly posed in the framework of the quantum theoretical program. Nevertheless, by the beginning of the 1920s a path to the attainment of an ideal of unity contained within the program gradually started to appear. What we are referring to is in essence the construction from elementary discrete elements such as electrons and light quanta a world of atoms and molecules by means of the concept of the "elementary quantum of action." The attainment of this unity appeared to be a very distant prospect, however, not only because numerous phenomena of the type mentioned above still remained unexplained, but also because it was still necessary to find many laws and a single theoretical basis of them. Thus, in contrast to the program of unified geometrized field theories, with its characteristic axiomatic and deductive approach to the problem, the quantum theoretical program had an inductive approach to realization of the ideal of unity.

As we have seen, the aims of the two programs were in many respects opposed:

Unified Field Program	Quantum Theoretical Program
Continuity (classical field)	Discreteness, atomism (concept of quanta)
Classical, unique causality	Probabilistic causality (or willingness to give up classical determinism)
High level of theorization, strongly mathematical structure	Theoretical eclecticism, predominance of experimental–empirical aspect
Geometrization (geometrical nature of fundamental entities)	"Material" nature of fundamental entities
Relativism, symmetry, invariance	Absence of well-developed approach based on invariance
Axiomatic–deductive method of solution of the problem of unity	Inductive path to unity

The Development of Weyl's Theory

In the previous chapter, we considered in detail the original form of Weyl's theory, the criticism of the theory by Einstein and Pauli, and certain improvements to it, made mainly by Weyl himself. Here we shall discuss a new form of the theory, the initial development of which was made in 1919 [164] and which already existed in its developed form in 1921 [165]. This form, without significant modifications, was included in the final, fifth edition of Weyl's book *Raum-Zeit-Materie* (1923) [64].

We start by summarizing the main difficulties of Weyl's theory (in its original form) as discussed in the previous chapter.

(1) The abandonment, under the criticism by Einstein, of a direct relationship between geometric characteristics ("the metric field") and the readings of measuring instruments. Weyl's theory then appeared, in the words of Pauli, "to have been robbed of its inherent convincing power from a physical point of view" ([63], p. 196).

(2) The absence of a physical basis for the geometrization of the electromagnetic field matching the equality of the inertial and gravitational masses (or the equivalence principle) in the case of the gravitational field.

(3) In Riemannian geometry, the "infinitesimal" aspect was associated only with the direction of a vector. In the geometry of Weyl, which he himself called "pure infinitesimal geometry" (*reine Infinitesimalgeometrie*), this aspect was developed further, being extended to the length of a vector. Even more radical would be the "infinitesimalization" of the angle between vectors or affine properties. There was no convincing justification of the level adopted in the theory for the infinitesimalization.

(4) The deep physical nature of the geometrization of the gravitational field was also manifested in the fact that the world lines of material points and light rays were geodesics of the Riemannian space. In Weyl's theory, this correspondence did not hold in the general case.

(5) There were several difficulties associated with the field equations. The method of obtaining the equations, based on a variational principle and the requirement of invariance, led to fourth-order equations that in the absence of an electromagnetic field did not reduce to the generally relativistic equations of gravitation, violating the correspondence principle. In addition, the choice of the Lagrangian was not unique (although the number of possible forms was significantly less than Mie's theory by virtue of the requirement of gauge invariance).

(6) The fourth order of the field equations led to great computational difficulties. New solutions were not obtained. As a result, the theory did not give new consequences capable of experimental testing. A not insignificant achievement was Pauli's proof that the Schwarzschild solution of the field equations of general relativity was also a solution of the field equations of Weyl's theory.

(7) Static and spherically symmetric solutions of the field equations that were everywhere nonsingular, i.e., solutions that could be interpreted as charged particles, were not found. In addition, the theory was, in principle, invariant with respect to a change in the sign of the charge and thus did not appear to be able to explain the actually existing asymmetry of electricity.

As we see, these difficulties can be divided into two groups. The first group is associated with a certain lack of connection between the physical and geometrical foundations of the theory, and the second is associated with the field equations and their solutions. Taking into account cosmological aspects, Weyl found, as it appeared to him, a way to modify the theory that, while preserving its geometrical basis would, in part at least, eliminate the shortcomings of both groups. This meant that the theory then became significantly more complicated, losing the simplicity and attraction of the original theory. We shall consider Weyl's construction in more detail.

It began with a discussion of the fact on which Einstein based his main objection against the first form of the theory, i.e., the fact of the constancy of the spectral frequencies of atoms or, in a different formulation, the fact of the constancy of the mass ratios of atoms, for example, hydrogen and oxygen, independent of the point of space-time at which the ratio is determined. In the first place, it postulated from the very beginning that atomic masses, the periods of clocks, and the lengths of rods did not behave in accordance with a certain inertial tendency (*Beharrungstendenz*) but in accordance with a tendency to adjustment (*Einstellungstendenz*):

> The preservation of this mass ratio must, thus, be based on the fact that each atomic mass separately, at each instant, is adjusted to have a definite ratio to the value of a known *field quantity* that has the dimensions of length (= mass). ([64], p. 299)

It would be natural to take as this quantity the radius of curvature of the space-time continuum.

Here Weyl used the field equations of general relativity with cosmological term and right-hand side in the form of the energy–momentum tensor of the electromagnetic field:

$$\left(R_{ik} - \tfrac{1}{2}g_{ik}R\right) - \lambda g_{ik} = -\kappa S_{ik}. \tag{1}$$

Because S_{ik} has vanishing trace, contraction of the left- and right-hand sides of this equation, respectively, gives the value

$$R = -4\lambda \tag{2}$$

for the scalar curvature. This expression was also used in one of the first form of a unified geometrized field theory developed by Einstein. Incidentally, as Einstein noted, it enabled one "to regard the world as spatially closed without recourse to an additional hypothesis" [93], i.e., the cosmological constant was now not introduced as an *ad hoc* hypothesis, acquiring an interpretation through the scalar curvature. Einstein himself and Pauli rejected this theory, but the idea of the corresponding interpretation of the cosmological constant was taken up by Weyl in the second form of his theory. The same relation (2) was one of the initial points in the rearrangement of the theory.

Taking as a basis "infinitesimal" geometry, or Weyl's geometry, one could assume that rods, clocks, and atomic masses were adjusted to the scalar curvature F of the Weyl space-time. This adjustment had the consequence that the original infinitesimal geometry became unobservable. The correspondence procedure of parallel transport of rods was, as Weyl noted, "falsified" by the readings of clocks, being replaced by "adjustment to the world curvature" ([64], p. 299). As a result, there was a doubling of geometry; as Weyl wrote:

> Thus, there is a difference between the original *geometry of the ether*[3] and the geometry constructed from the readings of measuring instruments, the so-called *natural geometry*.... This natural geometry arises from the former because the infinitesimal congruent displacement of intervals is replaced by their adjustment to the curvature. ([64], p. 299)

Eddington called the gauge expressed by the relation

$$F = -4\lambda \tag{3}$$

the "natural gauge." By this doubling of the geometry, Weyl succeeded in preserving the geometrical structure of the theory that so elegantly related the relativity of length and the scalar vector φ_i to the electromagnetic potential, and he was able to give a well-argued answer to the "Einstein objection" and explain the discrepancy between the physical and geometrical foundations of the theory.

Weyl then considered the problem of the field equations. Taking as basis the invariant relations of the "geometry of the ether," he chose as the most suitable expression for the action the "naturally measured" volume

$$\int \sqrt{g F^n} \, dx,$$

which in the four-dimensional case ($n = 4$) is equal to

$$W = \int \lambda^2 \sqrt{g} \, dx. \tag{4}$$

By simple transformations of the variational integral (4), Weyl reduced it to the form

$$\int \left[G + \lambda - \tfrac{3}{4}(\varphi_i \varphi^i) \right] \sqrt{g} \, dx \tag{5}$$

---

[3] At the beginning of the 1920s there was a certain rehabilitation of the ether concept. In 1921 Einstein wrote about the "ether of the general theory of relativity" ([166]). Weyl also made wide use of this concept, identifying it, essentially, with the concept of the space-time continuum.

(G is the Riemannian scalar curvature). Variation of this action with respect to g_{ik} gives the ordinary equations of gravitation with a cosmological term, while variation with respect to φ_i gives the equations

$$\varphi_i = 0. \tag{6}$$

Thus, the variational principle for the action (4) is consistent with the original "natural" gauge (2), but does not give the equations of the electromagnetic field. Therefore Weyl, having identified φ_i with the electromagnetic potential, and the scale curvature (*Streckenkrümmung*)

$$f_{ik} = \frac{\partial \varphi_k}{\partial x_i} - \frac{\partial \varphi_i}{\partial x_k}$$

with the electromagnetic field strengths, arrived at the need to add to the action (4) or (5) the Maxwell component with Lagrangian $\sqrt{g}M$,

$$\frac{1}{4} \int \sqrt{g} f_{ik} f^{ik} \, dx. \tag{7}$$

As a result, the basis of the theory was now the Lagrangian

$$\mathcal{L} = \left[G + \alpha M + \frac{1}{4} [1 - 3(\varphi_i \varphi^i)] \right] \sqrt{g}. \tag{8}$$

Here α is a certain numerical constant, and the cosmological constant is determined by the normalization $\lambda = \frac{1}{4}$, which means that the scales of measurement have "cosmic" order. If one goes over to a different gauge, in which the coordinates x_i' have, for example, the order of the dimensions of the human body ($x_i = 2\varepsilon x_i'$, where ε is a very small constant),

$$g_{ik}^i = g_{ik}, \quad \varphi_i' = 2\varepsilon\varphi_i, \quad \lambda' = -\varepsilon^2, \tag{9}$$

then the Lagrangian (8) can be rewritten in the form

$$\mathcal{L} = \left\{ (G + \alpha M) + \varepsilon^2 [1 - 3(\varphi_i \varphi^i)] \right\} \sqrt{g} \tag{8'}$$

(at the same time $(ds')^2 = \frac{1}{4}\varepsilon^2 (ds)^2$, and $\frac{1}{\varepsilon}$ is the radius of curvature of the world). Thus, from the expression (8') it can clearly be seen that for such a choice of the action we must, apart from very small terms of a cosmological nature having the order ε^2, obtain the field equations of gravitation and the electromagnetic field corresponding to the general theory of relativity and Maxwell's theory.

Despite the "doubling of the geometry" ("geometry of the ether" and "natural geometry"), the apparent abandonment of the more direct method of constructing the field action using a quadratic Lagrangian, and a certain artificiality of the new construction of the Lagrangian, Weyl believed that the basic idea of the theory could be preserved. Exactly repeating one of the main conclusions of the paper of 1921, he wrote in the fifth edition of his famous book:

> The extension of the world geometry due to the principle of gauge invariance leads in conjunction with the use of an action principle constructed in a simple and sensible manner from the quantities of state of the metric field to consequences that agree with experiment and make redundant the extra physical field assumed hitherto, in addition to the metric, in the form of the electromagnetic field. ([64], p. 304)

"Descartes's dream of a purely geometrical physics," noted Weyl somewhat earlier, "is realized in a remarkable manner, which, of course, was not foreseen by him ..." ([64], p. 302). Fully understanding that all the ingenious devices that made possible the geometrization of the electromagnetic field were due to the absence of an electromagnetic analog of the equivalence principle, he simultaneously attempted to give physical and mathematical arguments against the very existence of such an analog.[4]

These considerations essentially demonstrated that the geometrization of gravitation and electromagnetism were not on an equal footing, reducing the physical significance of the theory still further. First, Weyl emphasized that the electromagnetic field strengths f_{ik} did not depend on the gauge (in contrast to the gravitational field strengths, which depend on the frame of reference), and therefore the "ponderomotive [i.e., mechanical] effect of the electromagnetic field on charged bodies is in no way associated with the gauge" ([64],

---

[4]   Here, we still feel a certain dissatisfaction. In Einstein's case, the connection that maintains the direction of inertia and gravitation could be transparently and directly expressed by means of the equality of the inertial and gravitational masses, or the "equivalence principle." Is there a corresponding transparent basis for the connection formulated here between congruent transport and the electromagnetic field? It is generally assumed that there is no such transparent basis. However, we can at least explain why it must be absent. ([64], p. 305)

p. 305). Second, this difference was due to the different structures of the electromagnetic and gravitational Lagrangians. If, Weyl noted, the gravitational action were completely analogous to the electromagnetic action (for example, were expressed by $\int R_{ik}R^{ik}\sqrt{g}\,dx$), then in general relativity too there would appear forces that were not proportional to the mass, forces "whose connection with the guiding field could not be made convincing by means of the equivalence principle" ([64], p. 305). In the second form of the theory, electromagnetism was not associated with variability of the "metric connection" (i.e., the Weyl connection)—and in this it differed fundamentally from gravitation, interpreted as "variability of the affine connection"—but mainly with the structure of the action and was interpreted as a "concomitant phenomenon." The role of an analog of the gravitational field was taken over by a field generated by the "cosmological" component of the Lagrangian

$$\varepsilon^2 \sqrt{g}\left[1 - 3(\varphi_i\varphi^i)\right], \tag{10}$$

manifestations of which were characterized by the second order in ε and were therefore too small to obtain experimental confirmation. Finally, the entire construction of the theory associated with the doubling of geometry and the fundamental distinction of the processes of "congruent transport" and the real behavior of rods and clocks, namely, their adjustment to the radius of curvature, excluded an effect of the electromagnetic potentials on the scale behavior of bodies and, thus, a physical basis for an electromagnetic equivalence principle. The same considerations made it possible to understand the nongeodesic motion of charged particles.

According to Weyl, the justification for augmenting the principle of general covariance with the principle of relativity of magnitude (or the principle of scale relativity) was ultimately the possibility of achieving a unified geometrization of gravitation and electromagnetism. It is true that to achieve consistency of the theory with experimentation and the original field theories (general relativity and electrodynamics) it was necessary to complicate the entire structure to such an extent that its physical persuasive power was significantly undermined.

As we have seen the problem of the field equations was solved in the theory with much more care. Weyl abandoned a Lagrangian quadratic in the curvature and, thus, fourth-order equations. Correspondence with the field equations of general relativity and electrodynamics could then be readily and naturally established. The choice of the Lagrangian was determined by the need to ensure this correspondence and only partly by the geometrical invariance arguments. With regard to the four quadratic invariants in the

Lagrangian mentioned above, augmented subsequently by two more, Weyl now wrote: "In contrast to the opinion that I previously expressed, I am now almost convinced that they do not play any part in nature" ([64], p. 317). Thus, the numerous difficulties of a physical and mathematical nature associated with the fourth-order equations led to their abandonment. The choice of the geometrical part of the Lagrangian ("the naturally measured volume") was now determined rather naturally, but the addition of the electromagnetic part appeared artificial, suggesting an *ad hoc* structure. In fact, the only difference in the new field equations was replacement of the ordinary cosmological term $\varepsilon^2 \sqrt{g}$ by the expression (10),

$$\varepsilon^2 \left[1 - 3(\varphi_i \varphi^i) \right],$$

whose contribution to the observable consequences of the theory must be negligible in view of its extreme smallness. As a result, Weyl's theory in the new modification had even less hope of experimental confirmation going beyond the framework of general relativity and classical electrodynamics.

Weyl emphasized the possibility of reconciling the theory with the cosmologies of both Einstein ("cylindrical world") and De Sitter. At the same time, in the framework of the Einstein cosmology, one could speak of static centrally symmetric solutions with arbitrary masses and charges, which left no serious hope for a satisfactory solution to the problem of matter. In fact, by 1921 Weyl's continuum radicalism had diminished appreciably. Noting that the "theory gives no explanation for the unequal status of positive and negative electricity," he emphasized:

> However, this cannot be regarded as a defect of the theory, since this difference is without doubt based on the fact that in the two main particles of matter, the electron and the nucleus of the hydrogen atom, the positively charged particle is associated with one mass, and the negatively charged particle with another. This difference is due to the nature of matter rather than the field. ([64], p. 308)

The final sentence indicates that Weyl now allowed a certain dualism of field and matter and an irreducibility of particles to the field.

In a paper *"Feld und Materie"* ("Field and matter") published in 1921, he considered two possible mutual relationships of field and matter (understanding by "matter" charged elementary particles and the atoms and molecules associated with them): (1) the field conception of matter in the spirit of Mie's theory and the maximum program of unified geometrized field theories; (2) the dualistic conception, in accordance with which the field is uniquely determined by the matter. Weyl wrote in this paper:

From the first to the third edition of the book *Raum-Zeit-Materie*, I completely adhered to the first point of view, the reason for which was the beauty and unity of the purely field theory. In the fourth edition [prepared for publication at the end of 1920] I became convinced, however, of the incorrectness of the field theory of matter and gave preference to the second point of view. ([167], p. 242)

The fifth edition of Weyl's book ended with the following words:

Whoever looks back over the ground that has been traversed, leading from the Euclidean metrical structure to the mobile metrical field which depends on matter, and which includes the field phenomenon of gravitation and electromagnetism ... must be overwhelmed by a feeling of freedom won — the mind has cast off the fetters which have held it captive. Our ears have caught a few of the fundamental chords from that harmony of the spheres of which Pythagoras and Kepler once dreamed. We were not able to make our analysis of space and time without considering the details of the structure of matter. But here we stand before a mystery, the solution of which cannot be expected from the physics of the field. In the mist that still surrounds the problem of matter, the first ray of the rising sun is, perhaps, quantum theory. ([64], p. 317)

This fifth edition was the last. During the five years that had elapsed since the appearance of the first form of his unified theory, Weyl's views had changed considerably. At the beginning of the 1920s he evidently no longer had especially great hopes for the program of unified geometrized field theories (in its maximalistic aim) as the strategic high road of physics. In grappling with the problem of matter, his attention was drawn more and more to the rapidly progressing quantum theoretical program, although in the eyes of the majority of physicists he was still one of the leaders of the field program.

EDDINGTON'S THEORY

Eddington, one of the founders of modern astrophysics, was probably one of the first British scientists to study general relativity, and he immediately became a convinced supporter of it. He learned about the theory from the Dutch astronomer De Sitter, to whom in 1916 Einstein had sent his fundamental studies on the general theory of relativity.[5] De Sitter sent Eddington not only a copy of Einstein's papers, but also his own studies [169]. The classical

---

[5] The correspondence between Einstein and De Sitter, which occurred at just that time, 1916–1917, is extremely important for understanding the history of the development of relativistic cosmology [168].

investigations of De Sitter into relativistic cosmology were published in the *Monthly Notices of the Royal Astronomical Society*, whose secretary at that time was Eddington [170].

In 1918 he published his *Report on the Relativity Theory of Gravitation* especially for the London Physical Society [171]. This was probably the first extensive exposition of general relativity in the English language. The following year, two British expeditions, organized at his initiative and under his leadership, convincingly confirmed the relativistic deflection of light rays in the gravitational field of the sun during the eclipse of May 29, 1919 [172].

In 1920 Eddington published his brilliant popular science book *Space, Time, and Gravitation*, which was soon translated into many languages, including Russian (1923) [173]. Already in this book, as in subsequent editions of the *Report on the Relativity Theory of Gravitation*, Eddington gave a high estimation of Weyl's theory, although he did note some difficulties with it, for example, those associated with the choice of a Lagrangian that decomposes into gravitational and electromagnetic terms and the assertion that a necessary condition for applicability of the field equations of general relativity was the absence of electromagnetic fields.

On February 19, 1921 Eddington submitted to the *Proceedings of the Royal Society of London* a paper that contained a further generalization of Weyl's theory and opened up a new stage in the development of unified geometrized field theories. "I believe that Weyl's geometry," wrote Eddington in the introductory part of his paper, "far-reaching though it is, yet suffers as from an unnecessary and harmful restriction" ([174], p. 104). He explained:

> In passing beyond Euclidean geometry, gravitation makes its appearance;
> in passing beyond Riemannian geometry, electromagnetic force appears;
> what remains to be gained by further generalization?

"Clearly the non-Maxwellian binding forces which hold together an electron," was Eddington's answer to his own question, evidently hoping that the generalization undertaken by him would open up a path to the solution of the problem of the electron. It is true that he made a reservation here that "the present paper does not seek these unknown laws, but aims at consolidating the known laws." In other words, as yet he was going to consider only the development of a new geometrical scheme and its agreement with already known phenomena, above all general relativity and the theory of the electromagnetic field. Naturally, the central task was the unification of gravitation and electromagnetism on a unified geometrical basis. In this first paper, Eddington regarded the geometrical generalization he had thought up as "a new light . . . thrown on the origin of the fundamental laws of physics" ([174], p. 105) and not merely as a mathematical simplification of the whole scheme. In a more extended exposition of his theory, completed one and a half years later (in August 1922) [11], Eddington's position on this question had undergone certain changes, about which we shall speak later. The sequence of development of the mathematical skeleton of the theory and the main physical formulations as they were developed in the paper of 1921 are almost completely repeated (but with certain additions) in the book [11] just cited.[6]

In constructing Riemannian geometry and general relativity, and also his own unified field theory, Weyl made essential use of the concept of infinitesimal parallel transport, which is the basis of affine-connection geometry. But in both general relativity and Weyl's unified theory the basic geometrical quantities (identical to the basic physical variables characterizing gravitation and the electromagnetic field) were ultimately the metric characteristics g_{ik} (or g_{ik} and the four-vector φ_i).

In contrast, Eddington, without denying reality to the observed metric

$$ds^2 = g_{ik}\, dx_i\, dx_k,$$

showed that it can be constructed from the coefficients $\Gamma^\alpha_{\mu\nu}$ of the affine connection, which acquire their elementary geometrical meaning in the operation of infinitesimal parallel transport:

$$dA^\mu = -\Gamma^\mu_{\nu\alpha} A^\alpha\, dx_\nu, \tag{11}$$

where A^μ is the vector of an infinitesimal displacement transported parallel to itself from the point $P(x_\mu)$ to the infinitely close point $P'(x_\mu + dx_\mu)$. This is the most general continuous expression for the variation of A_μ, the smallness of A_μ and dx_μ making it possible to ignore the higher-order terms. The system of quantities $\Gamma^\mu_{\nu\alpha}$ is not a tensor and in the general case contains 64 components. Despite the nontensorial nature of $\Gamma^\alpha_{\mu\nu}$ and the too great number of them, it was these quantities that Eddington regarded as the fundamental geometrical characteristics and, simultaneously, physical variables, taking the relation (11) as the basis of his entire construction:

> We are going to build the theory afresh starting from this notion of infinitesimal parallel displacement.... Our fundamental axiom is that parallel displacement has some significance in regard to the ultimate structure of the world. ([11], p. 213)

---

[6] Thus, §§91–94 of the book completely reproduce the section of the paper entitled "Geometrical theory." The second section of the paper, "Physical theory," exactly corresponds to §§95 and 96 of the book, although the subsequent material is presented somewhat differently from §§97–101.

What was the basis for the fundamental nature of the relation (11), which, in Eddington's opinion, could not be confirmed experimentally? A natural, practical consideration from the point of view of an adherent of the program of unified geometrized field theories would be that an affine-connection geometry, permitting the development of the entire formalism of the theory of curvature, is the next stage of generalization in a series of four-dimensional geometries: Euclidean geometry → Riemannian geometry → Weyl's geometry → affine-connection geometry. Some difficulties and shortcomings of Weyl's theory could be due to the somewhat unfortunate, insufficiently general nature of the "world geometry." In 1921, affine-connection geometry could be regarded as the most general and, at the same time, sufficiently nontrivial geometrical scheme capable of accommodating not only gravitation but also electromagnetism and possibly other as yet unknown physical phenomena.

Eddington still wished to find arguments, possibly of a philosophical or methodological nature, for the generalization he had chosen. Such arguments were absent in the 1921 paper, but are contained in the book *The Mathematical Theory of Relativity,* published at the end of 1922 (§98). Despite his support for the program of unified geometrized field theories, Eddington assumed that physical reality has "*structure* and *substance*," and that as regards substance "we can only give a name to it.... (Any attempt to do more than give a name leads at once to the attribution of structure)," whereas "structure can be described to some extent" ([11], §98). Understood as a "complex of relations," relations that in principle are comparable (otherwise, as Eddington acutely noted, all parts of it would be "alike in their unlikeness," and there could not be "even the rudiments of a structure)", a structure of continuum type can be introduced in the simplest way by means of the relation (11).[7] "Thus," he summarized, "our axiom of parallel displacement is the geometrical garb of a principle which may be called 'the comparability of approximate relations' " [ibid.]. At the end of the book (§103), this thought is expressed as follows:

---

[7]

The axiom of parallel displacement is the expression of this comparability, and the comparability postulated seems to be almost the minimum conceivable. Only relations which are close together, i.e., interlocked in the relation-structure, are supposed to be comparable, and the conception of equivalence is applied only to one type of relation. This comparable relation is called displacement. By representing this relation graphically [or geometrically] we obtain the idea of location in space; the reason why it is natural for us to represent this particular relation [i.e., the expression (11)] graphically does not fall within the scope of physics. ([Ibid.])

The idea is that the affine connection is the most general structure coming within the scope of continuous analysis (a contention not fully demonstrated) and may therefore be used as a basis of prediction. ([11], note 15 to §103)

It should be admitted that a justification of this kind, even in the 1920s, did not sound too convincing, especially since soon there developed different generalized geometrical systems (for example, geometries involving projective and conformal connections).

Sensible geometrical arguments make it possible to require symmetry for the coefficients of the affine connection,

$$\Gamma^{\alpha}_{\mu\nu} = \Gamma^{\alpha}_{\nu\mu}, \tag{12}$$

which reduces the number of basic variables from 64 to 40. Actually, it is under this assumption that an affine-connection geometry is obtained.

They [i.e., these 40 quantities $\Gamma^{\alpha}_{\mu\nu}$] are descriptive of the relation-structure of the world, and should contain all that is relevant to physics. Our immediate problem is to show how the more familiar variables of physics can be extracted from this crude material. ([11], §91)

This problem is solved as follows. First, in the well-known manner, by considering parallel transport of a displacement along a small closed contour, an expression is derived for the curvature tensor of fourth rank (the affine analog of the Riemann–Christoffel tensor):

$$\delta A^{\mu} = -\tfrac{1}{2} \iint *B^{\mu}_{\varepsilon\nu\sigma} A^{\varepsilon} \, dS^{\nu\sigma}, \tag{13}$$

where

$$*B^{\mu}_{\varepsilon\nu\sigma} = -\frac{\partial}{\partial x_{\sigma}}\Gamma^{\mu}_{\nu\varepsilon} + \frac{\partial}{\partial x_{\nu}}\Gamma^{\mu}_{\sigma\varepsilon} + \Gamma^{\mu}_{\nu\alpha}\Gamma^{\alpha}_{\sigma\varepsilon} - \Gamma^{\mu}_{\sigma\alpha}\Gamma^{\alpha}_{\nu\varepsilon}, \tag{14}$$

the asterisk ($*$) to the left of the symbol of this tensor denoting the tensorial nature of this quantity with respect to gauge transformations (in Eddington's terminology, the *in-tensor* nature).[8] It is obvious that in the constructed theory an extremely important part must be played by the corresponding curvature tensor of second rank (the analog of the Ricci tensor):

$$*G_{\mu\nu} = *B^{\sigma}_{\mu\nu\sigma}. \tag{15}$$

---

[8] Naturally, in affine geometry all the basic tensors not associated with the use of the metric are in-tensors.

Contracting the tensor $*B^{\varepsilon}_{\mu\nu\sigma}$ with respect to the indices ε and μ, we obtain one further important antisymmetric second-rank tensor:

$$2F_{\nu\sigma} = \frac{\partial\Gamma_{\nu}}{\partial x_{\sigma}} - \frac{\partial\Gamma_{\sigma}}{\partial x_{\nu}}, \tag{16}$$

where $\Gamma_{\nu} \equiv \Gamma^{\alpha}_{\nu\alpha}$.

It can be seen from expression (15) that the tensor $2F_{\mu\nu}$ is the antisymmetric part of $*G_{\mu\nu}$:

$$*G_{\mu\nu} - *G_{\nu\mu} = 2F_{\mu\nu}. \tag{17}$$

Thus, as Eddington noted ([11], pp. 215–216), the tensors

> $*B^{\varepsilon}_{\mu\nu\sigma}$ and $*G_{\mu\nu}$ are the most fundamental measures of the intrinsic structure of the world.... They take precedence of the $g_{\mu\nu}$ [i.e., the metric], which are only found at a later stage in our theory.

In physics, we undoubtedly deal with metrical quantities, but the construction employed by Eddington is rather complicated. The transition to an operationally measured metric is made in two stages. First, a metric is introduced formally, but this is not yet the metric that is measured directly. "Length as here defined is not anything which has to be consistent with ordinary physical tests" ([11], p. 217). It is only subsequently that, introducing the so-called natural gauge, Eddington shows "how $g_{\mu\nu}$ must be chosen in order that conventional length [defined in the first stage] may obey the recognized physical tests and thereby become physical length" ([11], p. 217).

For the time being we formally introduce the symmetric second-rank tensor $g_{\mu\nu}$ and metric ds^2:

$$ds^2 = g_{\mu\nu}\,dx_{\mu}\,dx_{\nu}. \tag{18}$$

Now let A_{μ} be some interval (vector); its length is then expressed by the formula

$$l^2 = g_{\mu\nu}A_{\mu}A_{\nu}. \tag{19}$$

Under infinitesimal parallel transport, the square of this length changes as follows:

$$\delta(l^2) = \left(\frac{\partial g_{\mu\nu}}{\partial x_{\sigma}} - \Gamma_{\sigma\mu,\nu} - \Gamma_{\sigma\nu,\mu}\right)A^{\mu}A^{\nu}\,dx^{\sigma}. \tag{20}$$

It is easy to show that the quantity in the parentheses forms the tensor $2K_{\mu\nu,\sigma}$. Forming the following combination from $K_{\mu\nu,\sigma}$ and denoting it by $S_{\mu\nu,\sigma}$,

$$S_{\mu\nu,\sigma} = K_{\mu\nu,\sigma} - K_{\mu\sigma,\nu} - K_{\nu\sigma,\mu}, \tag{21}$$

we obtain the expressions

$$\Gamma_{\mu\nu}^{\sigma} = \{\mu\nu, \sigma\} + S_{\mu\nu}^{\sigma}, \tag{22}$$

for $\Gamma_{\mu\nu}^{\sigma}$, where $\{\mu\nu, \sigma\}$ are the Christoffel symbols.

In Weyl's geometry, in particular,

$$K_{\mu\nu,\sigma} = g_{\mu\nu}\varphi_{\sigma},$$

and the coefficients of the affine connection are correspondingly specialized. Thus, Weyl's geometry can, as Eddington notes, be regarded as a special case of generalized affine geometry.

If the vector $S_{\mu\alpha}^{\alpha}$ is denoted by $2\varphi_{\mu}$, then, as is readily shown,[9]

$$F_{\mu\nu} = \varphi_{\mu\nu} - \varphi_{\nu\mu}. \tag{23}$$

The fundamental tensor $*G_{\mu\nu}$ can be represented as a sum of its antisymmetric, $F_{\mu\nu}$, and symmetric, $R_{\mu\nu}$, components:

$$*G_{\mu\nu} = R_{\mu\nu} + F_{\mu\nu}. \tag{24}$$

Having developed the geometrical formalism of the theory, which suggests, after comparison with Weyl's theory, that the symmetric tensor $R_{\mu\nu}$ is responsible for gravitation and the antisymmetric tensor $F_{\mu\nu}$ (and the vector φ_{μ}) for the electromagnetic field, Eddington starts to fill this geometrical scheme with physical content. The main problem here, as in the second form of Weyl's theory, was to reconcile the ideal "geometry of the world," or "geometry of the ether," which is an affine geometry, with the real physical geometry of space-time that we find by means of measurement procedures.

This is achieved by the introduction of the so-called natural gauge of the world, taking the form of a specialization of the metric tensor $g_{\mu\nu}$, which has hitherto been arbitrary, in such a way "that the lengths of displacements agree with the lengths determined by measurements made with material and optical appliances" ([11], p. 219). In accordance with Eddington's idea, $g_{\mu\nu}$ must be constructed from the basic building material contained in the "world

---

[9] Eddington shows by simple calculation how $\Gamma_{\mu} = \Gamma_{\mu\alpha}^{\alpha}$ and φ_{μ} are related: $\Gamma_{\mu} = \frac{\partial}{\partial x_{\mu}}(\ln \sqrt{-g}) + 2\varphi_{\mu}$.

geometry." The tensor $*G_{\mu\nu}$ or rather its symmetric part $R_{\mu\nu}$, is the only such tensor. This justification for expressing the "natural length" by means of it:

$$l^2 = R_{\mu\nu}A^\mu A^\nu.$$

This leads, with allowance for expression (19), to a relation that is the basis of the "natural gauge":

$$\lambda g_{\mu\nu} = R_{\mu\nu}, \tag{25}$$

where λ is a certain universal constant.

Further, it is shown that measurements of length made not only by means of rigid rods and clocks but also by optical methods lead to relation (25). Thus, if the natural gauge is used, tensors $g_{\mu\nu}$ and $K^\sigma_{\mu\nu}$ are fundamental variables. In Eddington's opinion, however, the metric is the unique fundamental property of physical space, in terms of which the remaining geometrical characteristics must be expressed. Therefore, he concludes, "if $K^\sigma_{\mu\nu}$ does not vanish, then there is something else present not *recognized* [our italics] as a property of pure space," and thus, $K^\sigma_{\mu\nu}$ "must therefore be attributed to a 'thing' " ([11], p. 221).[10]

If $K^\sigma_{\mu\nu} = 0$, then $R_{\mu\nu}$ reduces to the ordinary Ricci tensor $G_{\mu\nu}$, and the equation of the natural gauge reduces to the equation of the gravitational field characteristic of general relativity (for empty space):

$$G_{\mu\nu} - \lambda g_{\mu\nu} = 0.$$

As a result, the need for a special action principle to derive the equations of gravitation disappears.

As we have seen, the idea of a natural gauge was advanced by Weyl in the second form of his theory in order to eliminate the difficulty associated with the Einstein objection. Weyl subsequently wrote,

> Eddington extended my idea that the radius of curvature of the world determines the gauge (*Eichmass*) to the curvature tensor R_{ij}, and this had the consequence that the world ... must be described, not by means of a metric, but by means of the 40 quantities $\Gamma^i_{kl} = \Gamma^i_{lk}$, the coefficients of the affine connection. ([18], p. 430)

---

[10] The tensor $K^\sigma_{\mu\nu}$ is a more general characteristic of the electromagnetic field than $F_{\mu\nu}$ or φ_μ. In the special case of Weyl's geometry,

$$K_{\mu\nu,\sigma} = g_{\mu\nu}\varphi_\sigma.$$

In the fifth edition of *Raum-Zeit-Materie*, in Appendix IV, which was specially devoted to Eddington's theory, Weyl pointed out that Eddington's procedure of "natural gauging" could be interpreted as adjustment to the curvature. As a result, in this theory one could say that a small measuring rod remained unchanged "not only when displaced to another position but also when rotated at a given place around the initial point" ([64], p. 323). Thus, in Eddington's theory, a scale was established, or adjusted, in each direction differently, namely, in accordance with the radius of curvature corresponding to the given point and given direction. This more sensitive (than Weyl's theory) adjustment of scales made it possible to reconcile the affine geometry, understood as an unobservable ether geometry, with the natural geometry found experimentally and given by general relativity, i.e., Riemannian geometry.

Eddington implemented the geometrization program on the basis of a "principle of identification" of the elements of the geometrical structure developed by him with fundamental physical quantities. He notes:

> If we can do this completely, we shall have constructed out of the primitive relation-structure a world of entities which behave in the same way and obey the same laws as the quantities recognized in physical experiments. ([11], p. 222)

The first step in this process of identification was the introduction of the metric $g_{\mu\nu}$ through the tensor $R_{\mu\nu}$. This in fact led to the gravitational equation of general relativity for empty space. It was then necessary to identify "things" (the energy and momentum of "matter," the electromagnetic field strengths, charges, and currents) with geometrical entities. These identifications were made in such a way that the experimentally known and theoretically verified equations and relations of the type of conservation laws, Maxwell's equations, etc., were satisfied.

Thus, the energy–momentum tensor of matter was identified with the Einstein tensor,

$$-8\pi T_{\mu\nu} = G_{\mu\nu} - \tfrac{1}{2} g_{\mu\nu}(G - 2\lambda),$$

and as a result equations of the gravitational field in the general case of the natural gauge were obtained that obviously agreed with equation (25). The tensor of the electromagnetic field was identified with the tensor $F_{\mu\nu}$, which by virtue of its geometrical properties satisfied (identically) the first half of Maxwell's equations:

$$\frac{\partial F_{\mu\nu}}{\partial x_\sigma} + \frac{\partial F_{\nu\delta}}{\partial x_\mu} + \frac{\partial F_{\sigma\mu}}{\partial x_\nu} = 0.$$

If, further, in accordance with the charge conservation law ($J^\mu_\mu = 0$), the four-dimensional current vector J^μ was identified with the divergence of an antisymmetric contravariant tensor, this tensor being taken to be $F^{\mu\nu}$ ($J^\mu = F^{\mu\nu}_\nu$), then the second pair of Maxwell's equations was also obtained.

Eddington emphasized that, due to the absence of clear ideas about the structure of the electron, it was not possible to introduce as naturally and readily into the geometrical scheme an equation of motion for an electron under the influence of the electromagnetic field and an expression for the force that then acts on it.[11] Thus fundamental physics, specifically the theory of gravitation and electromagnetism, acquired a unified geometrical description, and the field equations were satisfied automatically, as a result of the identification procedure that was developed. These equations did not differ from those that occurred in general relativity and classical electrodynamics, just as the geometry of physical space-time remained Riemannian, despite the use of the fundamental entities of affine geometry in the construction. The achieved geometrization differed from the one to which Weyl strove in the first form of his theory, and even from the weakened modification of it given in the second form of Weyl's theory.

Although quite definite geometrical characteristics of the affine space corresponded to the electromagnetic quantities, the new, extended geometry remained unobservable, and in this respect the geometrical status of the gravitational and electromagnetic fields was fundamentally different. At the end of his 1921 paper, Eddington wrote:

> It is a natural impression that in the light of the more general theory Einstein's work [i.e., general relativity] must be regarded as only a close approximation. The impression is wrong, for the present paper leads to the conclusion that Einstein's postulates and deductions are exact. The natural geometry of the world ... is the geometry of Riemann and Einstein, not Weyl's generalized geometry or mine. ([174], p. 121)

Weyl's idea of a doubling of geometry found its complete expression in Eddington's theory. On the one hand, complete (or almost complete) geometrization was achieved on the basis of an unobservable geometry of the ether (or world geometry), which was an affine-connection geometry and was described by the tensors $*B^\varepsilon_{\mu\nu\sigma}$ and $*G_{\mu\nu} = g_{\mu\nu} + F_{\mu\nu}$ (for $\lambda = 1$). On the other hand, the adjustment of the scales to the curvature ("natural gauge") had

---

[11] "We can only show in an imperfect way," notes Eddington, "that our tensors will conform to this law, because a complete proof would require more knowledge as to the structure of an electron" ([11], p. 223).

the consequence that the geometry of the real space continued to be Riemannian geometry completely described by the metric $g_{\mu\nu}$.

In the book *The Mathematical Theory of Relativity*, this point of view toward the geometrization of physics was expressed even more distinctly:

> Two possible ways of generalizing our geometrical outlook are open. It may be that the Riemannian geometry assigned to actual space is not exact; and that the true geometry is of a broader kind [for example, Weyl's geometry].... The alternative is to give all our variables...a suitable graphical representation in some new conceptual space—not actual space.... This generalized graphical scheme may or may not be helpful to the progress of our knowledge; we attempt it in the hope that it will render the interconnection of electromagnetic and gravitational phenomena more intelligible. ([11], p. 197)

As we have seen, the transition to Eddington's generalized geometry was associated with hopes of solving the problem of the electron. It appeared to Eddington that, despite the absence of any physical meaning, or, rather, measurement–operational justification, the increase in the number of fundamental variables (from 14 to 40) opened up new possibilities to include in the new geometry new physics associated with the study of the electron, quanta, and the atomic nucleus. In his opinion, the possibilities opened up by the transition from the tensor $*G_{\mu\nu}$ to the fourth-rank tensor $*B^{\varepsilon}_{\mu\nu\sigma}$ were no less rich.

If, however, it turned out that the extra variables $\Gamma^{\sigma}_{\mu\nu}$ or $B^{\varepsilon}_{\mu\nu\sigma}$ ("lumber," as Eddington called them) were not related to the problems of microscopic physics, then they would remain lumber and nothing could be learned about them by means of appropriate instruments. In this case, the lumber would be of no use, but it would do no harm if it was eliminated from the problems for which it was unimportant, for example, by introduction of the natural gauge.[12]

---

[12] Eddington continued these arguments with the following words, which are very interesting from the methodological point of view. The useless variables need not be eliminated from the theory or made to vanish, since

> we are only aware of a selection of the things which *exist* (in an extended meaning of the word), the selection being determined by the nature of the apparatus available for exploring nature.... Therefore, there is no need to insert, and puzzle over the cause of, special limitations on the world-structure, intended to eliminate everything which physics is unable to determine. The world-structure is clearly not the place in which the limitations arise." ([11], p. 228)

As we have seen, the basic equations of physics were obtained as a result of the identification procedure described above and the use of the geometrical properties of the identified tensors. In the main part of his theory, Eddington was actually able to dispense with the action principle. In *The Mathematical Theory of Relativity*, however, he considered, following Weyl, the possibility of using an action principle with generalized volume as Lagrangian:

$$\delta \int \sqrt{-\left|*G_{\mu\nu}\right|}\, d\tau = 0. \tag{26}$$

Calculating the Lagrangian in Galilean coordinates, and then, making use of its invariance, extending the obtained results to the case of an arbitrary coordinate system, he obtained for it the expression

$$\mathcal{L} = \tfrac{1}{4}(R_{\mu\nu}R^{\mu\nu} + F_{\mu\nu}F^{\mu\nu})\sqrt{-g} = \tfrac{1}{4}*G_{\mu\nu}*G^{\mu\nu}\sqrt{-g}. \tag{27}$$

This result was obtained with neglect of fourth powers of $F_{\mu\nu}$, but if terms of higher order are taken into account, then, as Eddington showed, one obtained a certain generalization of electrodynamics, in accordance with which the vacuum possesses definite values, different from 1, of a spurious inductive capacity and permeability, and even a spurious charge and current. "It has seemed worth while," summarized Eddington, "to show in some detail the kind of amendment to Maxwell's laws which may result from further progress of the theory . . . but the present proposals are not intended to be definitive" ([11], p. 235).

Eddington clearly understood that his generalized theory did not, in essence, come to grips with the cardinal problems of microscopic physics associated with the structure of the electron and quanta, although he did not rule out the possibility of a development of the theory in that direction. He did believe, however, that in his theory he had approached the very limits of what "analysis of continuous quantities" could give physics altogether.

Eddington's theory stimulated great interest among those who were concerned with the problem of constructing a unified geometrized field theory. We shall briefly consider the reaction of the chief authorities—Weyl and Einstein. Weyl wrote about Eddington's theory in the paper of 1921 cited earlier [165], and then in somewhat more detail in Appendix IV to the fifth edition of *Raum-Zeit-Materie* [64]. He emphasized that the theory was a successor to the second form of his theory, although he also pointed out the even greater discrepancy from experiment that arose in the new theory. He saw the main shortcoming of Eddington's scheme as the absence of any justification for nondegeneracy of the metric and constancy of the index of inertia of the metric quadratic form (a problem that, in Weyl's opinion, arose in any nonmetric theory). In metric theories, this problem could be solved by invoking group-theoretical arguments.

Weyl also believed that the abandonment of the primary status of the metric and the placing of the affine connection in the foreground did not have any experimental justification, and that Eddington's form of natural gauge associated with adjustment of scales and masses to the curvature differently in different directions did not appear physically convincing. Weyl subsequently wrote twice in historical reviews about Eddington's theory (in 1931 [17], and in 1950 [18]). He emphasized that inertial motion was more probably a projective rather than an affine property and that therefore it could not serve as a physical justification for giving preference to affine geometry, as in Eddington's theory. In the 1931 paper, however, he agreed with Eddington that the affine connection played an exceptional role in the formulation of physical laws. This was also emphasized by Einstein in his Nobel lecture (1923) [175], although he believed that Eddington had not succeeded in finding field equations corresponding to all 40 variables $\Gamma^\alpha_{\mu\nu}$ [176].

KALUZA'S THEORY

In 1921 there arose one more new direction in the development of unified geometrized field theories. This was to play an important role not only in the formation of the geometrized field program for the synthesis of physics but also in its subsequent development. We are referring to the five-dimensional unified theory of gravitation and the electromagnetic field of Kaluza. Whereas Weyl and Eddington in their attempts to construct unified geometrized field theories remained on the ground of a four-dimensional space of events, achieving generalizations of general relativity by giving up a Riemannian metric and going over to different forms of spaces with affine connection, the German mathematician Kaluza found a way to give a unified description of gravitation and electromagnetism in the framework of Riemannian geometry, but using five, rather than four, dimensions.

Theodor Franz Eduard Kaluza was born in Ratibor (today Raciborz in Poland) in 1885, the son of Max Kaluza, a great specialist of the English language and literature, and author of well-known works on the phonetics of the English language and the life and work of Chaucer. Theodor Kaluza entered the University of Königsberg at the age of 18, and after five years (in 1909) had successfully defended a dissertation of the Tschirnhaus transformation under the supervision of the Königsberg mathematician F.W.F. Meyer. The Tschirn-haus transformation was associated with one of the methods of studying the problem of the solvability of algebraic equations in radicals. Kaluza remained at the University of Königsberg as a privatdozent, but did not succeed in re-ceiving a professorship. One was awarded to him after he transferred to the University of Kiel in 1929, i.e., 20 years after the defense of his dissertation. In 1935, he transferred to Göttingen, where he continued to work until the end of his life in 1954 [177].

His interest extended far beyond the limits not only of algebra but also of mathematics. In 1910 he published a paper on the special theory of rela-tivity in the *Physikalische Zeitschrift* [178]. It appears that from then on he maintained a lively interest in the theory of relativity. He worked for several years on his five-dimensional generalization of general relativity. At the very least, it is known that the first sketch of his unified theory was contained in a letter he wrote to Einstein in March or April 1919 ([177], p. 212). Kaluza's famous paper, which initiated the investigations of five-dimensional unified field theories, was presented by Einstein at the meeting of the Berlin Acad-emy of Science on December 8, 1921 [179]. Essentially, it was this paper of Kaluza that made his name widely known among specialists in theoreti-cal physics. His other studies—on algebra, analysis, mathematical physics, methodological aspects of the theory of relativity, and calculations of nuclear models—did not become widely known. In 1938, he and Joos published a textbook on applied higher mathematics that had certain methodological advantages.

Anticipating somewhat, we mention that Einstein regarded Kaluza's idea very highly in the spring of 1919, after he had received the letter giving the sketch of the future theory. Einstein wrote (in a letter dated April 21, 1919):

> The idea to achieve this (the electric field strengths) through a five-dimen-sional cylinder world never occurred to me and is probably altogether new.
> My first reaction to your idea is extremely positive. ([177], p. 212)[12a]

[12a] This letter is published completely in the supplement to the following article: D. Laugwitz, "Theodor Kaluza 1885–1954," in *Jahrbuch Überblicke Mathematik*, 1986, Vol. 19, pp. 179–187. In the same place are listed all known Einstein–Kaluza letters, and are given facsimiles of three known postcards from Einstein to Kaluza. In this issue there is also an article on the modern theories of Kaluza–Klein: E. Mielke, "*Kaluza–Klein Theorien: Wege zur geometrischen Vereinheitlichung fundamentaler physikalischer Wechselwirkungen*," pp. 127–138. Fragments of two letters from Ein-stein to Kaluza (April 28, 1919 and October 14, 1921) are in another article on the modern theories of Kaluza–Klein: D.Z. Friedmann, P. van Nieuwenhuizen, "The Hid-den Dimensions of Space-Time." *Scientific American*, 1985, Vol. 252, No. 3, p. 62.

It is also possible that Einstein made some attempts to obtain a professorship for the Königsberg Privatdozent. In a November 1926 letter to one of his colleagues, he wrote that he highly regarded the creative talent (*schöpferische Begabung*) of Kaluza, who "works under very difficult circumstances" and "would be very pleased if he were to receive an appropriate post." For Kaluza, as for almost all of those who joined in the development of unified geometrized field theories at the beginning of the 1920s, Weyl's theory had decisive importance. Kaluza wrote at the start of his paper:

> Some years ago, Weyl took a remarkably bold step toward the solution of this problem[13]—one of the great ambitions of the human spirit (*den grossen Lieblingsideen des Menschengeistes*). He radically reexamined the foundations of geometry and introduced, alongside the tensor $g_{\mu\nu}$, a fundamental metric vector, interpreting it as the electromagnetic potential q_s. In his theory, this complete world metric is the common source of all the phenomena of nature (*gemeinsamer Quell alles Naturgeschehens*). Here, the same aim is adopted, but a different path is chosen. ([180], p. 966; [179], p. 529)

Kaluza regarded his method of solution as more adequate from the point of view of the principle of the unity of the physical picture of the world. Referring to his theory, he said that in it "a more complete realization of the idea of unity (*eine noch volkommenere Vewirklichung des Unitätsgedankens*)" was achieved, since "both the gravitational and the electromagnetic field are deduced from a *single* universal tensor."

In fact, Kaluza did not begin his construction with this tensor, namely the curvature tensor, or the corresponding metric tensor of the five-dimensional space, but with the coefficients of the affine connection, or Christoffel symbols. An important heuristic argument for him was the analogy, noted by many investigators, between the equations of the gravitational and electromagnetic fields. This analogy had already been noted in 1913 by Einstein, though he considered gravitation in the framework of a tensor (but not generally covariant) theory of gravitation. Einstein showed that if the velocities of the

---

[13] Kaluza characterized the problem as follows:

> To describe the world events in the general theory of relativity, one must use not only the tensor potential of gravitation—the fundamental metric tensor $g_{\mu\nu}$ of the four-dimensional manifold—but also the *electromagnetic four-potential* q_s. The *dualism* of gravitation and electromagnetism which then persists does not rob this theory of its captivating beauty but does throw down a challenge—to overcome it and give a completely *unified picture of the world*. ([180], p. 966; [179], p. 529)

masses were taken into account in the energy–momentum tensor only to the first power, then the equations of the gravitational field and the equation of motion of a material point in this field acquired a form completely analogous to the corresponding equations of electrodynamics. In this case, the gravitational potential was a four-vector. Mentioning that this analogy was important for him, Kaluza referred to a paper of Thirring, who had analyzed it in detail in the framework of general relativity [181].

The chain of arguments that led Kaluza to five dimensions was as follows. The analogy between the equations of the gravitational and electromagnetic fields, in particular a certain structural similarity between the tensor of the electromagnetic field strengths $F_{\mu\nu}$ and the Christoffel symbols, identified as the gravitational field strengths, suggested that the former were a certain degenerate form of the latter, but since in a four-dimensional space all co-efficients of the affine connection were identified with the gravitational field strengths and were combinations of derivatives of the gravitational potential, a way forward could consist in the use of a five-dimensional Riemannian space. These correspondences and analysis, Kaluza emphasized,

> suggest strongly the suspicion that the quantities $\frac{1}{2}F_{\kappa\lambda} = \frac{1}{2}(q_{\kappa\cdot\lambda} - q_{\lambda\cdot\kappa})$ could be certain "deformed" (*verstümmelte*) three-index quantities $\begin{bmatrix} i\lambda \\ \kappa \end{bmatrix} = \frac{1}{2}(g_{i\kappa\cdot\lambda} + g_{\kappa\lambda\cdot i} - g_{i\lambda\cdot\kappa})$.[14] Allowing such a thought, the investigator quite definitely finds himself on a path that initially appears rather uninviting. Namely, since in the four-dimensional world no other three-index quantities exist apart from those already used as components of the gravitational field strengths, the suggested view of the $F_{\kappa\lambda}$ can hardly lead to anything else but what must be a very offputting decision: to call on a new *fifth dimension of the world* for assistance (*wohl stark befremdenden Entschluss... eine neue fünfte Weltdimension zu Hilfe zu rufen*). ([180], p. 967; [179], p. 529)

Of course, as Kaluza noted further on, the real four-dimensional space of events can be regarded as embedded in the corresponding five-dimensional Riemannian space, just as two-dimensional formations are embedded in three-dimensional space. Then, bearing in mind that "all experience accumulated by physics gives no indications of such an additional world parameter," it is necessary to require that the derivatives of all characteristics of the state of systems with respect to the fifth coordinate must vanish. This requirement, which of course may also be satisfied approximately, Kaluza called the *cylinder condition* (*Zylinderbedingung*). Of course, this condition in no way meant that all the additional connection coefficients had to vanish.

---

[14] An index with a dot in front of it denotes differentiation with respect to the corresponding coordinate.

Kaluza may have been prompted to the use of five dimensions in the problem of geometrical unification of gravitation and electromagnetism by Nordström's earlier similar attempt. Nordström by 1914 had unified his scalar theory of gravitation with classical electrodynamics in the framework of an elegant five-dimensional scheme. As we saw in Chapter 1, he considered an antisymmetric tensor of second rank in a five-dimensional space (the physical meaning of the fifth coordinate remained obscure), six components of which were interpreted as electromagnetic field strengths, and the remaining four as the four-vector of the gravitational field strength for which a corresponding generalization of Maxwell's equations was given. His scheme also used an analog of the cylinder condition. Nordström emphasized the formal nature of the unification achieved by him, although he hoped that it "could have a deeper basis." We emphasize once more that this was essentially the first unified geometrized field theory that used a curved space-time, since his metric differed from a pseudo-Euclidean metric:

$$ds^2 = \Phi \sum_i dx_i^2.$$

Nordström's paper [72] was published in the well-known journal *Physikalische Zeitschrift*, in which, not long before this, Kaluza's own paper on relativity theory had been published [178]. Moreover, Nordström's generalization had not passed unnoticed. It was discussed in Abraham's review article on theories of gravitation (1914) [113], and also in *Raum-Zeit-Materie*. Although Kaluza in his paper refers only to Weyl's first paper on unified field theory (in the *Sitzungsberichte der Preussichen Akademie der Wissenschaften*) [86], which does not contain references to the papers of Nordström and Abraham, it is hard to believe that he was unfamiliar with Weyl's famous book, in which there are references. Of course, by 1919–1921, when scalar theories of gravitation had already ceased to be topical, Kaluza could have forgotten Nordström's paper, which would also explain the absence of a reference. However, the absence could also be due to the fact—if Kaluza was aware of Nordström's theory—that he regarded his construction as entirely different, and the gravitational part of Nordström's theory as completely incorrect.

We now turn to Kaluza's construction. He actually postulated a five-dimensional Riemannian space of events R_5 (in the system of notation adopted by him) with coordinates x^0, x^1, x^2, x^3, x^4 (the new, fifth coordinate was x^0), metric tensor g_{rs} (Latin indices take values from 0 to 4, Greek indices from 1 to 4), and Christoffel symbols $\left[{ik \atop l} \right] = -\Gamma_{ikl}$. Then, taking into account the cylinder condition, he wrote down the values of all Γ_{ikl} in terms of the derivatives of g_{rs} with respect to the coordinates. Besides the 40 "four-dimensional" $\Gamma_{\mu\nu\lambda}$

describing the gravitational field, there appear 35 new three-index quantities $\Gamma_{0\kappa\lambda}$, $\Gamma_{\kappa\lambda0}$, $\Gamma_{00\kappa}$, $\Gamma_{0\kappa0}$, Γ_{000} of the form

$$2\Gamma_{0\kappa\lambda} = g_{0\kappa\cdot\lambda} - g_{0\lambda\cdot\kappa}, \tag{28}$$

$$2\Gamma_{\kappa\lambda0} = -(g_{0\kappa\cdot\lambda} + g_{0\lambda\cdot\kappa}), \tag{29}$$

$$2\Gamma_{000} = 0. \tag{30}$$

Besides the sextuplet of quantities $\Gamma_{0\kappa\lambda}$, which could be identified with the components of the antisymmetric tensor of the electromagnetic field, there are two further types of quantity $\Gamma_{\kappa\lambda0}$ and two quartets of gradient quantities that differ in sign, $\Gamma_{00\kappa} = -\Gamma_{0\kappa0}$, "which in the proposed interpretation should also possess an electromagnetic nature" and therefore "threaten to become a hindrance." To ensure proportionality of $F_{\mu\nu}$ and $\Gamma_{0\mu\nu}$, it was natural to identify the four components of the electromagnetic potential q_μ with the four components $g_{0\mu}$ of the five-dimensional metric tensor:

$$g_{0\mu} = 2\alpha q_\mu,$$

where α is a certain constant whose value will be determined subsequently. The significance of the final, "corner" component, denoted g by Kaluza,

$$g_{00} = 2g, \tag{32}$$

remained as yet unexplained. Also unexplained is the significance of the "subsidiary" field (*Nebenfeld*)

$$\Sigma_{\mu\nu} = q_{\mu\cdot\nu} + q_{\nu\cdot\mu} = -\frac{1}{\alpha}\Gamma_{\mu\nu0}, \tag{33}$$

which by symmetry has ten components. In accordance with the spirit of general relativity, however, the field equations must be expressed in terms of the curvature tensor of second rank. In other words, the equations

$$R_{ik} = 0 \tag{34}$$

for the five-dimensional analog of the Ricci tensor must include both the gravitational equations and the equations of the electromagnetic field. Kaluza made the corresponding identification in the approximation of weak fields, which he called the first approximation:

$$g_{ik} = \delta_{ik} + \gamma_{ik}, \tag{35}$$

where δ_{ik} is a five-dimensional pseudo-Euclidean tensor, and γ_{ik} is an infinitesimally small symmetric second-rank tensor.

Calculation of the components of R_{ik} in the adopted approximation leads to the results

$$R_{\mu\nu} = \Gamma^{\rho}_{\mu\nu\cdot\rho} = 0,$$
$$R_{0\mu} = \alpha(\text{Div } F)_{\mu} = 0, \tag{36}$$
$$R_{00} = -\Box g = 0.$$

In the adopted approximation, the first equation is identical to the system of equations of the gravitational field. The second gives the divergence quartet of Maxwell's equations. Incidentally, the other pair of Maxwell's equations can be obtained from the identity for Γ_{ikl} with allowance for the cylinder condition. The subsidiary field $\Sigma_{\mu\nu}$ has disappeared somewhat surprisingly from the system of equations (36), but the final equation still requires an interpretation. The obtained result can, in Kaluza's words, be regarded as a "first confirmation of our assumption and a support for the hope of interpreting gravitation and electricity as manifestations of some universal field" ([180], p. 969; [179], p. 531).

It was still necessary first to consider the case of nonempty space (with matter-energy-momentum tensor $T_{ik} = 0$), and second to clarify the significance of the quantity g and the third equation of the system (36), and finally to consider the question of the equations of motion of charged particles in electromagnetic and gravitational fields. In solving the first problem, Kaluza considered a five-dimensional tensor T_{ik}—a generalization of the four-dimensional kinetic tensor of Minkowski:

$$T_{\rho\nu} = \mu_0 u_\rho u_\nu.$$

The zeroth components of T_{ik} were identified with the components of the density of the four-current:

$$T_{0\alpha} = \mu_0 u^0 u^\alpha = \frac{\alpha}{\kappa}\rho_0 v^\alpha, \tag{37}$$

where ρ_0 is the invariant charge density; κ is the gravitational constant; $v^\sigma = \mathrm{d}x^\sigma/\mathrm{d}\sigma$; $\mathrm{d}\sigma = g_{\mu\nu}\,\mathrm{d}x^\mu\,\mathrm{d}x^\nu$. "Thus," noted Kaluza, "the space-time energy tensor is, essentially, bordered by the current density." This tensor should now be substituted in the five-dimensional equation

$$R_{ik} - \tfrac{1}{2}g_{ik}R = -\kappa T_{ik}, \tag{38}$$

and calculations made in the approximation $g_{ik} = \delta_{ik} + \gamma_{ik}$. In the paper of Kaluza himself, the details of this calculation are not given, but they are reproduced, for example, in Beck's article on the general theory of relativity written in 1929 for Vol. IV of the *Handbuch der Physik* [15]. As is well known, Eq. (38) in the above approximation reduces to the following equations for γ_{ik}:

$$\tfrac{1}{2}\square(\gamma^{ik} - \tfrac{1}{2}\delta^{ik}\gamma_{rs}\delta^{rs}) = \kappa T^{ik}. \tag{39}$$

Since $\delta_{ik}\delta^{ik} = 5$, it is easy to reduce Eq. (39) to

$$\tfrac{1}{2}\square\gamma^{ik} = \kappa(T^{ik} - \tfrac{1}{3}\delta^{ik}T). \tag{40}$$

For $i = 0$, $k = 1, 2, 3, 4$, this equation takes the form of a wave equation for the electromagnetic potential. For $i, k = 1, 2, 3, 4$, the corresponding equations of the gravitational field are obtained (in the first approximation).

To clarify the significance of g (or γ^{00} in Beck's notation), Kaluza made one further assumption, which he called the second approximation. He assumed that the components of the five-velocity $u^r = dx^r/ds$ satisfied the following conditions of smallness:

$$u^0, u^1, u^2, u^3 \ll 1, \quad u^4 \sim 1. \tag{41}$$

From these conditions there follow not only smallness of the velocities but also smallness of the specific charge of a moving object. Indeed, from relation (37) and the relations

$$d\sigma^2 \sim ds^2, \quad v^\mu \sim u^\mu,$$

which hold under conditions (41), we obtain

$$\rho_0 = \frac{\kappa}{\alpha}\mu_0 u^0 = 2\alpha\mu_0 u^0 \ll \mu_0. \tag{42}$$

Here it is assumed that $\alpha = \sqrt{\kappa/2}$, a choice that, as we shall see, will be justified by the corresponding interpretation of the geodesic equation as the equation of motion of a charged particle. As Kaluza noted, relation (42) makes it possible to interpret the electric charge as the fifth component of momentum. Since in the second approximation $T_{00}, T_{11}, T_{22}, T_{33} \sim 0$, it follows that $T = -\mu_0$, and we find that

$$R_{00} = -R_{44} = \frac{\kappa}{2}\mu_0,$$

from which it follows, in accordance with the third equation of the system (36), that the "corner" potential g reduces to the gravitational potential with opposite sign. Naturally, the same result can be obtained from Eq. (40) for γ^{00}:

$$\Box \gamma^{00} = -\tfrac{2}{3} \kappa \delta^{00} \mu_0 c^2,$$

which means that γ^{00} and γ^{44} are proportional. Thus, the difficulties associated with the appearance of superfluous quantities, namely, the subsidiary field and "corner" potential, are (in the considered approximations) unimportant.

Further, Kaluza considered the equation of a geodesic in the five-dimensional Riemannian space,

$$\frac{du^l}{ds} - \Gamma^l_{rs} u^r u^s = 0,$$

and showed that in the second approximation it can be interpreted as the equation of motion of a charged particle in the gravitational and electromagnetic fields. Indeed,

$$\frac{dv^\lambda}{d\sigma} = \Gamma^\lambda_{\rho\sigma} v^\rho v^\sigma + 2\alpha F^\lambda_\kappa u^0 v^\kappa - g_{.\lambda} u^{0^2}. \tag{43}$$

It is here assumed that $d\sigma \sim ds$, and the following identifications are also used:

$$\Gamma_{0\kappa\lambda} = \alpha F_{\kappa\lambda}, \quad \Gamma_{00\kappa} = -\Gamma_{0\kappa0} = g_{0\kappa}.$$

From this, in view of the smallness of u^{0^2}, the density of the force ("ponderomotive force") is

$$\pi^\lambda = \Gamma^\lambda_{\rho\sigma} T^{\rho\sigma} + F^\lambda_\kappa j^\kappa, \tag{44}$$

where j is the current density. Incidentally, the expression of α in terms of the gravitational constant, $\alpha = \sqrt{\kappa/2}$, is justified by this formula. The zeroth component of the geodesic equation (for $l = 0$) ensures the law of conservation for the density of the electric charge when the adopted approximations are taken into account (i.e., in the quasistatic case). Kaluza also emphasized that the subsidiary field (Nebenfeld) was not manifested in the equations of motion.

Nevertheless, the problem of unifying the gravitational and electromagnetic fields was only solved approximately. In particular, Kaluza himself emphasized that for an electron, for example, the condition of smallness of ρ_0/μ_0 is not satisfied, and therefore everything obtained in the second approximation of the theory applied only to macroscopic phenomena and "there arises the radical problem of its applicability to the primordial particles (Urteilchen) [i.e., microscopic particles]." Thus, he believed that a significant modification of the theory was needed to describe microscopic phenomena.

Although in the paper there was no explicit formulation of the maximum problem of the program of unified geometrized field theories—the obtaining of particle–like solutions of one kind or another from the field equations—the somewhat ambiguous ending of the paper suggests that Kaluza also thought of this possibility. He also clearly saw the difficulty of reconciling his theory, and in fact all other unified geometrized field theories, with the quantum conception: "Quite generally, any hypothesis that lays claim to universal significance is threatened by the sphinx of modern physics—quantum theory." Noting the remarkable "formal unity" (*formale Einheitlichkeit*) of the two field phenomena, Kaluza doubted whether this was "merely the capricious game of deceiving chance." He hoped that behind it there was "something more than empty formalism."

Naturally, the first who responded to Kaluza's paper was Einstein. About a month after the paper was presented to the Berlin Academy of Sciences he and the talented mathematician Grommer[15] completed a paper devoted to proving the absence in Kaluza's theory of an everywhere-regular centrally symmetric solution that could be identified with the electron [183]. Kaluza's theory is evaluated in the following way: "Recently T. Kaluza presented to the Academy of Sciences at Berlin the project of a theory that eliminates all these shortcomings [the shortcomings of Weyl's theory: the gap between the physical and geometrical foundations; the representation of the Lagrangian as a sum of electromagnetic and gravitational components; the fourth order of the field equations]" ([183]).

Einstein and Grommer described the formal structure of Kaluza's theory and noted that the "corner" component (g_{55} or g_{00}), which did not have any physical interpretation, could be connected with the Poincaré pressure of classical electron theory. Then they pointed out an important advantage of this theory: "We have a possibility to construct the physical picture of world on the basis of a unified Hamilton function which is not the sum of the heterogeneous

---

[15] Jakob Grommer studied mathematics at Göttingen under Toeplitz and wrote a paper that, in the opinion of Toeplitz, and then Hilbert, went significantly beyond the requirements of a doctoral dissertation. The granting of a degree was hindered, however, by the absence of Grommer's diploma for completion of the gymnasium. Hilbert finally succeeded in seeing that Grommer received a doctoral degree [106]. Interesting material about Grommer, who in the last years of his life was Professor of Mathematics at the Belorussian University in Minsk, can be found in [182].

parts" ([183], p. 131). It was natural that Einstein wanted to go further along the path opened up by Kaluza, and he and Grommer noted:

> The final aim is a pure field theory in which the field variables represent both the field of "empty space" and electrically charged elementary particles forming "matter." [183]

As Einstein and Grommer showed, however, the hope that Kaluza's theory would make it possible to solve this problem was not justified. In fact, they also pointed out a fundamental defect of the theory, which was subsequently regarded as the main theoretical shortcoming of all five-dimensional theories:

> It is not possible to pass over in silence the essentially weak pont of Kaluza's idea. In the general theory of relativity, which deals with a four-dimensional continuum, the form
>
> $$ds^2 = g_{\mu\nu}\, dx_\mu\, dx_\nu$$
>
> represents something that can be directly measured by means of rods and clocks in a locally inertial frame, whereas in the five-dimensional continuum of Kaluza's theory ds^2 is a pure abstraction, devoid, apparently, of direct metrical significance. Therefore, from the physical pont of view, the requirement of general covariance of all equations in the five-dimensional continuum appears entirely unjustified. In addition, there arises a dubious asymmetry when one dimension is distinguished from all the others by the cylinder condition, whereas in the structure of the equations all five dimensions must be on an equal footing. [183]

As we recall, for almost two years Pauli occupied himself with great enthusiasm with Weyl's theory. When Kaluza put forward his theory, Pauli had already moved away from unified field theories and was entirely immersed in the problems of quantum theory. About 15 years later, his attention was attracted to a projective formulation of the idea of five dimensions developed at the beginning of the 1930s by Veblen and Hoffmann, and also Schouten and van Dantzig. In the English edition of his encyclopedia article on relativity theory (1956), Pauli included a supplement devoted to cosmology and unified field theories. In this supplement, he summarized his criticism of five-dimensional theories (including projective ones)[16] as follows:

---

[16] At the beginning of the 1940s, Bergmann showed that the projective reformulations of Kaluza's theory were essentially equivalent to the original form of the theory. It should be mentioned, however, that Kaluza's theory was here considered in the form given it in 1926 by O. Klein. In this form, the geometrical unification of gravitation and electromagnetism was not approximate but exact [12].

> There is, however, no justification for the particular choice of the five-dimensional curvature scalar as integrand of the action integral, from the standpoint of the restricted group of the cylindrical metric. ([2], p. 230)

In other words, the field equations are not a consequence of the geometrical structure, since the covariance group of the theory is smaller than the group of general covariance of a five-dimensional Riemannian space.

Thus, Einstein (together with Grommer) correctly identified the main weaknesses of Kaluza's theory. Nevertheless, they tested it for the existence of centrally symmetric everywhere-regular solutions and concluded that they did not exist, this being, in Einstein's eyes, one indication of a merely formal nature of the unification. The nondecomposability of the Lagrangian, which appeared to Einstein and Grommer to be an important advantage of the theory, was illusory. As O. Klein showed, the five-dimensional scalar curvature could be represented as a sum of gravitational and electromagnetic components

$$P = R + \tfrac{1}{4} f_{ik} f^{ik}.$$

Nevertheless, the idea of five dimensions determined one of the main paths of realization of the program of unified geometrized field theories for many years. There were subsequently two very significant bursts of activity in the development of this direction—in 1926–1927 (O. Klein, Mandel, Fock, Gamow and Ivanenko, Einstein, de Broglie, Frederiks, de Donder, Ehrenfest and Uhlenbeck, and others) and 1931–1933 (Veblen and Hoffmann, Veblen, Schouten and van Dantzig, van Datzig, Pauli, Einstein and Mayer, etc.). The first was associated with attempts to unify in a five-dimensional scheme not only gravitation and electromagnetism but also quantum mechanics. The second was associated on the one hand with a projective generalization of the Kaluza–Klein theory and, on the other, with the attempt of Einstein and Mayer to make the five-dimensional scheme more physically sensible. There were lesser peaks of activity in the field of five dimensions at the end of the 1930s and beginning of the 1940s (Einstein and Bergmann; Einstein, Bergmann, and Bargmann; and others) and at the end of the 1940s and beginning of the 1950s (Jordan, Ludwig, Bergmann, Thiry, Rumer, Ikeda, and others). One was due to papers of Einstein (and collaborators) in which attempts were made to give a physical or geometrical justification for the cylinder condition and the taking of g_{55} (or γ_{55}) equal to unity. The other activity peak, which was much more significant, had several thematic focuses:

(1) The inclusion of Dirac's idea of variability of the gravitational constant in the five-dimensional conception (Jordan, Thiry, Fierz, and others);

(2) The five-optics of Rumer, in which the fifth coordinate was ascribed the meaning of action, and where there was also a periodic nature of the variation with period equal to Planck's constant h;

(3) The study of various mathematical and physical aspects of five dimensions (cosmology, spinor calculus, development of the projective formalism, variational principles).

We have rushed far ahead, however, and we must return to the beginning of the 1920s to consider in more detail the question of the extent to which Einstein himself participated in the creation of the scientific research program of unified geometrized field theories. Of course, in discussing his reaction to the appearance of the unified theories of Hilbert, Weyl, Eddington, and Kaluza, we have already broached this question. In the following section, it will be considered more systematically.

EINSTEIN'S THEORIES OF 1919–1923

In Chapter 1, we considered Einstein's early attempts to construct a unified field description of the electromagnetic field, its quantum properties, and the electron. Einstein sought a way of generalizing Maxwell's equations that would make it possible to obtain particle-like solutions and relate them to the electron and light quanta. In this project, he was guided by the electromagnetic field program, the unified field being understood as a generalized classical electrodynamic field. The generalization itself was to be realized by guessing a suitable mathematical structure of the field equations. Important additional restrictions were also the requirement of Lorentz covariance, a correspondence principle, simplicity, and certain other principles. As yet, there was no intention to include gravitation in the generalized field equations. Although in Einstein's searches the leading aim was the mathematical construction of field equations (he tried a nonlinear generalization and raising the order of the equations), physical considerations and transparent ideas were most important to him.

Einstein did not succeed in constructing a unified theory and switched his efforts to the problem of the gravitational field, but he did not simply abandon the acquired experience. In his work on general relativity, he made use not only of general concepts and methods, for example the concept of a classical field and the mathematical hypothesis method, but also of the idea of the nonlinear nature of the field equations and the idea of obtaining equations of motion from field equations. He had acquired certain experience in applying the methodological principles of physics in the construction of a new theory, above all in the finding of the basic equations of the theory. This episode indicates Einstein's early interest in the idea of a field synthesis of physics, and the presence in his thought of a thematic coupling of two themes (in Holton's sense)—continuity and unity.

In our consideration in Chapters 2 and 3 of the unified theories of Hilbert and Weyl, we devoted no little space to discussing Einstein's attitude to these theories. As we saw, it was very negative. We are referring mainly to the events of 1915–1919. In those years, he achieved outstanding successes in the development of quantum theory (the paper of 1916) and especially general relativity: the completion of the foundations of the theory, the prediction of gravitational waves, the solution to the problem of energy–momentum conservation, the creation of the foundations of relativistic cosmology, etc. During this period, Einstein thought neither of including the electromagnetic field in a unified geometrized scheme nor of obtaining particles and quanta from field equations like those of the gravitational field. Moreover, in one of the papers of 1916 [184], he emphasized the need for a subsequent quantum generalization of general relativity, the classical nature of which he clearly understood (see Chapter 2).

The successful development of the fundamental physical aspects of general relativity and the weakness of the unified field theories of Hilbert and Weyl could have been for Einstein a very weighty argument for a restricted, purely gravitational understanding of the general theory of relativity, and accordingly the inclusion only of gravitation in the geometrical scheme. Both these attempts at a synthesis of gravitation and electromagnetism on the basis of general relativity had a Göttingen origin, and thus were associated with the Göttingen tradition of mathematical physics, about which Einstein was very skeptical. The synthetic tendency, which was always present in Einstein, was embodied to some extent at this period in the relativistic cosmology.

It is a fact that after the appearance of Weyl's paper, Einstein, for all his doubts, often returned to it in his thoughts. This is particularly noticeable in his correspondence with Besso, Sommerfeld, Ehrenfest, Weyl himself, and Born. Einstein had a high opinion of Weyl's book *Raum-Zeit-Materie* [185]. At the same time, in his lecture "Motives of scientific research," given in 1918 in honor of Planck's 60th birthday, he considered two ideas that may indicate a process of reorientation with respect to the concept of unified field theories. First, he objected to the idea of equal validity of alternative systems of theoretical physics and emphasized the uniqueness of a theory that corresponded to physical reality:

History had shown that, at a given time, among all conceivable constructions only one is dominant.... The theoretical system is more or less uniquely determined by the world of observations, although no logical path leads from observations to the fundamental principles of the theory. This is the essence of what Leibniz aptly called "pre-established harmony." [186]

Second, as the "most important physical problem of our time" he named the problem of unifying quantum mechanics, electrodynamics, and mechanics in a logically elegant system.[17]

In one of his letters to Weyl (on December 16, 1918) Einstein wrote: "All my remarks, from a mathematical standpoint, are made with the greatest respect for your theory and . . . I also admire it as an intellectual construction" ([57], p. 169). All this could be regarded as definite indications of Einstein's increasing interest in the concept of unified field theories.

Indeed, on April 10, 1919 Einstein presented to a session of the Prussian Academy of Sciences his paper "Do gravitational fields play an essential part in the structure of the elementary particles of matter?" [93], which initiated a new series of Einsteinian investigations in unified field theories. It is true that in this paper the main problem for Einstein was not the geometrical unification of the gravitational and electromagnetic fields but the proof of the possibility of constructing elementary particles of matter on a purely field basis.

Having briefly mentioned how this problem was solved in the preceding unified theories of Mie, Hilbert, and Weyl, and having placed his theory in the framework of the corresponding program, Einstein formulated the main task—to show that charged particles preserve their stability only by virtue of gravitation: "It will be shown in the following pages that there are reasons for thinking that the elementary formulations which go to make up the atom are held together by gravitational forces" ([93], p. 191 of the English translation) (". . . *die Bausteine der Atome bildenden elektrischen Elementargebilden durch Gravitationskräfte zusammehgehalten werden*" [187], p. 349). By contrast, in the earlier unified theories the balancing forces within charged particles were also assumed to have an electric origin. In connection with these theories, he noted, without going into a detailed discussion:

> On the one hand the multiplicity of possibilities is discouraging, and on the other hand those additional terms [explaining the equilibrium of charged

----

[17] May the love of science continue to adorn his [i.e., Planck's] life and lead him to the solution of the most important physical problem of our time, proposed and significantly advanced by himself. May he succeed in unifying quantum mechanics, electrodynamics, and mechanics in a logically elegant system. [186]

particles] have not as yet allowed themselves to be framed in such a simple form that the solution could be satisfactory. ([93], p. 191)

Einstein assumed that, as in general relativity, the Maxwell energy–momentum tensor

$$T_{ik} = \tfrac{1}{4} g_{ik} \varphi_{\alpha\beta} \varphi^{\alpha\beta} - \varphi_{i\alpha} \varphi_{k\beta} g^{\alpha\beta} \tag{45}$$

must be proportional to a differential expression of second order constructed from g_{ik}, but if, as such an expression, one were to take the left-hand side of the gravitational field equation of general relativity,

$$R_{ik} - \tfrac{1}{2} g_{ik} R = -\kappa T_{ik},$$

then nothing new would be obtained. These field equations were consistent with the energy–momentum conservation law for the energy–momentum tensor of matter in the presence of gravitation,

$$\frac{\partial(\sqrt{-g} T_i^{\sigma})}{\partial x_{\sigma}} + \tfrac{1}{2}(\sqrt{-g} T_{\sigma\tau}) \frac{\partial g^{\sigma\tau}}{\partial x_i} = 0,$$

which on the transition to the constant limiting values of g_{ik} characteristic of the special theory of relativity becomes identical to the corresponding expression for the energy–momentum conservation law when gravitation is ignored.

Einstein noted that

> if gravitational fields do play an essential part in the structure of the particles of matter, the transition to the limiting case of constant $g_{\mu\nu}$ would, for them, lose its justification, for indeed, with constant $g_{\mu\nu}$ there could not be any particles of matter. ([93], p. 192)

Requiring the vanishing of the covariant derivative of $T_{\mu\nu}$ thus would become meaningless, and the arguments for including the term $-g_{ik} R/2$ on the left-hand side of the field equations would also lose their force. As a result, it would be necessary to give up the standard equations of gravitation. The requirement of general covariance did not leave, however, too large a choice for the required second-order differential expression in g_{ik}: it must have the form

$$R_{ik} + a R g_{ik}. \tag{46}$$

Since, however, the energy–momentum tensor T_{ik} of matter has zero trace (for particles possessing rest mass must be definite constructions from the gravitational and electromagnetic fields), the trace of expression (46) must also be zero. This immediately gives the value $-\frac{1}{4}$ for a, and the equations of gravitation in this purely field world are somewhat different from the corresponding equations of general relativity:

$$R_{ik} - \frac{1}{4} g_{ik} R = -\kappa T_{ik}. \tag{47}$$

Of course, to these equations it is necessary to add the standard Maxwell equations

$$\frac{\partial(\sqrt{-g}\,\varphi_{\sigma\tau}\,g^{\sigma\alpha}g^{\tau\beta})}{\partial x_\beta} = \sqrt{-g}\,J^\alpha, \qquad \frac{\partial\varphi_{\mu\nu}}{\partial x_\rho} + \frac{\partial\varphi_{\nu\rho}}{\partial x_\mu} + \frac{\partial\varphi_{\rho\mu}}{\partial x_\nu} = 0. \tag{48}$$

A simple calculation shows that there are precisely 12 independent equations (47) and (48) for determining the 16 variables $g_{\mu\nu}$ and $\varphi_{\mu\nu}$, in agreement with the requirement of general covariance. In the framework of general relativity, as is readily shown, the relation

$$\varphi_{i\alpha}(\sqrt{-g}\,J^\alpha) = 0,$$

holds, but in the new theory the formation of the divergence of the right- and left-hand sided of Eq. (47) gives the relation

$$\varphi_{\sigma\alpha}J^\alpha + \frac{1}{4\kappa}\frac{\partial R}{\partial x_\sigma} = 0, \tag{49}$$

which can be interpreted as a balancing of the electric forces by gravitational pressure.

If at the same time one sets $J^\sigma = \rho\frac{dx^\sigma}{ds}$ and multiplies (49) by J^σ, then, $\varphi_{\sigma\alpha}$ being antisymmetric, it is found that

$$\frac{\partial R}{\partial x_\sigma}\frac{dx^\sigma}{ds} = 0, \tag{50}$$

from which there follows constancy of the scalar curvature along each world line of electric charge. In a space without charges, the scalar curvature also remains constant by virtue of relation (49):

$$R = \text{const} = R_0. \tag{51}$$

Equation (49) is interpreted as follows:

The scalar of curvature R plays the part of a negative pressure which, outside of the electric corpuscles, has a constant value R_0. In the interior of every corpuscle there subsists a negative pressure (positive $R - R_0$), the fall of which balances the electodynamic force (*dessen Gefälle der elektrodynamischen Kraft das Gleichgewicht leistet*). ([187], p. 352)

The expression $\frac{1}{4\kappa}R$ is, in fact, the potential energy of gravitation, the forces of which counter the electrostatic forces of repulsion.

Having compared the new equations (47) with the field equations containing the cosmological constant λ—and these equations, as Einstein assumed, must retain their importance, since they led to a satisfactory cosmological solution, namely a spatially closed world—he obtained the value

$$\lambda = -\frac{R_0}{4} \tag{52}$$

for the cosmological constant. Einstein saw in this result an important achievement of the new theory:

The new formulation had this great advantage, that the quantity λ appears in the fundamental equations as a constant of integration, and no longer as a universal constant peculiar to the fundamental law. ([93], p. 196)

If we wish to retain the gravitational equations of general relativity (with the term $-\frac{1}{2}g_{ik}R$), then as energy–momentum tensor of the matter ("tensor of the gravitation mass," *der gravitierenden Masse*) it is necessary to take the tensor

$$T_{ik}^* = T_{ik} + \frac{1}{4\kappa}g_{ik}(R - R_0). \tag{53}$$

Einstein then showed that a closed world with constant density of rest mass was a solution of Eqs. (47), the energy of this world consisting of three quarters energy of the electromagnetic field and one quarter energy of the gravitational field.

Having emphasized in his conclusions the above advantages of his theory (the possibility of constructing charged particles "out of gravitational field and electromagnetic field alone, without the introduction of hypothetical supplementary terms on the lines of Mie's theory" and without "the necessity of introducing a special constant λ for the solution of the cosmological problem" [93], p. 198), Einstein himself drew attention to a decisive difficulty of the theory, leading it into a blind alley:

On the other hand, there is a peculiar difficulty. For, if we specialize (1) [there is here obviously a misprint, and Eq. (1a), i.e., (47), is meant] for the spherically symmetrical static case we obtain one equation too few for defining the $g_{\mu\nu}$ and $\varphi_{\mu\nu}$, with the result that *any spherically symmetrical distribution* of electricity appears capable of remaining in equilibrium. Thus the problem of the constitution of the elementary quanta cannot yet be solved on the immediate basis of the given field equations. [Ibid.]

In the encyclopedia article on relativity theory, which appeared in September 1921, Pauli associated himself with Einstein's conclusion: "However satisfactory the foundations of this theory may be, it, too, is not capable of providing an answer to the problem of the structure of matter" ([63], p. 205). We recall that Einstein's idea of relating the cosmological constant to the scalar curvature was then used both in the second form of Weyl's theory and in Eddington's theory.

As he prepared this paper for publication, Einstein already knew that his new theory was incorrect, but as a model it was very instructive and it very convincingly demonstrated deep possible connections between general relativity and the problem of the structure of matter. Therefore, he did publish the paper. In fact, a year and a half later, Pauli too had a high opinion of Einstein's theory, devoting an entire section (§66) to it in *The Theory of Relativity* and noting, in the quotation just given, the satisfactory nature of its foundations. Although, in this theory, the electromagnetic field was not geometrized, the theory was in its essence close to unified geometrized field theories, but, paradoxically posing the maximum problem, it appeared to postpone for a time the solution of the minimum problem. In fact, in 1919 Einstein may have believed—and for this there were fairly serious grounds—that only the gravitational field could be geometrized. At the same time, belief in the power and universality of the classical field concept forced him to seek ways of reducing particles to a field.

A second attempt at the construction of a unified field theory was made by Einstein in March 1921, when there arose several different directions in the development of unified geometrized field theories. This birth of multiple theories enables one to say that the program of unified geometrized field theories arose precisely in 1921. With his theory of 1919, Einstein had already initiated his investigations in that direction. During the subsequent two years, he did not cease to ponder the problem; the thought of a geometrical unification of gravitation and electromagnetism no longer appeared unacceptable to him, although Weyl's realization of the idea still appeared incorrect.

In April 1919, Kaluza entered into correspondence with Einstein about his five-dimensional unified field theory, and the latter approved and supported the Königsberg mathematician (see the beginning of the previous section). It is entirely possible that Kaluza's investigations already had a strong influence on Einstein's change in position with regard to the possibility of a geometrical synthesis of gravitation and electromagnetism.

The measurements of the deflection of light at the edge of the solar disk made by two British expeditions under the leadership of Eddington at the time of the solar eclipse of May 29, 1929, confirmed the prediction of general relativity, and this was taken as a triumph of the new theory of gravitation. Einstein learned of this on September 22nd in a telegram from Lorentz. This event not only made general relativity and its author universally known,[18] but also decisively strengthened the position of general relativity in its struggle with competing theories of gravitation (especially the scalar, Lorentz-covariant theory of Nordström), and also the authority of the geometrical interpretation of physical fields.

As we have mentioned, in 1919 Pauli published two papers on Weyl's theory. These papers are often mentioned in Einstein's correspondence at the end of 1919 and the beginning of 1920, either with the hope that they could refute Weyl's theory [for example, in the letter from Sommerfeld to Einstein of October 24, 1919 ([68], p. 207)] or with the conviction of an error in Weyl's theory associated with, as proved by Pauli, the absence in it of static solutions for nonvanishing electromagnetic potentials—for example, in the letter from Einstein to Ehrenfest on September 4, 1919 ([57], p. 168 of the English translation) or in the letter from Einstein to Besso on December 12, 1919 ([56], p. 93).

Einstein's criticism of Weyl's theory no longer signified a negative attitude toward the concept of geometrical unification of the gravitational and elec- tromagnetic fields and a field solution of the problem of particles and quanta

---

[18] Seelig, for example, wrote: "The confirmation of a completely abstract thought through a real phenomenon in the universe finally made Einstein's name a household word in the Anglo-Saxon countries. He was inundated by an army of autograph hunters, editors, reporters and brainless worshippers of transitory renown." Seelig goes on to give a very expressive extract of a letter from Einstein to his former collaborator Hopf on February 2, 1920:

Since the flood of newspaper articles, I have been so swamped with ques- tions, invitation, challenges, that I dream I am burning in Hell and that the postman is the Devil eternally roaring at me, throwing new bundles of letters at my head because I have not yet answered the old ones. In addition to this, at home I have my mother who is on her deathbed and I have to spend the best part of this "great occasion" in countless meetings, etc. ([57], pp. 162–163 of the English translation)

based on this. The corresponding reorientation of Einstein occurred, apparently, no later than the spring of 1919. Indeed, on February 2, 1920 he wrote to Holland (Seelig, quoting an extract from this letter, does not say to whom it was addressed):

> The electromagnetic field, despite all my efforts, and a consistency in the differential equations... have proved unsuccessful [Einstein probably means inclusion of the electromagnetic field in a single geometric structure together with gravitation]. But I firmly believe that this is the way to real inner progress. ([57], p. 159 of the English translation)

Incidentally, this does not mean that at that time Einstein gave up attempts to solve the maximum problem (obtaining particles and quanta on the basis of a field theory) and switched his efforts to the solution of the more restricted minimum problem (geometrical unification of gravitation and electromagnetism). In a letter to Born on March 3, 1920 he wrote:

> In my spare time I always brood about the problem of the quanta from the point of view of relativity. I do not think the theory can work without the continuum. But I do not seem to be able to give tangible form to my pet idea, which is to understand the structure of the quanta by redundancy in determination, using differential equations. ([89], p. 26)

He wrote to Ehrenfest on much the same topic on April 17, 1920: "I have made no further progress in the general relativity theory. The electric field still remains there in the air.... Nor have I solved any questions to do with electrons" ([57], p. 159 of the English translation).

On May 5, 1920 Einstein gave a lecture at the University of Leyden with the title "The ether and the theory of relativity." In this lecture for the first time he judged as very important the problem of unifying gravitation and electromagnetism in the framework of some unified field theory:

> It would be a great step forward to unify in a single picture the gravitational and electromagnetic fields. Then there would be a worthy completion of the epoch of theoretical physics begun by Faraday and Maxwell; the contrast between the ether [it is clear from the preceding that by the "ether" Einstein understood curved space-time] and matter would be smoothed away, and all physics would become a closed theory like the general theory of relativity, encompassing geometry, kinematics, and the theory of gravitation. [166]

Einstein further mentioned Weyl's theory as a concrete example of a unification of such type but immediately rejected it, regarding it as not capable of comparison with experiment. Thus, already at the beginning of 1920 Einstein had clearly delineated the ideal of a closed unified field theory "like the general theory of relativity." It is true that he had some sobering reservations:

> In thinking about the immediate future of theoretical physics, we undoubt-edly cannot deny the possibility of encountering unexpected frontiers for field theory, which may be imposed by facts encompassed by quantum theory.

It is possible that this reservation was the result of the first unsuccessful attack on the problem of understanding "the structure of the quanta ... using differ-ential equations," although, as he wrote at approximately the same time to one of his Dutch colleagues, he continued to believe firmly "that this way [i.e., the way associated with realization of the program of unified geometrized field theories] will become the way of genuine progress."

In the correspondence of Einstein and Besso in the second half of 1920, Weyl's theory was still at the center of attention. Einstein did not change his attitude to it; indeed, he found more and more new arguments against it, although he remarked that "Weyl always remains a deep and clear thinker, and it is always a pleasure to read what he has written ... " (from the letter of July 26, 1920) ([56], p. 93).

In the second half of 1920, the anti-Einstein campaign, in which the No-bel laureate P. Lenard played an unattractive role, reached a climax ([188], pp. 156–163). At the end of September, at the 86th Congress of German Nat-ural Scientists at Bad Nauheim, Lenard led the anti-relativists, who mustered very sharp attacks against Einstein and the theory of relativity. At the same congress, Weyl gave an exposition of his theory [144].

Einstein once again formulated his arguments against the theory. Pauli noted the failures of field theories, including Einstein's theory of 1919, in the attempts to solve the problem of matter, and pointing out the absence of an operational measuring procedure for defining the concept of field strength within the electron, thereby emphasized the limitations of "theories of con-tinuum type." Einstein's answer to Pauli's equation of whether one should adhere to theories of continuum type in the solution of the problem of matter or whether it was necessary to seek a solution by modifying the classical field and space-time conception "in the sense of atomism" was not entirely definite, but it appeared that he must have chosen the first alternative.

Einstein gave a clearer answer to Pauli's question in his lecture "Geometry and experience," read at a ceremonial meeting of the Prussian Academy of Sciences in Berlin on January 27, 1921 [189]. Emphasizing that the notion of a Riemannian structure of space-time had deep experimental foundations (in contrast to Weyl's geometry), he commented:[19]

> The physical interpretation of geometry presented here fails, it is true, when applied directly to spaces of submolecular size. Nevertheless, it retains part of its significance in questions of the construction of elementary particles. For one can attempt to ascribe a physical significance to the field concepts which have been defined for the description of the geometrical behavior of bodies large compared with molecules even when one is considering the description of the electric elementary particles that constitute matter. Only success can decide the justification of such an attempt to ascribe physical reality to the basic concept of Riemannian geometry beyond their physical domain of definition. It may turn out that this extrapolation is as inappropriate as that of applying the concept of temperature to parts of a body of the size of molecules. [190]

In January 1921 the discussion was only of an attempt to use the classical, macroscopic concepts of field and space-time to study the structure of microscopic particles, which already exhibited manifestly nonclassical features in their behavior and structure. Einstein himself emphasized that "only success can decide the justification of such an attempt." In the period when the program was born, its "parents" and adherents had doubts about its justification and prospects; they expressed caution in predictions, and in the event of failure a willingness to abandon it.

---

[19] Die hier vertretene physikalische Interpretation der Geometrie versagt zwar bei ihrer unmittelbaren Anwendung auf Räume von submolekularer Grössenordnung. Einen Teil ihrer Bedeutung behält sie indessen auch noch die Fragen der Konstruktion der Elementarteilchen gegenüber. Denn man kann versuchen denjenigen Feldbegriffen, welche man zur Beschreibung des geometrischen Verhalten von gegen das Molekül grossen Körpern physikalisch definiert hat, auch dann physikalische Bedeutung zuzuschreiben, wenn er sich um die Beschreibung elektrischen Elementarteilchen handelt, die die Materie konstruieren. Nur der Erfolg kann über die Berechtigung eines solches Versuches entscheiden, der den Grundbegriffen der Riemannschen Geometrie über ihren physikalischen Definitionsbereich hinaus physikalische Realität zuspricht. Möglichweise könnte es sich zeigen, dass diese Extrapolation ebensowenig angezeigt ist wie diejenige der Temperaturbegriffs auf Teile eines Körpers van molekularer Grössenordnung. ([190], p. 6)

We have given this extract in the original because of its extreme importance for understanding the process of formation of the aims of the program of unified geometrized field theories in the case of Einstein himself.

In his "A brief outline of the development of the theory of relativity," published at the beginning of 1921 in *Nature*, Einstein formulated four main problems of general relativity "which are awaiting solution at the present time" [191]. Two of them related to cosmology, and the other two to unified field theory (the problem of the formal unification of gravitation and electromagnetism and the part played by gravitation in the structure of the elementary particles of matter). Thus, despite Einstein's critical attitude to the specific unified field theories of Hilbert and Weyl, at the start of the 1920s he himself proclaimed as basic problems of general relativity the creation of a unified geometrized field theory capable of solving the problem of the structure of matter (expansion of general relativity inward), and at the same time the development of cosmology (expansion of general relativity outward). By 1921 the latter direction had achieved much greater success than the former. The program of generally relativistic cosmology progressed rapidly during the 1920s and 1930s (especially in connection with the work of De Sitter, Friedmann, Lemaître, Eddington, Tolman, Hubble, and Einstein himself.) By contrast, the unification program in this period did not lead, as we shall see, to real achievements.

Nevertheless, until 1921 Einstein was less interested in the problem of unifying gravitation and electromagnetism than in the problem of the field structure of material particles, which was the main subject of discussion in his theory of 1919. Before 1921 he did not make attempts to develop a form of geometrical synthesis of gravitation and electromagnetism of his own and did not associate himself with any of the known attempts, above all Weyl's. Einstein's correspondence shows, however, that he continually pondered Weyl's theory and was still delighted by its mathematical elegance, but this theory, as we have seen, was unacceptable for him on physical grounds. Then, in March 1921, Einstein attempted a generalization of Weyl's geometry that, while preserving all the advantages of the mathematical scheme, would eliminate from the theory the defects associated with the physical interpretation of the foundations of the theory.[20]

---

[20] Here Einstein once more summarized his criticisms of Weyl's theory:

> For all my admiration for the unity (*Einheitlichkeit*) and beauty (*Schönheit*) of Weyl's conceptual framework, it does not appear to me to meet the requirements of physical reality. We do not know of any natural object suitable for making measurements whose relative extension (*Ausdehnung*) depends on their prehistory. In addition, the shortest line introduced by Weyl together with the electric potentials that appear explicitly in this and the other equations of Weyl's theory do not appear to have any direct physical meaning. ([193], p. 262)

He reported his attempt to Sommerfeld on March 9:

> I have found a kind of addition to the foundations of the general theory of relativity, related to Weyl's but different from it in that the φ_v appear only as electric field strengths $\left(\frac{\partial \varphi_\mu}{\partial x_v} - \frac{\partial \varphi_v}{\partial x_\mu}\right)$; however, it is doubtful whether it is physically acceptable. The Lord does what he wants and will not be dictated to. ([68], p. 219)

In his answering letter Sommerfeld wrote: "It is good that you assimilate Weyl (and, hence, Mie too) in a modified form" ([68], p. 220).

This paper of Einstein's was dated March 17, 1921. He first presented the basic ideas of Weyl's theory, which warranted "great interest at the least by virtue of its logical and daring mathematical structure" (*"die [Theorie] schon ihres folgerichtigen und kühnen mathematischen Aufbaues wegen ein hohes Interesse verdient"*) ([192], p. 104; [193], p. 261). Essentially, Einstein presented two arguments as the basis of the theory. First, there was a proposal to replace the interval ds^2, as the fundamental characteristic of the space-time continuum, by the equation $ds^2 = 0$ of the local light cone. The physical justification for this replacement could be that metrical relations should be formulated on the basis of operations with light signals, in which only ratios of the $g_{\mu v}$ occur.

The second basic proposition of Weyl's theory was the assumption that the length of a transported measuring rod depended on the integral $\int \varphi_\mu \, dx_\mu$ taken along the path of this displacement, the vector φ_μ, like $g_{\mu v}$, making a definite contribution to the metric (and connection) and being identified with the electromagnetic potential.

On the basis of these two considerations, Weyl's entire theory breaks down. In Einstein's opinion, the difficulties of Weyl's theory, listed in the footnote, were mainly associated with the second proposition. Einstein noted that Riemannian geometry was based on two assumptions: the existence of transportable measuring rods, and their independence of the path of transport. In Weyl's theory, only the second assumption was rejected. If we were to reject the first assumption as well, then we would arrive at a theory more general than Weyl's. The fundamental assumption of this generalized theory could be the equation of the local light cone, determining the covariance group of the theory:

$$ds^2 = g_{\mu v} \, dx_\mu \, dx_v = 0. \tag{54}$$

Neither the concept of distance nor the concept of rods and clocks would be used in such a theory. The geometrical structure of the corresponding space was not entirely clear; if condition (54) was satisfied, one would be dealing with a curved space with conformal geometry. Einstein considered—and took advice on this from the Viennese mathematician Wirtinger—a generalization of the equation of a geodesic in which only ratios of the components $g_{\mu\nu}$ would play a significant role. This led him to consider the possibility of the following generalization:

$$\delta \int d\sigma = 0, \tag{55}$$

where as $d\sigma$ one must take

$$d\sigma = J g_{\mu\nu} \, dx_\mu \, dx_\nu, \tag{56}$$

and J is an invariant of weight -1 formed from the Weyl tensor,

$$J = \sqrt{H}, \tag{57}$$

$$H = H_{iklm} H^{iklm}, \tag{58}$$

the Weyl tensor itself being expressed in terms of the Riemann tensor:

$$
\begin{aligned}
H_{iklm} = R_{iklm} & \\
& - \frac{1}{d-2} \left(g_{il} R_{km} + g_{km} R_{il} - g_{im} R_{kl} - g_{kl} R_{im} \right) \\
& + \frac{1}{(d-1)(d-2)} \left(g_{il} g_{km} - g_{im} g_{kl} \right) R.
\end{aligned}
\tag{59}
$$

Here, R is the scalar curvature, and d is the dimension of space.

In such a geometry, giving up the concept of metrical connection but using the concept of curvature, Einstein considered a unification of gravitation and electromagnetism that would avoid introducing transportable rods and an explicit operation with electromagnetic potentials. It is probable that systematic development of this geometrical scheme would necessarily have led to the geometry of conformal connection developed in 1923 by Cartan [194]. From the beginning, however, it faced serious physical and mathematical difficulties associated with the interpretation of the electromagnetic field as a geometrical phenomenon, the construction of an action functional, etc. At the end of the paper, Einstein wrote, fully understanding the sketchy and model nature of the scheme he had proposed: "Our aim here was merely to present a logical possibility that warrants publication irrespective of whether or not it will be used in physics" ([192], p. 101; [193]). The German mathematician Bach attempted to develop a similar scheme at approximately the same time [195]. In the fifth edition of *Raum-Zeit-Materie*, Weyl judged these attempts as having no prospects, since the invariance possibilities in the scheme did not make it possible to obtain a sensible expression for the action functional ([64], p. 324).

Einstein took up the active development of unified geometrized field theories; in fact, the theory of 1919 was already fully in the spirit of such theories. It did not pose the problem of geometrizing electromagnetism and, in essence, did not invoke a new geometry more general than Riemannian geometry. In the conformal theory of 1921, however, a geometrical scheme that was even more general than in Weyl's theory was put forward. Incidentally, the nature of the geometrization of the electromagnetic field remained obscure. This direction was not subsequently developed.

From 1920, Einstein began his series of foreign journeys, in which he gave reports and lectures on the theory of relativity. In 1920 he was in Leyden and Copenhagen (for about five weeks altogether). In 1921 he spent the whole of February in Vienna, Prague, and Amsterdam, and then from April to June he was in the United Stated and Britain; in October, he was in Italy. In 1922, he was abroad for almost six months, visiting France, spending the summer in Holland, and passing November to March 1923 in Japan, Palestine, and Spain. These journeys not only helped to popularize the theory of relativity in the world, but also had a larger, even political, significance, helping to restore the scientific ties broken by the war and raising the prestige of German science.

At Princeton in May 1921, Einstein gave four famous lectures, soon published as a separate book under the title *The Meaning of Relativity*, which contained a masterly exposition of the foundations of the special and general theory of relativity [196]. The Princeton lectures contained the bare bones, or foundations, of the theory of relativity and were not actually concerned with unified field theories. Just a single remark indicates the change in Einstein's position with regard to the unification problem. Discussing the question of the equations of electrodynamics in the framework of general relativity, he emphasized that the standard way of taking into account the electromagnetic field in the equations of gravitation, by using the corresponding energy–momentum tensor "has been considered arbitrary and unsatisfactory by many theoreticians." Einstein continued

> Nor can we in this way conceive of the equilibrium of the electricity which constitutes the elementary electrically charged particles. A theory in which the gravitational field and the electromagnetic field enter as an essential unity would be much preferable. ([196], p. 108)

Mentioning the results of Weyl and Kaluza (although Kaluza's paper was published only in December 1921), Einstein noted that he was "convinced that they [i.e., these results] do not bring us nearer to the true solution of the fundamental problem." He also did not mention his conformal theory.

In the fall (on September 19) the volume of the *Encyklopädie der mathematischen Wissenschaften* with Pauli's paper on the theory of relativity, which contained a detailed review and critical analysis of the theories of Mie and Weyl and Einstein's 1919 theory, was published [10, 63]. At the same time, Einstein began to be interested in the idea of an experiment capable of determining unambiguously whether light had a wave or corpuscular nature. On December 8, along with Kaluza's paper, he presented a paper of his own in which he proposed an experiment with radiation of light by canal rays and reported that he was about to prepare the experiment with Geiger. On December 30, Einstein reported the successful completion of the experiment and the conclusion in favor of quanta to Born:

> The experiment on light emission had now been completed, thanks to Geiger and Bothe's splendid cooperation. The result: The light emitted by moving particles of canal rays is strictly monochromatic while, according to wave theory, the color of the elementary emission should be different in the different directions. It is thus clearly proved that the wave field does not really exist, and that the Bohr emission is an instantaneous process in the true sense. This had been my most impressive scientific experience in years. ([89], p. 65)

This experiment had the most direct relation to investigations in unified field theory. The successes of quantum theory and the failures of Hilbert, Weyl, Eddington, and Einstein himself in the attempts to unify gravitation and electromagnetism (in its classical, Maxwellian form) could indicate that the electromagnetic field in reality had a quantum structure and that the unification must take this into account from the very beginning. The experiment with the canal rays confirmed this point of view. As a result, the idea of an essentially classical continuum structure of the electromagnetic field, the previous basis of all attempts at geometrical synthesis of gravitation and electromagnetism, must be shattered. This is why Einstein wrote that the result of the experiment had been for him the "most impressive scientific experience in years."

Less than a month later, however, Einstein, largely under the pressure of criticism by von Laue and Ehrenfest, understood the error of his conclusion. The instructive history of this error is considered in detail in a paper by M. Klein [152]. On January 28, 1922 Einstein wrote to Sommerfeld: "I must tell you that the experiment on which I placed such hopes proves nothing. A more rigorous examination has shown that the wave theory leads to the same consequences as the quantum theory . . . " ([68], p. 231). In a paper presented to the Berlin Academy of Sciences on February 2, 1922 Einstein recognized the mistake of his original idea [197].[21]

Considering Kaluza's theory, we have already mentioned the joint paper of Einstein and Grommer, which was received by the editors of the Jerusalem journal on January 10, 1922 and was devoted to proving the nonexistence of an everywhere-regular centrally symmetric solution in this theory. Einstein now firmly adopted the strategy of the program of unified geometrized field theories. The paper begins,

> At the present time, the most important question in the general theory of relativity is that of the unified nature of the gravitational and electromagnetic fields. Although the unified nature of these two forms of field does not in any way follow *a priori*, the overcoming of this dualism would undoubtedly be a great success of theory. [183]

Having criticized the first attempt in this direction, namely, Weyl's theory, the authors of the paper highly rated the new five-dimensional theory of Kaluza ("which eliminates all these shortcomings [i.e., the shortcomings of Weyl's theory] and is distinguished by a remarkable formal simplicity") and briefly

---

[21] In February 1922, Einstein evidently wrote to Born (the letter is undated): "I too committed a monumental blunder (*ein monumental Bock geschossen*) some time ago (my experiment on the emission of light with positive rays), but one must not take it too seriously. Death alone can save one from making blunders" ([89], p. 71). Born's commentary on these words of Einstein, written in 1965, appears very appropriate:

> Here Einstein admits that the considerations which led him to the positive-ray experiments were wrong: "a monumental blunder *(ein kapitalen Bock)*". I should add that now (1965), when I read through the old letters again, I could not understand Einstein's observation at all and found it untenable before I had finished reading [we recall, however, that Born and Franck, in answer to Einstein's letter of December 30, 1921 wrote that they were "very much shaken by the contents of [the letter]" and that they could not "reconstruct the set-up of the positive-ray experiment for ourselves ..." [89], p. 65)]. This is, of course, quite simply because we have learned a good many things about the propagation of light during the intervening forty-odd years. The same is true of the idea that the laws of the propagation of light in transparent media have nothing to do with quanta but are correctly described by the wave theory.... It is quite possible that Laue had already realized this at that time, and used it in argument against Einstein's ideas. Einstein understood this and agreed that it was a blunder. [Ibid.]

We note that the earliest mention of the idea of the experiment with canal rays is probably to be found in Einstein's letter to Sommerfeld on September 27, 1921 ([68], p. 226). About a month later he wrote about it to Besso ([56], p. 102).

considered its foundations. They noted the "fundamentally weak point of Kaluza's theory" associated with the absence of a direct operational significance of ds^2 in the five-dimensional space, and also the "questionable asymmetry" associated with the distinguishing of the fifth dimension by the cylinder condition.

In view of the recent result on the nature of light, Einstein could conclude that all attempts to unify gravitation and the electromagnetic field (in his classical manner) could hardly lead to success. As we know, for Einstein one of the main criteria for the promise (or fruitfulness) of a unified field theory was an affirmative answer to the question of the existence in the theory of an everywhere-regular centrally symmetric solution that could be interpreted as an electron (or, more generally, as a charged elementary particle). The absence of such solutions was one of the main reasons why Einstein rejected both Weyl's theory and his own theory of 1919. The calculation made by Einstein (with Grommer) showed that "in Kaluza's theory there is no centrally symmetric solution that depends only on the $g_{\mu\nu}$ that could be identified with a (nonsingular) electron" [183].

The authors proceeded from a variational formulation of the theory with Lagrangian

$$H = g^{\mu\nu}\Gamma^{\alpha}_{\mu\beta}\Gamma^{\beta}_{\nu\alpha}, \tag{60}$$

the field equations of which in the first approximation take the form

$$\frac{\partial\Gamma^{\sigma}_{\mu\nu}}{\partial x_{\sigma}} = 0. \tag{61}$$

They then formulated conditions for the central symmetry of a static solution (the vanishing of g_{14}, g_{24}, g_{34}, g_{15}, g_{25}, g_{35}, the possibility of representing the purely spatial $g_{\mu\nu}$ in the form $g_{\alpha\beta} = \lambda\delta_{\alpha\beta} + \mu x_{\alpha}x_{\beta}$, where λ and μ are functions of $r = \sqrt{x_1^2 + x_2^2 + x_3^2}$; g_{44}, g_{45}, g_{55} must also be functions of r alone).

Varying the action with the Lagrangian (60) calculated under these conditions with respect to the variables g_{44}, g_{45}, g_{55}, they obtain the relations

$$g_{44}/g_{45} = \text{const}, \quad g_{55}/g_{45} = \text{const}. \tag{62}$$

Noting that at infinity the space-time manifold must be flat and that the electrostatic potential must vanish, the authors noted the vanishing of the ratios (62) as well. This proved the nonexistence of a spatially inhomogeneous electric potential, and thus the nonexistence of a corresponding solution.

In a letter to Weyl on June 6, 1922, Einstein gave a somewhat pessimistic estimate of the situation with regard to unified geometrized field theories, giving, incidentally, preference to Kaluza's theory:

I do not believe in your connection between the electric field and interval curvature (*Streckenkrümmung*) [he is here referring to Weyl's theory]. I find the Eddington argument to have this in common with Mie's theory: it is a fine frame, but one cannot see how it can be filled. Have you been through Kaluza's research? He seems to me to have come closest to reality, even though he, too, fails to provide the singularity-free electron. To admit singularities does not seem to me the right way. I think that in order to make real progress we must once more find a general principle conforming more truly to nature. ([57], p. 177 of the English translation)

Thus, at the beginning of the 1920s Einstein had passed from a clearly negative attitude to the concept of geometrical unification of the gravitational and electromagnetic fields (1915–1918) to the recognition of this idea as the basic strategy for the development of fundamental physical theory (1919–1921).

Although he sharply criticized each of the actual schemes proposed for the geometrical synthesis of the fields—the theories of Hilbert, Weyl, Eddington, and Kaluza—the idea itself, grandiose in design, took an ever stronger hold on Einstein. From time to time he attempted to develop first one form and then another. To a certain degree, Einstein's theory of 1919 was related to Hilbert's theory; Einstein's conformal theory of 1921 was a direct development of Weyl's theory; the joint paper with Grommer was devoted to Kaluza's theory; and beginning in 1923 as we shall see, Einstein attempted to modify Eddington's theory.

Einstein regarded as one of the main criteria of variability of the different forms of unified geometrized field theories the existence in the theory of everywhere-regular static spherically symmetric solutions that could be interpreted as charged particles (above all the electron). Thus, from the very beginning he associated the searches for a unified geometrical description of fields with the maximum problem of every unified theory—the explanation of the existence of particles on its basis. In the theory of 1919, it was this maximum problem that was in the foreground, while the problem of the geometrical unification of gravitation and electromagnetism was not even posed.

As we have attempted to show, Einstein's investigations in the unification program and his attitude toward it may have been influenced by his rare (for that time) conviction of the real existence of light quanta, which he hoped to confirm in 1921–1922 by a direct experiment. The idea behind this experiment, which had the aim of precluding the possibility of a wave interpretation of the structure of radiation, was incorrect. The idea of wave–corpuscle dualism of light, of which he was, of course, one of the main authors, was frequently questioned by Einstein himself; indeed, in 1921 the continuum advocate Einstein appeared to give preference to a purely corpuscular point of view.

Summary

If we regard a research program as a matrix that generates theories of a definite type, then the main indicator of the appearance of such a program is the beginning of multiple production of these theories. Such a situation was created in 1921 with regard to unified geometrized field theories. Although the center of attention was still Weyl's theory, which in fact at this time underwent significant changes (compared with the original form of 1918), in the course of one year there arose several essentially different unified theories (or projects of such theories)—Eddington's affine theory, Kaluza's five-dimensional theory, and Einstein's project of a conformal theory. These were unified by a common aim and mathematical formalism. The unity was expressed in the search for a generalization of Riemannian space-time structure, the basis of general relativity, that would make it possible to interpret not only the gravitational but also the electromagnetic field as a manifestation of the geometry of the space-time continuum. The characteristic features of this geometrical approach, with certain specific features manifested in the considered theories, were listed at the beginning of this chapter. They were geometrization of physical interactions, continuity of the basic structures (a classical field identified with the space-time continuum), classical causality, and dominance of the mathematical structure and the axiomatic and deductive aspects.

Subsequently, Einstein himself and other physicists who had been concerned with unified geometrized field theories (Pauli, Bergmann, Rumer, and others) emphasized that, essentially, there were two main directions in the construction of these theories: (1) a generalization going beyond the framework of Riemannian geometry (mainly by some generalization of the concept of connection—in the theories of Weyl and Eddington and in Einstein's conformal theory) and (2) an increase in the number of the dimensions of the four-dimensional Riemannian space-time (Kaluza's five-dimensional theory).[22] Both of these directions arose around 1921.

---

[22] For example, in 1938 Einstein wrote (in a joint paper with Bergmann): "So far two fairly simple and natural attempts to connect gravitation and electricity by a unitary field theory have been made, one by Weyl, the other by Kaluza" ([198], p. 683). Bergmann, in one of his first books (1942), in which particular attention was devoted to unified theories, also considered Weyl's theory, Kaluza's theory, and generalizations of Kaluza's theory [12]. Pauli, reissuing his encyclopedia article in English in 1955, added a supplement devoted to unified theories that had appeared after 1920. He divided them into two large groups: (a) theories with asymmetric g_{ik} and Γ^l_{ik} (an example of which is Eddington's theory) and (b) five-dimensional and projective theories (initiated by Kaluza's theory) ([2], p. 225 ff).

The considered unified theories encountered difficulties of the same kind whatever the theory (Weyl's, Eddington's, or Kaluza's): the absence of sufficient physical grounds for the geometrization of the electromagnetic field and, in contrast to general relativity, the gap between the physical and geometrical foundations of the theory; the non-uniqueness in the choice of the Lagrangian of the theory; the absence of everywhere-regular static spherically symmetric solutions; the absence of new effects amenable to measurement; etc. However, the triumph of general relativity, the grandeur and mathematical depth of the unified theories, and their level of theorization, contrasting with the empiricism and eclecticism of the quantum schemes at the beginning of the 1920s, attracted many physicists and mathematicians to the unification program, generating considerable hopes that a solution to the problem along these lines was an entirely realistic prospect and not beyond the moon. The eminent names of the pioneers of the program (Einstein, Hilbert, Weyl, Eddington, Pauli, and others) was a warrant of the depth, seriousness, and good prospects of the direction.

In its most consistently developed form the program of unified geometrized field theories was, as was emphasized at the beginning of this chapter, opposed to the quantum theoretical program, but in 1921 this opposition was not entirely clear. Some of the authoritative supporters of unified theories regarded the existence of field–matter dualism as inescapable and did not even pose the maximum problem of the program associated with a field interpretation of particles and also did not attempt to deduce quantum laws on a purely field geometrical basis (Weyl, Eddington, and others).

Even before the unification program was fully formulated, some supporters of Weyl's theory, who had made a significant contribution to its development, saw in it fundamental difficulties of a physical nature and switched their efforts to the field of quantum theory (Pauli). Other supporters, and at times, founders of the direction, for example Weyl, gradually lost interest in the unification program after 1921. This was due both to the absence of real physical progress as well as to the significant mathematical difficulties in the path of its development. The program took an even stronger hold of Einstein: even after the development of quantum mechanics, which opened up an epoch of new theorization, when the interests of the overwhelming majority of theoreticians were turned to the sphere of the quantum theoretical program, Einstein con-

---

Rumer, the author of the article on unified field theories in the *Fizicheskii éntsiklopedicheskii slovar* (*Physics Encyclopedia*, 1962) also identified two classes of such theories: non-Riemannian four-dimensional (with Weyl's theory as paradigm) and the five-dimensional direction (with Kaluza's theory as paradigm) ([199], pp. 5–6).

tinued with even greater energy to work on the program of unified geometrized field theories. He tried out more and more new geometrization schemes and returned to old forms, modifying them. Sometimes, partly due to his authority, he succeeded in attracting to the work on unified field theories young researchers, mainly mathematicians (Grommer, Mayer, Rosen, Hoffmann, Bergmann, and others). Sometimes there were also brief bursts of interest in these theories among physicists who had successfully worked in the field of quantum theory. For example, in 1926–1927 there were definite hopes of establishing deep connections between five dimensions and quantum mechanics (work by O. Klein, de Donder, Fock, de Broglie, Gamow and Ivanenko, Ehrenfest and Uhlenbeck, London, and others), and at the beginning of the 1930s the projective formulation of the five-dimensional approach attracted the interest not only of mathematicians (Veblen, van Dantzig, Schouten, and others) but also physicists (Pauli).

Only a few, above all Einstein, did not depart from the framework of the unification program over the following years, so there is particular interest in the history of unified field theories in considering the "world line" of Einstein in this period. In the present chapter, we have considered one of the most important periods in the history of the unification program (1921), and we have attempted to consider all the main theoretical schemes advanced at that time by different investigators. In the following chapter, choosing the main figure in this history, Einstein, we attempt to follow the evolution of the unified theories, beginning with 1922–1923 and ending with 1932–1933, doing this on the basis of his papers. Thus, the approaches used in these two chapters (the present one and the following one) bear approximately the same relationship as the representations of Euler and Lagrange in hydrodynamics.[23]

It is true that 1921 was not an ordinary year in the history of unified field theories, and Einstein was certainly not an ordinary figure in this history. In addition, the year 1921 has been taken with a certain latitude (1919–1922), and the evolution of Einstein's investigations will be considered against a larger background of studies in this field.

---

[23] In the Eulerian representation, a certain point is identified in space, and one considers what happens there as different particles (infinitesimal volumes) of fluid pass through it; the behavior of the fluid velocity at the point is studied. In the Lagrangian representation, one identifies a certain fluid particle (element of fluid volume) and consideres the trajectory of its motion from its initial position to the present time; the velocity of this particle is expressed as a function of the time and the coordinates of its initial position. The first method is sometimes called statistical, since to describe the liquid as a whole it is necessary to average over the points of space. The second is called historical, because a "history" is followed in it ([200], p. 539).

CHAPTER 5

Einstein's Decade of Hopes and Disappointments

Several circumstances justify a closer examination of Einstein's "world line" in the stream of investigations on unified field theories in the period from 1923 to the beginning of the 1930s. At this time he became the recognized leader of the investigations, taking over, as it were, the baton from Weyl, who had been the leading authority for the previous five years. In Einstein's studies, the basic features in the program of unified geometrized field theories are most clearly and consistently expressed. In particular, he never lost sight of the maximum problem that confronted him: interpretation of the corpuscular and quantum aspects of the structure of matter.

Einstein also carefully followed the work of other investigators, picking up the ideas that appeared promising. He worked on almost all the main forms of geometrical unification of fields, returning at times to theories advanced earlier; the most durable were the Weyl–Eddington and Kaluza directions considered in Chapter 4.

During the period 1922–1923 the fundamental principles of the classification of generalized spaces going beyond the framework of Riemannian geometry were developed in studies by Cartan, Schouten, Veblen, and others, using the work of Levi-Civita and Weyl as a basis. Except for the five-dimensional Riemannian schemes, the overwhelming majority of the geometries used to construct unified geometrized field theories were special cases of the geometry of four-dimensional affine spaces. The classification of spaces of this type given by Cartan in 1923 was distinguished by its simplicity and transparency [194]. During 1920–1930, however, neither Einstein nor the other adherents of the unification program attempted to use geometrical schemes obtained by systematic selection in accordance with Cartan's classification, although by the beginning of the 1930s almost all possible forms of four-dimensional non-Riemannian affine spaces had been tried in one way or another.

We shall briefly describe Cartan's classification, which was also used in the single systematic survey of unified field theories: Tonnelat's book [22] (see also [301]). According to Cartan, an arbitrary space with affine connection can be described by two forms of curvature (rotation, or ordinary Riemannian, curvature and homothetic, or segmental, curvature) and by one form of torsion. Cartan considered the change of the increments of an arbitrary vector and of unit vectors under parallel transport around an infinitesimal closed loop: $\oint d\mathbf{m}$ and $\oint d\mathbf{e_i}$.[1] In an affine space, in particular in a Euclidean space, both these integrals vanish. In Riemannian space, which is a special case of an affine space, the first integral is also equal to zero, and this leads to a symmetric affine connection. The second integral $\oint d\mathbf{e_i}$ is nonzero, however, and this leads to the presence at each point of a nonvanishing rotation curvature. Finally, in an arbitrary affine space the integral $\oint d\mathbf{m}$ is also nonzero, and this leads to nonvanishing torsion. If the increment $\oint dl$ of the length of a vector in this case (we have in mind the case of an arbitrary metric connection) is nonvanishing, then there exists a further form of curvature—homothetic curvature. In the case of a Riemannian space, $\oint dl$, and the homothetic curvature is also zero. If $ds^{\mu\nu}$ is the area of the closed loop, these three quantities can be expressed in terms of the curvature tensor $R^{\nu}_{\mu\rho\sigma}$ and the coefficients $\Gamma^{\kappa}_{\mu\nu}$ of the affine connection as follows:

$$\Omega^{\nu}_{\mu} \text{ (rotation curvature)} = -R^{\nu}_{\mu\rho\sigma}\, ds^{\rho\sigma}, \tag{1}$$

$$\Omega^{\rho} \text{ (torsion)} = -(\Gamma^{\rho}_{\mu\nu} - \Gamma^{\rho}_{\nu\mu})\, ds^{\mu\nu}, \tag{2}$$

$$\Omega \text{ (homothetic curvature)} = \Omega^{\mu}_{\mu} = -R^{\mu}_{\mu\rho\sigma}\, ds^{\rho\sigma}. \tag{3}$$

This immediately leads to the following classification of spaces with an affine (more precisely, metric) connection if the homothetic curvature is taken into account. Without torsion, we have

$$\Omega^{\mu}_{\nu} = 0, \ \Omega^{\mu} = 0, \ \Omega = 0 \ \text{ for Euclidean space;}$$

$$\Omega^{\mu}_{\nu} \neq 0, \ \Omega^{\mu} = 0, \ \Omega = 0 \ \text{ for Riemannian space;}$$

$$\Omega^{\mu}_{\nu} \neq 0, \ \Omega^{\mu} = 0, \ \Omega \neq 0 \ \text{ for affine space;}$$

---

[1] More precisely, the projections of the corresponding vectors and their increments onto the tangent affine space at some arbitrary point of the manifold are considered.

These cases were used in the unified theories of Weyl and Eddington and in Einstein's theories from 1923–1925, which will be considered in the present chapter. The affine spaces with torsion correspond to

$$\Omega^\mu_\nu \neq 0, \ \Omega^\mu \neq 0, \ \Omega = 0,$$
$$\Omega^\mu_\nu = 0, \ \Omega^\mu \neq 0, \ \Omega = 0,$$
$$\Omega^\mu_\nu \neq 0, \ \Omega^\mu \neq 0, \ \Omega \neq 0.$$

These spaces were used in various forms of unified gravitational field theories by Einstein, Infeld, and others. As far as we know, the two remaining cases ($\Omega^\mu_\nu = 0, \Omega^\mu = 0, \Omega \neq 0$ and $\Omega^\mu_\nu = 0, \Omega^\mu \neq 0, \Omega \neq 0$) were not considered.

At that time, Schouten [202] developed a similar classification. In condensed form, it is given in the well-known book of Schouten and Struik published in 1935 [203]. Schouten showed that an arbitrary affine connection can be expressed by the relation

$$\Gamma^\kappa_{\mu\lambda} = \left\{ {\kappa \atop \mu\lambda} \right\} + S^{\cdot\cdot\kappa}_{\mu\lambda} - S^{\cdot\kappa}_{\lambda\cdot\mu} - S^{\cdot\kappa}_{\mu\cdot\lambda} + \tfrac{1}{2}(Q^{\cdot\cdot\kappa}_{\mu\lambda} + Q^{\cdot\kappa}_{\lambda\cdot\mu} - Q^{\kappa}_{\cdot\mu\lambda}), \qquad (4)$$

where $S^{\cdot\cdot\kappa}_{\mu\lambda} = \tfrac{1}{2}(\Gamma^\kappa_{\mu\lambda} - \Gamma^\kappa_{\lambda\mu})$ expresses the degree of symmetry of the connection (transport), $\left\{ {\kappa \atop \mu\lambda} \right\}$ are Christoffel symbols, and $Q^{\cdot\kappa\lambda}_\mu = \Delta_\mu g^{\kappa\lambda}$ are the components of the covariant derivative of the second-rank fundamental tensor $g_{\mu\nu}$ (introduced formally in the affine geometry). Then if $S^{\cdot\cdot\kappa}_{\mu\lambda}$ and $Q^{\cdot\cdot\kappa}_{\mu\lambda} = 0$, the connection is called symmetric and metric (Riemannian geometry); but if $S \neq 0$, then the connection is asymmetric (this being equivalent to the presence of torsion); the case $S = 0$ and $Q^{\cdot\kappa\lambda}_\mu = Q_\mu a^{\kappa\lambda}$, which Schouten called symmetric and semimetric, leads to Weyl's geometry. Such a classification, which in essence is equivalent to Cartan's, was used to classify unified field theories by, for example, Schmutzer [23] and Rodichev [204].

In fact, Schouten's classification took into account the possibility of different types of connection for the contravariant and covariant vectors ($\Gamma^\lambda_{\mu\nu}$ and $\Gamma'^\lambda_{\mu\nu}$), the measure of their difference being determined by the quantities $C^\lambda_{\mu\nu} = \Gamma^\lambda_{\mu\nu} - \Gamma'^\lambda_{\mu\nu}$. Then all possible affine geometries (of which there are $3^3 = 27$) are determined by different combinations of these entities (tensors) $C^i_{kl}, S^i_{kl}, Q_{ikl}$, which Kagan (in his brilliant review paper given at the All-Russian Congress of Mathematicians at Moscow in May 1927) called Schouten characteristics [205]. As far as we know, spaces with nonvanishing C^j_{kl} have also not been used to construct unified field theories.

Although from 1922 mathematics had already given a systematic principle for testing various approaches to the solution of the problem of geometrical synthesis of gravitation and electromagnetism, the adherents to the program of unified geometrized field theories relied on their own intuition and heuristic physical arguments in choosing a geometrical scheme.[2] In addition, there were in essence no physical arguments for choosing affine geometry. During 1923–1924, a start was made, by Cartan above all, on the development of the theory of spaces of projective and conformal connections, which are constructed locally as projective and conformal spaces and make it possible to establish a correspondence between local spaces by means of projective and, respectively, conformal transformations [194]. These geometries also could have been taken as the basis of a geometrical unification of fields. On the other hand, after Kaluza's paper it became possible to use manifolds with their number of dimensions greater than four.

Other classifications of unified theories are also possible. We consider the classification ideas of Weyl and Pauli formulated at the beginning of the 1950s. Weyl distinguished three classes of theories: (1) theories based on the use of gauge invariance, (2) theories of the affine type, and (3) theories that do not require symmetry of the metric tensor or of the affine-connection coefficients [18].

In this characteristic classification, the basic principle is the method used to introduce the electromagnetic field. Thus, the first class includes both Weyl's theory and Kaluza's five-dimensional theory. Both theories are associated with an extension of the group of arbitrary continuous coordinate transformations that, in one way or another, induces a gauge transformation of the electromagnetic potential. The projective versions of the five-dimensional scheme are also included in this class.

In the affine theories, the fundamental quantities are the coefficients of the affine connection. In them, there exists considerable freedom in the choice of

---

[2] This feature of the development of unified geometrized field theories in the 1920s was noted by Tonnelat:

> In fact, the majority of authors originally did not pay attention to the existence of a number of quantities (two forms of curvature and torsion) that characterize the structure of a space and which could be used. It was for this reason that for a long time unified theories were somewhat arbitrary and based on a geometry that could allow artificial modifications. However, if in the construction of unified theories one remains within the framework of an affine manifold, then there is actually a possible subchoice of only three elements characterizing the structure of the manifold: two types of curvature and torsion. ([201], pp. 372–373)

invariants, making it possible, through the action principle, to obtain equations of both the gravitational and the electromagnetic field. The first purely affine theory was Eddington's theory. During 1923–25, Einstein actively developed this direction.

The third class of theories includes various schemes based on giving up the requirement of the conditions of symmetry for either the metric tensor or the coefficients of the affine connection. For example, if symmetry of g_{ik} is not required, this tensor can be decomposed into a symmetric part, which is identified with the gravitational field, and into an antisymmetric part, which is identified with the electromagnetic field. Actually, a sharp distinction between second and third class theories does not exist. Theories of the third class can be either purely affine or affine-metric theories. In theories of the second class, the electromagnetic field is often related to the antisymmetric part of the second-rank curvature tensor (Ricci tensor).

From this point of view, Pauli's classification, introduced in the Supplement of the English edition of his *The Theory of Relativity* (1956) ([2, 63], p. 225ff), appears more natural. He divided all unified geometrized theories into two groups: theories with asymmetric g_{ik} and Γ^l_{ik}, and five-dimensional and projective theories. In this Supplement, Weyl's theory was not considered, since it had already been treated in the first edition of Pauli's book. It is not entirely clear whether it should be included in the second group, or whether it must form a special group. In accordance with Weyl's ideas, it should probably be included in the second group.

Finally, Tonnelat regarded it as helpful (and we actually adhered to this point of view in previous chapters) to divide all theories in to those that solve only the minimum problem (in other words, theories in which electromagnetism and gravitation are combined into a single whole on a geometrical basis) and so-called nondualistic theories, in which the maximum problem is at the forefront (reduction of the field sources to the field itself) [22, 201]. Naturally, a consequential unified field theory of the first kind must also be ultimately nondualistic. To be nondualistic, unified field theories constructed on the basis of general relativity must be nonlinear.

In conclusion, we note that Cartan's classification (and the closely related classification of Schouten) is the most consistent, detailed and systematic, especially for consideration of non-Riemannian four-dimensional schemes. This is the reason it was used in various reviews of unified gravitational field theories, above all in the studies quoted above of Tonnelat, and also reviews by Schmutzer, Rodichev, and others. From the physical point of view, however, the Pauli–Weyl classification also appears fully justified. The majority of the unified gravitational field theories considered earlier and in this chapter are theories that solve only the minimum problem, although the possibility of solving the maximum problem within their framework was also often mentioned during their development.

Returning to Einstein's searches for a field synthesis and anticipating what is to follow, we emphasize once more that, first, Einstein tried almost all possible forms of geometrical unification: five-dimensional, including projective schemes, and four-dimensional non-Riemannian schemes with different combinations of Ω_{μ}^{ν}, Ω^{μ}, and Ω. Second, Einstein never lost sight of the maximum problem. Some of his studies were specially devoted to such nondualistic schemes; in others, he discussed the possibilities for solving the maximum problem in the framework of specific unified theories (the question of the existence of nonsingular static centrally symmetric solutions); in yet others he sought different possibilities for a field description of the sources (particles).

AFFINE THEORY: 1923–1924

In the summer of 1922, as we have seen, Einstein had the greatest hopes for the five-dimensional approach, despite the fact that six months earlier he had, together with Grommer, established the nonexistence in the five-dimensional theory of nonsingular static centrally symmetric solutions. He still regarded Weyl's theory as incorrect, while Eddington's theory appeared to him too broad and indefinite, although attractive in its basic idea ("it is a fine frame, but one cannot see how it can be filled" [57], p. 177).

From November 1922 to the spring of 1923, Einstein made a long and most attractive journey. He was in China and Japan, then Palestine and Spain. In Japan Einstein learned that he had been awarded the Nobel Prize for physics for 1921. While there, he also wrote the foreword to his first collected works, prepared by a Japanese publishing house (December 27) [206]. In a paper "On the present crisis of theoretical physics" published in December 1922 in the Japanese newspaper *Kaizo* he said nothing about unified theories but concentrated mainly on the crisis of the fundamental concepts of physics brought about by the discovery of quantum properties of the microscopic world: "It must be expected," he wrote, "that the progress of science will give rise to a revolution in its foundations no less deep than the one associated with field theory" ([207], p. 57). The paper ended with an expression of doubt about the ability of the mathematical formalism of differential equations to provide a basis of quantum theory:

> In order to find a true basis of quantum relations, it appears that we need
> a new mathematical language. At the least, the expression of the laws of

nature in the form of a combination of differential equations and integral conditions, as we do today, contradicts common sense. ([207], p. 60)

Despite this, Einstein continued to work in the direction of unified geometrized field theories, relying essentially on continuum concepts of mathematics, above all differential geometry and the theory of differential equations. On February 15, Planck presented Einstein's paper "On the general theory of relativity" to the Prussian Academy of Sciences. This paper had been completed in January 1923 in Japan and sent to Berlin [176]. It contains a first sketch of a unified geometrized field theory which followed the lead of Eddington's theory and which Einstein called affine field theory.

Having adopted Eddington's approach and taken as basis the law of "parallel transport,"

$$\delta A^\mu = -\Gamma^\mu_{\alpha\beta} A^\alpha \, dx^\beta, \tag{5}$$

and the coefficients of the symmetric affine connection $\Gamma^\mu_{\alpha\beta}$ as variables of the theory, Einstein decomposed the Ricci tensor into symmetric and antisymmetric parts,

$$R_{kl} = g_{kl} + \varphi_{kl}, \tag{6}$$

which he identified with the metric tensor g_{ik} and the tensor φ_{kl} of the electromagnetic field respectively.

Einstein then sought a suitable Lagrangian that, through the variational principle, would make it possible to obtain the equations of the gravitational and electromagnetic fields. He noted as the most promising possibility

$$\mathcal{L}\sqrt{-g} = 2\sqrt{-|R_{ik}|}, \tag{7}$$

i.e., an expression that has the meaning of the tensor density of the volume element and which "is formed from the R_{kl} without decomposition into symmetric and antisymmetric parts" ([176], p. 34). Einstein saw a significant quality of the theory in this absence of decomposition:

> If this Hamilton function is usable, then the theory will arrive in an ideal manner at unification of gravitation and electricity in a single concept; for not only will both forms of field be determined by the Γ alone, but also the Hamilton function will be unified, whereas hitherto it has consisted of terms logically independent of each other. [Ibid.]

Einstein further replaces the expression (6) by

$$\lambda^2 R_{kl} = g_{kl} + \varphi_{kl},$$

which was appropriate if the invariant $g_{ik} dx_i dx_k$ was to be on the order of "human dimensions" (λ must be a very large number). Then

$$\frac{1}{\lambda^2} g_{kl} = -\frac{\partial \Gamma^\alpha_{kl}}{\partial x_\alpha} + \frac{1}{2}\left(\frac{\partial \Gamma^\alpha_{k\alpha}}{\partial x_l} + \frac{\partial \Gamma^\alpha_{l\alpha}}{\partial x_k}\right) + \Gamma^\alpha_{k\beta}\Gamma^\beta_{l\alpha} - \Gamma^\alpha_{kl}\Gamma^\beta_{\alpha\beta}, \qquad (8)$$

$$\frac{1}{\lambda^2}\varphi_{kl} = \left(\frac{\partial \Gamma^\alpha_{k\alpha}}{\partial x_l} - \frac{\partial \Gamma^\alpha_{l\alpha}}{\partial x_k}\right). \qquad (9)$$

Then variation of the action with the Lagrangian (7) makes it possible to obtain 40 equations, from which all the $\Gamma^\alpha_{\mu\nu}$ can be determined. At the same time, symmetric and antisymmetric tensor densities $s^{ik}\sqrt{-g}$ and $f^{ik}\sqrt{-g}$, associated with the gravitational and electromagnetic fields, respectively, make their appearance. Einstein called $\sqrt{g} i^i = \frac{\partial(\sqrt{-g} f^{ik})}{\partial x_k}$ the current density. With allowance for the electromagnetic field, the coefficients of the affine connection can be expressed in the form

$$\Gamma^\alpha_{kl} = \frac{1}{2} s^{\alpha\beta}\left(\frac{\partial s_{k\beta}}{\partial x_l} + \frac{\partial s_{l\beta}}{\partial x_k} - \frac{\partial s_{kl}}{\partial x_\beta}\right) - \frac{1}{2} s_{kl} i^\alpha + \frac{1}{6}\delta^\alpha_k i_l + \frac{1}{6}\delta^\alpha_l i_k. \qquad (10)$$

Einstein showed further that in the absence of an electromagnetic field the coefficients of the affine connection were identical to the ordinary Christoffel symbols.

Substituting expression (10) for $\Gamma^\lambda_{\mu\nu}$ in Eq. (9), he obtained the equation

$$\frac{1}{\lambda^2}\varphi_{ik} = \frac{1}{6}\left(\frac{\partial i_i}{\partial x_k} - \frac{\partial i_k}{\partial x_i}\right). \qquad (11)$$

Since the value of λ^2 is very large, finite values of φ_{kl} are possible for very low current densities. If, in accordance with this, the current density everywhere are set equal to zero (except for singular regions), then Maxwell's equations without sources are obtained.

Thus, the proposed affine scheme gave a unified description of the gravitational and electromagnetic fields, a description, moreover, that was in agreement with the correspondence principle, which Einstein always gave great heuristic importance. It was also necessary to investigate the question of the existence of nonsingular particle-like solutions—he wrote:

> It will only be possible to decide whether our theory also encompasses electrical elementary formations after rigorous examination of the case of a centrally symmetric static field. At the least, Eq. (25) [our Eq. (11)] shows that finite values for the current density i^l are possible only if simultaneously i_l becomes a small quantity of order $1/\lambda^2$; thus, the existence of electrons without singularities is not ruled out. [176]

It also appeared to Einstein that in the new theory there was a possibility of explaining the charge asymmetry that exists in nature, and he optimistically concluded:

> This investigation shows that Eddington's general idea in conjunction with Hamilton's principle *leads to a theory almost free of arbitrary elements that correctly reproduces our present knowledge about gravitation and electricity and unifies the two types of field in a truly perfect manner* [our italics] [176].[3]

After he returned to Berlin, Einstein presented a short paper with some refinements of his affine theory to the Prussian Academy of Sciences. This note was published on May 15, 1923 [209]. It repeated in a concise form the results of the first paper, but contained a denial of his remark about charge asymmetry:

> In fact τ, and accordingly, $\left|\mathfrak{r}^{kl}\right|$, is an even function of the antisymmetric part $\mathfrak{f}$ of $\mathfrak{r}^{kl}$, and the term in Eq. (14) [i.e., in the equation
>
> $$\delta\left\{\int\left[-2\sqrt{-r} + \mathfrak{R} - \frac{1}{6}s^{\alpha\beta}i_{\alpha}i_{\beta}\right]d\tau\right\} = 0 \quad]$$
>
> depends quadratically on the current density. Therefore, the theory cannot take into account the difference between the masses of positive and negative electrons. [209]

Einstein presented this theory in a somewhat different, clearer form at the session of the Prussian Academy of Sciences on May 31, 1923 [210]. The only difference was in the choice of the Lagrangian. Einstein now took it to be

$$\mathcal{L}\sqrt{-g} = 2\alpha\sqrt{-g} - \frac{\beta}{2}f_{\mu\nu}(f^{\mu\nu}\sqrt{-g}), \tag{12}$$

which leads to the system of field equations

$$R_{\mu\nu} - \alpha g_{\mu\nu} = -\left[\beta\left(-f_{\mu\sigma}f_{\nu}^{\sigma} + \frac{1}{4}g_{\mu\nu}f_{\sigma\tau}f^{\sigma\tau}\right) + \frac{1}{6}i_{\mu}i_{\nu}\right], \tag{13}$$

$$-\beta f_{\mu\nu} = \frac{1}{6}\left(\frac{\partial i_{\mu}}{\partial x_{\nu}} - \frac{\partial i_{\nu}}{\partial x_{\mu}}\right); \tag{14}$$

---

[3] The original is as follows:

> Die vorstehende Untersuchung zeigt, dass Eddingtons allgemeine Gedanke in Verbindung mit dem Hamiltonschen Prinzip zu einer von Willkühr fast freien Theorie führt, welche bisherigen Wissen über Gravitation und Elektrizität gerecht wird und beide Feldarten in wahrhaft vollendeten Weise vereinigt. ([208], p. 38)

(here, α has the meaning of a cosmological constant, and β is a very small constant that must be set equal to zero in the absence of charges).

At this stage, Einstein noted that Eqs. (13) and (14) ultimately lead to the same results as the equations of Weyl's theory (for a certain special choice of the Lagrangian). Simultaneously, he understood that these "equations do not lead to an electron as a singularity-free solution" [210].[4] At the end of May, he wrote to Weyl:

> I shall soon be sending you the proofs of the paper on the generalized Hamiltonian function.... By and large I am once more in a fairly resigned mood as regards the whole problem. Mathematics are all well and good but Nature keeps dragging us around by the nose. Moreover, there is a comical aspect. I wanted to go in the opposite direction from you and yet I have arrived at the same equations as you with your special action principle.... I wanted to discard the potentials, but they crept back again through the back door.[5] The whole line of thought must be worked out and is of remarkable beauty. But above it stands the marble smile of implacable Nature which has endowed us with more longing than intellect ([57], p. 177 of the English translation).

In the summer of 1923, Einstein continued to regard the affine approach as promising. He wrote a paper "The theory of the affine field" for the British journal *Nature* [212], which almost exactly repeated the earlier paper. Einstein emphasized that his point of departure was "Eddington's idea... of basing 'field physics' mathematically on the theory of the affine connection." The only difference from the earlier paper was the abandonment of the cosmological

---

[4] In May 1923, Einstein worked at Leyden. From there, Ehrenfest reported to A.F. Ioffe (in a letter of May 16, 1923):

> Einstein has already been here for 14 days—he is completely possessed by the "electricity–gravitation" problem. It appeared that he had already found equations describing huge gravitational forces within the electron, as result of which the electron would not disintegrate, but again he had a failure. He is very intensively occupied with this." ([211], p. 167)

The failure about which Ehrenfest wrote was presumably that the theory did not permit an interpretation of the electron as a nonsingular solution of the field equations.

[5] In this theory, Einstein attempted from the very beginning to work solely with the electromagnetic field strengths, but in the last paper (presented on May 31) he found it necessary "for the physical interpretation of the field equation" to return to the potentials $f_\mu = -\frac{1}{\beta} i_\mu$.

term.[6] The field equations of the theory were identical to Eqs. (13) with $\alpha = 0$ and $i^\mu = -\gamma g^{\mu\sigma} f_\sigma$ (here $\gamma \equiv \beta$). Einstein summarized:

> The theory supplies us, in a natural manner, with the hitherto known laws of the gravitational field and of the electromagnetic field, as well as with a connection... of the two kinds of field; but it brings us no enlightenment on the structure of electrons. ([212], p. 449)

In July, Einstein went to Sweden to receive the Nobel Prize. He did not give the traditional Nobel address. The collection of Nobel addresses included a lecture that he had given at approximately the same time (July 11) at Göteborg at a congress of natural scientists of the Nordic countries. The lecture was entitled "Basic ideas and problems of the theory of relativity" and was mainly devoted to basic problems of the special and general theories of relativity [214]. In the first part of his address, however, Einstein touched on the problem of unified field theory:

> Minds are now especially concerned with the problem of the unified nature of the gravitational and electromagnetic fields. The thought that strives to unity of theory cannot accept the existence of two fields having a nature quite independent of one another. Therefore, attempts are made to construct a mathematically unified field theory in which the gravitational and electromagnetic fields are considered merely as different components of one and the same field; moreover, its equations should, as far as possible, not consist of terms that are logically independent of each other. [214]

Einstein noted further that a natural way to achieve such a synthesis was to generalize four-dimensional Riemannian geometry and, thus, geometrize the electromagnetic field. The main difficulty in this approach was, in Einstein's opinion, the absence of an experimental, physical basis for the geometrization of electromagnetism:

> Unfortunately, in this attempt [geometrization of the electromagnetic field], we cannot rely on experimental facts as in the construction of the theory of gravitation (equality of the inertial and gravitational masses) but are forced to restrict ourselves to the criterion of mathematical simplicity, which is not free of arbitrariness. [214]

He mentioned the ideas of Levi-Civita, Weyl, and Eddington and said that the affine approach that he had developed in the spring of 1923 in the spirit of Eddington was the most promising way of solving the problem:

---

[6] This was probably due to Einstein's recognition of his error in Friedmann's paper, which proved the possibility of a dynamic centrally symmetric solution of the equations of the gravitational field without cosmological term. A corresponding note was received by the *Zeitschrift für Physik* on May 31, 1923 [213].

> Finding the simplest differential equations that the affine connection can satisfy, we are justified in hoping that we shall find a generalization of the equation of gravitation that will also contain the laws of the electromagnetic field.... This hope was indeed justified, but we do not know if the formal connection obtained in this manner can be regarded as a real enrichment of physics as long as no new physical connections have been obtained. In particular, a field theory can be recognized as satisfactory, in my opinion, only when it permits the description of elementary electric particles by means of solutions that do not contain singularities. [214]

Thus, from the formal point of view Einstein regarded the affine theory as entirely acceptable, agreeing with the "criterion of mathematical simplicity," but he saw that it had not yet led to any physical progress. Moreover, he could not regard it as satisfactory, since it led to the conclusion that electrons could not be described in its framework by means of nonsingular solutions of the field equations. In this connection, Einstein also mentioned the need to bear in mind the quantum aspects of the behavior of particles and the electromagnetic field, although "the face of this deepest physical problem of the present day [i.e., the quantum problem] the theory of relativity has also proved to be powerless as yet" [ibid.].

Although there is no explicit mention in Einstein's paper of 1923 on unified field theory of the possibility that these theories could lead to observable effects, there is some evidence of searches by Einstein for such a development. Recalling his scientific contacts with Einstein in 1922–23, Gerlach wrote:

> At that time, Einstein proposed that I should take up a quite different problem [before this Gerlach and Stern had made the famous experiment that confirmed spatial quantization in a magnetic field], which, incidentally, had already been worked out by Faraday, namely, the problem of creating a magnetic field by moving matter (or changing the state of its motion). The idea was to make measurements near streams of water or waterfalls. ([215], p. 98)

Despite a long discussion of these experiments and the insistence of Einstein, Gerlach decided not to make them.[7]

In a letter to Born and his wife, Einstein wrote on July 22, 1923:

> Scientifically, I have at present a most interesting question, connected with the affine field theory. There are prospects now of understanding the earth's magnetic field and the electrostatic economy of the earth, and examining the concept experimentally. But we will have to wait for the experiment.

---

[7] I am grateful to B.E. Yavelov for drawing my attention to this evidence.

In a postscript, he reported that

> Franck [obviously, J. Franck from Göttingen], who had just been here, told me that according to the results of measurements of ionized gases already made, the effect I am looking for cannot exist. There is to be no understanding the earth's magnetic field. ([89], p. 80)

Despite this, Einstein did not abandon searched for experimental confirmation of his affine field theory.

He discussed possible consequences with the Brussels physicist A. Piccard in April 1924 ([216], p. 51). In particular, he was concerned with magnetic fields of cosmological origin and the possibility of measuring them. In this connection, Einstein was interested in possible experimental consequences of charge asymmetry reflected in a very small difference between the contributions to the electromagnetic field strengths from positive and negative charges (an idea related to the conception of Mossotti and Zöllner) [38]. Very accurate experiments made in 1925 by Piccard did not confirm the presence of this difference. At the time of his next visit to Leyden in October 1924, Einstein (together with Ehrenfest) considered making a delicate electrostatic experiment that would make it possible to confirm consequences of the affine theory that went beyond Maxwell's electrodynamics. Ehrenfest wrote to Ioffe on October 9,

> Einstein and I immerse ourselves for many hours everyday in an experimental study to establish whether there exists a completely crazy electrostatic effect assumed by him. At the present time, we have not yet sorted out the caprices of the string electrometer. ([211], p. 181)

Ehrenfest was obviously very skeptically inclined toward these experiments. Ten days later (on October 19) he wrote to Ioffe:

> The experiment that I have been making with Einstein has, to my joy, been abandoned. It turned out that the deviations associated with the ionization of air are much greater than the mystical effect itself that he sought. ([211], p. 184)

In a paper "On the ether" written in 1924, Einstein especially distinguished the problem of explaining the magnetic field of the earth and the sun [217]. It appeared to Einstein that on the basis of the notions of Maxwell's electrodynamics it was not possible to explain the rather large strengths of these fields.[8] He wrote:

> It appears more probable that the magnetic fields arise as a result of rotational motion of neutral masses. Such generation of fields is predicted neither by Maxwell's theory in its original form nor Maxwell's theory generalized in the sense of the general theory of relativity. It appears that here nature indicates to us a fundamental fact not yet explained by theory. [217]

In the facts of cosmic magnetism, above all the magnetic fields of the earth and the sun, which did not fit the generally accepted ideas of the time, based ultimately on Maxwell's electrodynamics, Einstein tried to see a direct physical, even experimental basis for an extension of that theory. He sought an electrodynamic analog of the fundamental fact of the equality of the inertial and gravitational masses. The development of the equivalence principle on the basis of the fact and the subsequent extension of the principle to the case of a rotating disk had played an important part in the genesis of the tensor-geometrical conception of gravitation [223]. In this occasion, too, the

---

[8]   The problem of the connection of a magnetic field with rotation of neutral massive bodies had interested Einstein since 1915, when in collaboration with de Haas he undertook an experimental proof of the existence of Ampere currents (the Einstein–de Haas effect) [218]. Having established proportionality of the magnetic and mechanical moments for the electron (the gyromagnetic ratio), Einstein and de Haas extended this conclusion to the rotation of the earth as well. They wrote:

> It is also found that to the rotation of the earth there corresponds a magnetic ponderomotive field that is parallel to the earth's axis and is directed from north to south, having a strength of order 10^{-11}. It is possible that this is the reason for the approximate coincidence of the magnetic axis and the rotation axis. [219]

The American physicist S.J. Barnett had studied the problem of a "rotational" origin of the magnetic field of the earth, having found a corresponding effect when an iron cylinder rotates (Barnett effect, first communication in 1909) ([218], pp. 73–74).

The prehistory of magnetomechanical investigations of this kind and their connection with cosmic magnetism and their subsequent development are briefly considered in the cited book [218] by Frenkel' and Yavelov and also in studies by Blackett [220] and Idlis ([221], pp. 52–75). At the present time, there is no generally accepted theory of the origin of magnetic fields of cosmic bodies, but Barnett's idea, which Einstein also shared, of a rotational origin of cosmic magnetism has been recognized to be wrong, I am grateful to B.E. Yavelov and G.M. Idlis for additional information on this question.

rotation of massive bodies appeared to indicate that it had a relationship to electromagnetic phenomena.[9]

During the 1930s and 1940s, the question of the magnetism of rotating masses was raised several times in connection with unified geometrized field theories. There is interesting material on this, for example, in the recollections of M.A. Markov about S.I. Vavilov, who took a lively interest in this problem [224]. In subsequent modifications of asymmetric affine field theory there arose similar magnetomechanical effects, which, however, were too weak to be detected experimentally ([201], pp. 374–375). In the 1940s and 1950s, this question had much less interest for Einstein than it did in the 1920s, as one can deduce from his papers. Thus, in 1923–1924 Einstein very insistently sought possibilities to relate his affine field theory to physical effects, above all macroscopic electromagnetic phenomena. He attempted to interest the experimentalists Gerlach, Franck, Piccard, and others, and also his friend Ehrenfest, in these ideas. None of these attempts gave results, but nevertheless they did not undermine Einstein's belief in the promise of the affine direction, which he continues to develop during the whole of 1925.

Let us return to the end of 1923, however, when Einstein, concluding that the affine theory did not contain nonsingular solutions that could be interpreted as the electron, attempted to deduce the quantum behavior of the electron on the basis of the overdetermination of a system of generally relativistic field equations.

THE CONCEPT OF OVERDETERMINATION OF FIELD EQUATIONS AND QUANTA, 1923

Broadly speaking, Einstein did not associate the problem of explaining the quantum behavior of the electromagnetic field and the electron with the maximum problem in the first stage of its solution, which consisted merely of proving that the field equations had a nonsingular static centrally symmetric solution that could be interpreted as the electron. Subsequently, however, there was no doubt that the quantum features must also be explained on the basis of the field equations of a unified theory.

It had become clear by the end of 1923 that none of the unified theories proposed since 1918 had such nonsingular solutions. On the other hand, already in the period of development of his first, nongeometrized unified field

---

[9] In this connection, B.E. Yavelov (private communication) has suggested that the nature of rotational motion was one of the fundamental Einsteinian themes (understood, for example, in the sense of Holton's "thematism").

theory during 1908–1910, Einstein had considered the possibility of inter-preting the elementary electric and light quanta as certain singular points of generalized field equations [55]. Moreover, the proximity in order of magni-tude of Planck's constant and the constant e^2/c (where e is the electron charge and c the speed of light) was for him a serious argument in support of the view that "a modification of the theory that gives as a consequence the elementary quantum [i.e., the electron charge] will also contain the quantum structure of radiation" [52].

At the meeting of the Berlin Academy of Sciences on December 13, 1923, Einstein presented his paper "Does field theory offer a possibility for solv-ing the quantum problem?" in which, without relation to a particular form of unification of gravitation and electromagnetism, the problem of the quantum behavior of particles was discussed for the first time in the framework of the program of unified geometrized field theories [225]. Hitherto, Einstein had either restricted himself to the minimum problem (the problem of unifying the gravitational and electromagnetic fields), or, on the basis of such a unification, had attempted to find the key to the solution of the maximum problem (mainly by finding from the equations of the unified field nonsingular centrally sym-metric static solutions that can be interpreted as charged elementary particles), or had sought such possibilities (i.e., the possibility of a field description of particles) on the basis of the equation of general relativity or certain general-izations of them not directly associated with a unified geometrical description of fields.

Einstein began by comparing the unification program to the quantum the-oretical program. He saw a weakness in the latter in its inadequate level of theorization: "The great successes achieved by quantum theory during the almost quarter of a century since its birth cannot hide from us the fact that *a logical basis of this theory does not yet exist* [our italics]" [225]. The failures of the classical method for describing microscopic processes on the basis of the idea of a four-dimensional space-time continuum and unique classical causal-ity, and also the associated mathematical formalism of differential equations, led to doubts about the universality of these assumptions. "From the point of view of epistemology," noted Einstein, "all these doubts are valid and, in the light of the existing deep difficulties, are fully comprehensible" [225].

Nevertheless, the successes of the classical field conception associated with the formalism of partial differential equations during the last half century were, in Einstein's opinion, so impressive that to abandon it in the absence of a sufficiently powerful theoretical alternative for the solution of the quantum problem could be decidedly premature. The approach to particles as concen-trations of field, which had been most fully developed in the work of Mie, had not led to success. Einstein's closely related approach to the solution of the maximum problem in the framework of the unification program also had not produced results.

An important feature of the theory, which cast doubt on the possibility of using a classical method of description, was, in Einstein's opinion, the fact that the "initial state of an electron moving around the hydrogen nucleus cannot be chosen arbitrarily," since "this choice must be consistent with the quantum conditions" (in other words, "not only the evolution in time but also the initial states satisfy certain laws") [225]. This property appeared to be difficult to reconcile with classical causality, which allowed an arbitrary choice of the initial state of the system, which, after this choice, developed in time in accordance with the differential equations that were the basis of the theory. "Can this property of the processes of nature, which we must, apparently, accord universal significance, be described by a theory based on partial differential equations?" asked Einstein [ibid.]. At the same time, he did not yet associate the classical continuum method of description with the geometrization program. Einstein saw a possibility of reconciling the identified property of the processes of nature with a dynamical law in the form of differential equations in the idea of "overdetermination," according to which "the number of differential equations must be greater than the number of field variables determined by them" ([225], p. 458; [226], p. 360).

He illustrated this idea by the example of Euclidean geometry, regarded as a special case of Riemannian geometry. A necessary and sufficient condition for a Riemannian space to be Euclidean is the vanishing of the fourth-rank curvature tensor $R_{ik,lm}$. The need for any initial conditions disappeared altogether. As a result, 20 algebraically independent equations for the ten components of the metric tensor g_{ik} were obtained.

He then sketched a field theory with an overdetermined system of equations that had the possibility of including a description of quantum behavior. The required system of equations must be generally covariant and it must contain only the gravitational potentials g_{ik} and the electromagnetic field strengths φ_{ik}. In accordance with the correspondence principle, it must also contain the equations of gravitation and electrodynamics, in other words, the equations

$$R_{ik} = -\kappa T_{ik}, \tag{15}$$

where $T_{ik} = -\varphi_{i\alpha}\varphi_k^\alpha + \frac{1}{4} g_{ik}\varphi_{\alpha\beta}\varphi^{\alpha\beta}$ is the energy–momentum tensor of the electromagnetic field. Finally, the required overdetermined system of equations must admit a static centrally symmetric solution that could be interpreted as a "positive or negative electron." What was now considered was a singular solution.

If these conditions were satisfied, one could hope that by means of the overdetermined system of equations "the mechanical behavior of the singular points (electrons) will also be determined because the initial states of the field and of the singular points will also satisfy restrictive conditions" ([226], p. 361; [225], p. 458). As a result, the quantum conditions would be obtained as a consequence of the overdetermined system of equations.

As a model system of this kind, Einstein chose, by analogy with Euclidean geometry, the system of equations

$$R_{ik,lm} = \Psi_{ik,lm}, \tag{16}$$

where $R_{ik,lm}$ is the Riemann–Christoffel tensor and Ψ is a certain fourth-rank tensor, homogeneous and of second degree in the tensor $\varphi_{\mu\nu}$ of the electromagnetic field strengths and possessing the same symmetry properties as the tensor $R_{ik,lm}$. These conditions of $\Psi_{ik,lm}$ make it possible to express it in the form

$$\Psi_{ik,lm} = A'\Phi'_{ik,lm} + A''\Phi''_{ik,lm} + A'''\Phi'''_{ik,lm}, \tag{17}$$

where the tensors $\Phi'_{ik,lm}$, $\Phi''_{ik,lm}$, and $\Phi'''_{ik,lm}$ are expressed as follows:

$$\Phi'_{ik,lm} = \varphi_{ik}\varphi_{lm} + \tfrac{1}{2}(\varphi_{il}\varphi_{km} - \varphi_{im}\varphi_{kl}),$$
$$\Phi''_{ik,lm} = g_{il}\Phi'_{km} + g_{km}\Phi'_{il} - g_{im}\Phi'_{kl} - g_{kl}\Phi'_{im},$$
$$\Phi'''_{ik,lm} = (g_{il}g_{km} - g_{im}g_{kl})\Phi',$$
$$\Phi'_{il} = g^{km}\Phi'_{ik,lm}, \qquad \Phi = g_{ik}\Phi'_{ik}.$$

If A', A'', A''' are taken to have the values -2, $-2/3$, and $-1/6$, respectively, and we multiply Eq. (16) by g^{il} and then sum over these indices, we obtain the equations of gravitation with the electromagnetic energy–momentum tensor. These equations, as Reissner and Weyl had shown earlier, admit a centrally symmetric static solution $L(m, \varepsilon)$ possessing a singular point determined by two constants—the mass and the charge—and could be interpreted as a positive or negative electron [64]. Therefore, the overdetermined system too must possess the solution $L(m, \varepsilon)$. The system (16) is not suitable, however, because in the absence of an electric field it leads to a Euclidean metric, i.e., the solution $L(m, 0)$ does not satisfy the system. In contrast, as Einstein could show, an electron without mass, $L(0, \varepsilon)$, is contained in the system (16). Therefore, Einstein assumed, the system of equations (16) is close to the one needed. He saw one of the ways of overdetermining Eqs. (15) in augmenting them with the relations

$$\Psi_{ik,lm;n} + \Psi_{ik,mn;l} + \Psi_{ik,nl;m} = 0, \tag{18}$$

obtained from the Bianchi identities for the tensor $R_{ik,lm}$.

Einstein was unable, "due to the great complexity of the calculations," to show that the solution $L(m, \varepsilon)$ satisfied the system of equations (18). The hope that this system nevertheless contained $L(m, \varepsilon)$ was based on the inclusion of $L(0, \varepsilon)$ and $L(m, 0)$ among the solutions of the system (18). "Thus," summarized Einstein, "there is a certain possibility that combination of the system (12) [i.e., (18) here] with Eqs. (8) [i.e., Eqs. (15)] leads to the required overdetermination of the complete field" ([225], p. 461; [226]).

In fact, basic questions remained unanswered: Is $L(m, \varepsilon)$ a solution of this overdetermined system of equations? Does this system determine the mechanical and quantum properties of the singularities? In a note added in proof, Einstein remarked that Grommer had succeeded in answering the first question in the affirmative. It appeared to Einstein that finding the answers to the second and third questions depended only on mathematical skill. "The two last questions," he wrote in the conclusions, "present great problems for a mathematician wishing to solve them; it is necessary to invent approximate methods for solution of the problem of motion" [ibid.]. Einstein did not regard Eqs. (16) and (18) as in any way final. Rather, they were an illustration of the idea of an overdetermined system of differential equations, which, in his opinion, was the only possible way of overcoming purely classical causality and formulating quantum properties in the language of differential equations:

> In conclusion, it must be emphasized once more that the most important thing for me in this communication is the idea of overdetermination; I readily grant that Eqs. (12) [i.e., Eqs. (16) and (18) here] have not been derived with the rigor that one would wish. ([225], p. 462; [226])[10]

Without insisting on a specific way in which the idea of an overdetermined system of field equations was to be realized, Einstein regarded the idea itself as very deep and promising. It appeared to open up a way to the theoretical comprehension of the quantum world on the basis of the classical field program, which for him was the ideal, the norm of a physical theory:

> The circumstances that there does appear here to be a possibility of a truly scientific foundation of quantum theory justifies great efforts (*Der Umstand, dass hier eine Möglichkeit zu einer wirklich wissenschaftlichen Fundierung*

---

[10] In the general introduction to the paper, he wrote: "If it is possible at all to solve the quantum problem by means of differential equations, one can hope to arrive at the goal in this way [i.e., by means of the idea of an overdetermined system of equations]. Below, I shall present my attempts in this direction, though I cannot assert that the equations established by me do indeed have a physical meaning." ([225], p. 459; [226])

der Quantentheorie vorzuliegen scheint, rechfertigt grosse Anstrengungen).
([226], p. 364)

It seemed to Einstein that the theoretical breakthrough he outlined in the field of quantum physics mainly required the overcoming of mathematical difficulties. Therefore, he attempted in this problem to attract the interest of mathematicians. "My efforts," he wrote, "will achieve their aim already if they interest mathematicians and convince them that the path proposed here can be taken and that it must necessarily be followed to the end" ([225], p. 459; [226]).

The basic idea of the paper was expressed very clearly, although very briefly, in a letter from Einstein to Besso on January 5, 1924:[11]

> The idea with which I struggle concerns the understanding of the essence of quanta, and the idea is overdetermination of the laws by differential equations, the number of which exceeds the number of field variables. In this way one can understand the absence of arbitrariness of the initial conditions without giving up field theory. Of course, this approach may be quite wrong, but it is logically possible and must therefore be tested. We completely give up equations of motion of material points (electrons); the mechanical behavior of them is determined by the laws of the field. ([227], p. 5)

It is worth saying a few words about the subsequent fate of the idea of overdetermination of the differential field equations. It passed the test of time and was crucial in solving the problem of the derivation of the equations of motion of matter from the field equations. In 1927, Einstein still believed that overdetermined field equations contained not only the law of motion of singularities (which he was able to prove in two papers in 1927, the first of them written with Grommer [228]), but also the quantum properties of the motion of particles. Einstein and Grommer wrote:

> The achieved success is ... that it has been shown for the first time that field theory can contain within it the theory of the mechanical motion of particles of matter. This may be important for the theory of matter, for example, for quantum theory. [228]

In the second paper of 1927, it was stated:

---

[11] Unfortunately, in the Russian translation, the dates of this and several subsequent letters are given incorrectly. Instead of January 5, 1924 the letter is dated November 5, 1924, and the dating of letters Nos. 67 and 68 (in the numbering of the Russian translation) have been moved forward a year (1925 is given instead of 1924) ([227], pp. 5–7).

> The majority of physicists are now convinced that the existence of quanta
> precludes field theory in the usual sense. However, this conviction is based
> on an inadequate knowledge of the consequences of field theory. Therefore,
> further investigation of the conclusions of field theory with regard to the
> motion of singularities appears to me justified despite the great number of
> quantitative results obtained by quantum mechanics. [229]

The investigations of the problem of the equations of motion did not, however, reveal any connection with quantum theory. At the end of this paper,
Einstein recognized that "the investigation ... has not given anything for the
understanding of quantum phenomena" [229].

In the middle of the 1930s, he again returned to the idea of eliminating
singularities from the theory, assuming that the principle of eliminating singularities was related to the rules of quantization [230]. In a classical paper
of 1938, written in collaboration with Infeld and Hoffmann and containing a
fairly general solution to the problem of deriving equations of motion from
field equations, the idea of this derivation was explicitly based on the concept
of overdetermination:

> The gravitational equations are non-linear and, because of the necessary
> freedom of choice of the coordinate system, are such that four differential
> relations exist between them so that they form an overdetermined system
> of equations. The overdetermination is responsible for the existence of
> equations of motion.... ([231], p. 65)

On this occasion, though, nothing was said about obtaining the quantum properties of particles.

At the beginning of the 1950s, in the second appendix to the third edition
of his Princeton lectures *The Meaning of Relativity*, Einstein also discussed the
fundamental significance of the concept of overdetermination. Having noted
that in general relativity there were ten field equations for six g_{ik}, Einstein
emphasized:

> The system of equations is indeed "overdetermined," but due to the existence
> of the identities [the Bianchi identities] it is overdetermined in such way
> that its compatibility is not lost.... The fact that the equations of gravitation
> imply the law of motion for the masses is intimately connected with this
> (permissible) overdetermination. ([232], p. 16)

In conclusion, we mention that overdetermination of the field equations
associated with general covariance was already used in Hilbert's unified field
theory. Maxwell's equations were ultimately obtained as a consequence of the
equations of the gravitational field. In the 1923 paper, Einstein sought a more
radical overdetermination in order to obtain not only the equations of motion

but also quantum properties. Although this direction was not unfruitful, it did not lead, as we see, to a decisive success in the development of a nondualistic unified field theory capable of encompassing quantum phenomena. It is true that for a number of years the idea of overdetermination sustained Einstein in his efforts to solve the quantum problem on the basis of the program of unified geometrized field theories. The real program in this direction was achieved in connection with the development of the problem of deriving the law of motion of matter from the equations of the gravitational field.

Affine–Metric Theory, 1925–1926

In quantum theory great events associated with the discovery of quantum mechanics came to a head in 1925–1926. Heisenberg's classical paper, which opened the new epoch of physics, was completed on July 29, 1925 [233]. Einstein participated in this development not only through his earlier studies [234, 235], but with later ones as well. On January 8, 1925 he presented to the meeting of the physics and mathematics sections of the Prussian Academy of Sciences his paper "Quantum theory of a monatomic ideal gas. Second communication" [236], which played an important part in Schrödinger's investigations of wave mechanics. Having this paper in mind, Schrödinger, in answer to a letter of April 16, in which Einstein gave a high estimation of Schrödinger's work, wrote to him on April 23:

> Your and Planck's approval means more to me than the approval of half the world. In fact, none of this would have happened now or ever (I am referring to my own participation) if you in your second paper on the quantum theory of gases had not made me aware of the importance of the ideas of de Broglie. ([237], p. 331)

During 1925 (and earlier) he was often in Kiel, where he worked with H. Anshütz-Kaempfe on a gyrocompass [218], but it was not this work, nor even the fundamental papers on quantum theory mentioned above, that were the most important for Einstein. As before, he was tortured by the problem of unified field theory. In connection with the translation of Eddington's book *The Mathematical Theory of Relativity* into German, the author and Courant suggested to Einstein that he should write a special supplement to the German edition devoted to the application of Hamilton's principle to the affine unified field theory whose foundations had been laid by Eddington. Einstein wrote the supplement "Eddington's theory and Hamilton's principle," which, in fact, contained only a clearer and more consistent exposition of his affine theory of 1923 [238]. This supplement was preceded (already in the second

English edition* of 1924) by an author's note, actually the supplementary §104, called "Einstein's new theory" and containing a consideration of Einstein's affine theory of 1923 [239]. It is interesting that Einstein called his theory Eddington's, and Eddington called his Einstein's.

As we have already noted, in his affine theory of 1923, Einstein essentially arrived at field equations characteristic of Weyl's theory (in its second form), as he wrote in the letter to Weyl that we quoted earlier. ("I wanted to go in the opposite direction from you and yet I have arrived at the same equations as you with your special action principle.") Eddington also emphasized the relation: "This development can be regarded as a form . . . of the theory of Weyl's action" ([239], p. 448 of the Russian translation). Neither Einstein nor Eddington was particularly enthusiastic about Einstein's generalization of Eddington's theory. Both were very skeptical about its possibilities. Eddington wrote,

> Einstein's theory is very formal . . . and I cannot free myself from the sus-
> picion that its mathematical elegance is achieved perhaps by devices that
> will not lead along the direct path of true physical progress. [Ibid.]

Einstein noted that the theory led, in a fairly good approximation, to the already known equations of gravitation and electromagnetism but, at the same time, did not lead to a nonsingular solution that could be interpreted as an electron. He also noted the absence of any experimental confirmation of the conclusion deduced from the exact theory of generation of a four-current by the electromagnetic field, and he gave the following summary with regret:

> The result we have obtained creates in me the impression that the deepening
> of the geometrical foundations undertaken by Weyl and Eddington has not
> been able to lead to progress in physical knowledge. ([238], p. 166 of the
> Russian translation)

It is true, however, that Eddington and Einstein still hoped that subsequent work in this direction could lead to success.[12]

---

* *Translator's note*: I have not been able to find this in any copy of the second edition.

[12] Eddington noted:

> it would be unreasonable to ignore entirely the possibility of advance on
> some new path; therefore, we give in what follows Einstein's results . . .
> ([239], p. 448 of the Russian translation)

Einstein himself concluded the above estimate of the theory with these words: "One can only hope that the future development of the theory will demonstrate the incorrectness of this pessimistic opinion" ([238], p. 166 of the Russian translation).

On his return at the beginning of June from South America, he wrote to Besso (in a letter of June 5, 1925):

> I am firmly convinced that the entire chain of ideas of Weyl, Eddington, and Schouten will not lead to anything physically useful, and I have now followed a different track that is physically more justified. ([227], p. 9)

The name of Schouten, the eminent Dutch mathematician, did not appear here by chance. In 1922, the *Mathematische Zeitschrift* published his program paper "*Ueber die verschiedenen Arten der Uebertragung*" ("On different forms of transport") [202], in which he classified spaces with affine connection of the basis of three tensor characteristics, which were subsequently called by Kagan the Schouten characteristics: C_{kl}^i, S_{kl}^i, and Q_{ikl} (see the beginning of this chapter) ([205], pp. 504–506). All three tensors vanishing gives Riemannian geometry; the first two Schouten characteristics vanishing gives a large class of spaces with symmetric affine connection (in particular, for $Q_{ikl} = Q_i g_{kl}$ Weyl's geometry is obtained); S_{kl}^i not vanishing leads to the presence of torsion. Einstein probably thought of the possibility of using the different Schouten spaces; at the least, this possibility fit very well into the general scheme of the affine approach developed by Eddington and Einstein. The new "physically more justified" direction about which Einstein wrote was evidently the affine-metric direction based on the use not only of the coefficients Γ_{ik}^l of the affine connection but also of a second-rank metric tensor g_{ik} not related to Γ_{ik}^l (in contrast to the Christoffel symbols of Riemannian geometry) as independent variables.

Before we turn to the consideration of the affine-metric approach, we mention one puzzling passage in the quoted letter: "It seems to me that the quantum problem requires the introduction of something like a special scalar, and I have now found an acceptable way of doing this" ([227], p. 9). The concept of overdetermination of field equations, which, as appeared to Einstein, opened up a way to obtain particle-like solutions, derive the equations of motion of particles, and describe their quantum properties, also did not lead to success in 1925. It appears to us that in the quoted phrase one can recognize an anticipation of the Schrödinger wave function, associated with the ideas of de Broglie rather than those of geometrization. We recall that in the second paper on the quantum theory of a monatomic ideal gas, published in February 1925, Einstein wrote: "The manner in which a material particle can be associated with a (scalar) wave field was shown by L. de Broglie in his paper, which warrants every attention" ([236], p. 496).

Einstein presented his paper on the affine-metric theory to a general session of the Prussian Academy of Sciences in Berlin on July 9, 1925. Having rejected the purely affine approach, Einstein expressed a firm conviction in the promise of the new direction: "I now believe that after two years of continuous searching we have succeeded in obtaining the true solution, which is presented below" [240].

Besides the asymmetric affine connection $\Gamma^{\lambda}_{\mu\nu}$ (64 variables), Einstein also introduced as independent variables 16 components of a contravariant second-rank tensor density (or corresponding second-rank tensor $g^{\mu\nu}$). The fundamental invariant characteristics of the continuum were curvature tensors of second and fourth rank. As Lagrangian of the theory, Einstein chose the scalar density

$$\sqrt{-g}\,H = \sqrt{-g}\,g^{\mu\nu}R_{\mu\nu}. \tag{19}$$

Then variation of the corresponding action with respect to $g^{\mu\nu}$ and $\Gamma^{\lambda}_{\mu\nu}$ gave the 16 equations

$$R_{\mu\nu} = 0, \tag{20}$$

and the 64 equations (these last as result of variation with respect to $\Gamma^{\lambda}_{\mu\nu}$)

$$\frac{\partial(\sqrt{-g}\,g^{\mu\nu})}{\partial x_{\alpha}} + \sqrt{-g}\,g^{\beta\nu}\Gamma^{\mu}_{\alpha\beta} + \sqrt{-g}\,g^{\mu\beta}\Gamma^{\nu}_{\alpha\beta}$$
$$- \delta^{\nu}_{\alpha}\left(\frac{\partial(\sqrt{-g}\,g^{\mu\beta})}{\partial x_{\beta}} + \sqrt{-g}\,g^{\sigma\beta}\Gamma^{\mu}_{\sigma\beta}\right) - \sqrt{-g}\,g^{\mu\nu}\Gamma^{\beta}_{\alpha\beta} = 0. \tag{21}$$

Incidentally, as was correctly noted by Ferraris, Francaviglia, and Reina [241], it was at this point that the variational principle in Palatini's form, which is associated precisely with independent variation with respect to $g_{\mu\nu}$ and $\Gamma^{\lambda}_{\mu\nu}$, was first used. The Palatini method is widely used in modern variational formulations of general relativity, although Palatini himself, using variations of the Christoffel symbols, did not go beyond the metric approach in his 1919 paper [242]. Thus, Einstein's paper was an important step forward in the development of the variational formulations of general relativity.

Simplification of system (21) leads to the equations

$$\begin{cases} \dfrac{\partial(\sqrt{-g}\,g^{\nu\alpha})}{\partial x_{\alpha}} - \dfrac{\partial(\sqrt{-g}\,g^{\alpha\nu})}{\partial x_{\alpha}} = 0, \\[3mm] -\dfrac{\partial g_{\mu\nu}}{\partial x_{\alpha}} + g_{\sigma\nu}\Gamma^{\sigma}_{\mu\alpha} + g_{\mu\sigma}\Gamma^{\sigma}_{\alpha\nu} + g_{\mu\nu}\varphi_{\alpha} + g_{\mu\alpha}\varphi_{\nu} = 0, \end{cases} \tag{22}$$

where φ_{μ} is a certain covariant vector. A complete system of field equations in agreement with the already known laws of gravitation and electromagnetism (with the symmetric part of $g_{\mu\nu}$ identified with the metric tensor, and the antisymmetric part of $g_{\mu\nu}$ with the tensor of the electromagnetic field) is thus provided by the system of equations (20) and (22), augmented by the condition of identical vanishing of the vector φ_τ. This yields an overdetermined system: $16 + 64 + 4$ algebraically independent equations for the $16 + 64$ variables $g_{\mu\nu}$ and $\Gamma^\alpha_{\mu\nu}$. In Einstein's opinion, this overdetermination offered a hope of deriving equations of the motion of particles from the field equations and a description of their quantum properties. At the same time, there was also a different possibility, associated with giving up the requirement of identical vanishing of the vector φ_τ; this would be entirely consistent with the requirements of Hamilton's principle.

The restriction to symmetric $g_{\mu\nu}$ leads to the symmetry of $\Gamma^\alpha_{\mu\nu}$ and to their representation in the form of the Christoffel symbols, and this gives general relativity in conjunction with equations (20). For this, the vanishing of φ_τ is unavoidable. In the case of an arbitrary (nonsymmetric) tensor $g_{\mu\nu}$, the analysis in the general case is too complicated, and to show that the system (20) and (22) contains the equations of electrodynamics Einstein made a restriction to the first approximation:

$$g_{\mu\nu} = -\delta_{\mu\nu} + \gamma_{\mu\nu} + \varphi_{\mu\nu}, \tag{23}$$

where $\gamma_{\mu\nu}$ and $\varphi_{\mu\nu}$ are, respectively, symmetric and antisymmetric infinitesimal tensors of first order. In addition, the condition $\varphi_\tau = 0$ is imposed.

As in the earlier attempts to construct unified field theories, Einstein regarded the main criterion of viability of a theory the presence in it of nonsingular centrally symmetric solutions that could be interpreted as charged elementary particles. At the end of the paper he reported that, with Grommer, he had begun to work on this problem. In a July 28 letter to Besso, Einstein described his new theory as follows:

> With pleasure I wanted to present to you personally the egg that I recently laid, but I now do this by letter. We introduce independently of each other an affine connection ($\Gamma^\alpha_{\mu\nu}$) and tensor ($g_{\nu u}$) (respectively $g^{u\nu}$ and $\mathfrak{g}^{u\nu}$) and require for arbitrary (independent) variation of $\mathfrak{g}^{u\nu}$ and $\Gamma^\alpha_{\mu\nu}$ fulfillment of the variational principle
>
> $$\delta\left(\int \mathfrak{g}^{u\nu} R_{u\nu}\, d\tau\right) = 0 \ldots$$
>
> If it is assumed that $\mathfrak{g}^{u\nu}$ and $\Gamma^\alpha_{\mu\nu}$ are symmetric, then we obtain the old law of gravitation for empty space. If we give up the assumption of symmetry, then in the first approximation we obtain the laws of gravitation and Maxwell's field laws for empty space, the antisymmetric part of $\mathfrak{g}^{u\nu}$ representing the

electromagnetic field. This is a remarkable possibility that may correspond to reality. The question now is whether such a field theory is compatible with the existence of atoms and quanta. I do not doubt its validity for the macroscopic world. If only the calculation of concrete problems were simpler! But all this as yet preliminary. ([227], p. 12; [56], p. 209, German original)

We have already emphasized more than once that there was not, in fact, an insurmountable barrier between theoreticians working on the quantum field program and those working on the unification program. Einstein took a lively interest in quantum theory, actively participating in its development (we mention, for example, his papers of 1924–1925 on the quantum theory of a monatomic ideal gas). Schrödinger, de Broglie, O. Klein, Fock, and other leading "quantists" attempted to relate unified geometrized theories to quantum theory (it is true that the new wave of such attempts arose after the appearance of quantum mechanics). In this respect, the following extract from a letter of Born to Einstein on July 15, 1925 is very instructive:

> I am tremendously pleased with your view that the unification of gravitation with electrodynamics has at long last been successful; the action principle you give looks so simple. As we have time, Jordan and I are going to try some variations of it. But we would be most grateful if you could send us your paper on this subject as soon as possible. This kind of thing is much deeper than our petty efforts. I would never dare tackle it. ([89], p. 85)

Incidentally, in the same letter Born discussed the state of affairs in quantum theory in considerable detail. In particular, he reported the new paper of Heisenberg that laid the foundations of quantum mechanics in its matrix form (only two weeks after this letter, Heisenberg submitted his paper to the journal *Zeitschrift für Physik*). Recalling those distant years, Born wrote in his commentary on this letter: "I think that my enthusiasm about the success of Einstein's idea was quite genuine. In those days we all thought that his objective . . . was attainable and also very important" ([89], pp. 87–88). Judging from his correspondence with Einstein, Sommerfeld also had similar views, at least in the middle of the 1920s.

In 1923, discussing his affine field theory, Einstein concluded that the theory could not take into account the difference between the masses of positive and negative electrons, i.e., it was charge symmetric. He regarded this circumstance as a serious drawback to the theory, since, as a consequence, it did not reflect the charge asymmetry that existed in nature. This circumstance had been discussed earlier by Weyl and, especially, Pauli, who in 1919 emphasized that the charge asymmetry required by experience contradicted the general covariance of the field equations. General covariance also includes inversion of time, under which, in the static case, a solution (φ, g_{ik}) goes over into the solution ($-\varphi$, g_{ik}), which is tantamount to charge symmetry.

At the end of the paper on affine-metric field theory, Einstein made an interesting comment that the theory requirement of invariance with respect to inversion (reflection) of the time ($t' = -t$) made it possible, instead of the usual identification of the components φ_{14}, φ_{24}, φ_{34} of the electromagnetic field tensor with the electric field strengths, to associate these quantities with the magnetic field strengths, and the components φ_{23}, φ_{31}, φ_{12} with the components of the electric field. Pondering the problems of the discrete space-time symmetries of the theory and the problem of including the experimentally observed charge asymmetry in the framework of the affine-metric theory, Einstein generalized Pauli's result in a paper in 1925 [243] and arrived at a fundamental conclusion that is now evaluated as a theoretical prediction of the charge symmetry of matter, i.e., the prediction of the existence of antiparticles [244, 245].

It is important to note that Einstein did not give his proof in the framework of some particular form of unified field theory but under the conditions of the most general situation valid in any generally covariant field theory in which the electromagnetic field is described by an antisymmetric tensor of second rank. Under these conditions, Einstein found that there do not exist covariant fields equations that: "(1) have a solution corresponding to a negative electron, (2) have no solution corresponding to a positive electron of equal mass" [243].

The proof is not complicated, and the idea is similar to Pauli's argument mentioned above. It differs from Pauli's argument in using the tensor of the electromagnetic field strengths to characterize the field, and not the potential. Therefore, Einstein's result agreed with the requirement of gauge invariance. In addition, Einstein proved specifically that the mass of a particle-like solution was unchanged under spatial or time reflections, which are used for his proof. Assuming the existence of a solution corresponding to an electron with charge ε and mass μ and characterized by certain metric ($g_{\mu\nu}$) and electromagnetic ($f_{\mu\nu}$) tensors, and using the conditions of general covariance of the theory, he considered the solution that could be obtained by the operation of time reversal. By virtue of the static and centrally symmetric nature of the solution, all time and nondiagonal components of $g_{\mu\nu}$ are zero, while the spatial components g_{ik} ($i = k$) are unchanged. The time components of f_{ik}, identified with the electric field, have their signs reversed ($f'_{14} = -f_{14}$, $f'_{24} = -f_{24}$, $f'_{34} = -f_{34}$), while the components of the magnetic field are equal to zero. A change in the sign of the electric field leads, as a consequence of Maxwell's equations, to a change in the sign of the charge, whereas the gravitational field and mass remain unchanged.

Naturally, Einstein regarded his result as a serious difficulty in the development of unified field theories. "Our searches for a way out of the difficulty," he wrote, concluding the proof of charge symmetry, "were unsuccessful . . ." ([243], p. 168). If, following the previous paper, the electric field is not identified with (f_{14}, f_{24}, f_{34}) but with (f_{23}, f_{31}, f_{12}), and the first triplet is identified with the magnetic field, then, as Einstein showed, spatial reflection, and not time reversal, leads to charge symmetry.

In a note added in proof to this paper, Einstein made some interesting comments aimed at a deeper understanding of the charge symmetry of unified theories. He noted that since the charge density is defined as the square root of the square of the current density,

$$\rho = \sqrt{g_{\mu\nu}i^\mu i^\nu}, \qquad i^\mu = \partial f^{\mu\nu}/\partial x_\nu, \tag{24}$$

the sign of the charge is not determined in the theory. A solution to the problem could be to break the symmetry of the theory with respect to time reversal. "It appears to us important," Einstein concluded his note, "that an explanation of the inequivalence of the two forms of electricity is possible only if time is ascribed a direction of flow . . ." [243].

Einstein's important result (the proof of charge asymmetry of any generally covariant field theory containing electrodynamics, and the prediction of antiparticles) did not attract attention and was not recalled even after an analogous result obtained by Dirac in 1930 in the framework of relativistic quantum mechanics. The main reason for the underestimation of Einstein's paper was evidently the fact that Einstein himself regarded the result as a serious hindrance in the path of the program of unified geometrized field theories. Subsequently, as we shall see, Einstein sought ways to break the symmetry he had established, but Einstein's 1925 paper "The electron and the general theory of relativity" was seen as a theoretical prediction of the charge symmetry of all field theories and, thus, the existence of antiparticles [244, 245].[13]

The affine-metric theory, like the previous unified theories, encountered the difficulty that Einstein appears to have regarded as the greatest—the absence in the theory of nonsingular centrally symmetric static solutions that could be interpreted as charged elementary particles. In addition, Einstein came to the firm conviction that this theory (and, in all probability, all unified field theories of such kind) was incapable of explaining the experimentally observed charge asymmetry of the elementary particles. The conclusion drawn from the

---

[13] See also the commentary in the Russian translation of the paper [243], p. 170.

charge symmetry theory he had proved was very pessimistic: "Therefore, all attempts to fuse into a single entity electrodynamics and the laws of gravitation appear to us insufficiently well founded" [243]. It was also not possible to arrive at experimental consequences that went beyond the framework of general relativity and classical electrodynamics. Einstein's letter to Besso on December 25, 1925 gave a negative summary of the period of development of affine and affine-metric field theories: "I was forced to abandon my work in the spirit of Eddington. Quite generally, I am now convinced that, unfortunately, nothing can be done with the complex of Weyl–Eddington ideas" ([227], p. 15).

It can be seen from this letter that at the end of 1925 and the beginning of 1926 (and, apparently, throughout almost the whole of 1926) Einstein, disappointed with the affine and affine-metric directions, returned to the more realistic theory of 1919 with the field equations

$$R_{ik} - {}^1\!/_4\, R g_{ik} = -\kappa T_{ik},$$ (25)

where T_{ik} is the energy–momentum tensor of the electromagnetic field.[14] Einstein's return to Eqs. (25) is also indicated by the paper "On the formal relation of the Riemann curvature tensor to the equations of the gravitational field," which he completed in January 1926 [246]. In this paper, he again justified Eqs. (25), using cosmological arguments associated with the need to introduce a cosmological term (the requirement of finiteness of spacelike sections of the world due to the desire to ascribe to the world a finite mean matter density). The hope of obtaining electrons and their equations of motion from the field equations (25) was not justified on this occasion, however: "Unfortunately, recent investigations have shown that in this manner it is not possible to arrive at a satisfactory theory of electrons" ([246], p. 184). It appeared to Einstein at this time that the high road of the program of unified geometrized field theories had reached a dead end:

> All out efforts have had the aim, following the approach proposed by Weyl and Eddington or something analogous, to arrive at a theory that unifies the

---

14 In the letter to Besso quoted above, Einstein wrote:

I regard the equation
$$R_{ik} - {}^1\!/_4\, g_{ik} R = -T_{ik}$$
(electromagnetically) as almost the best among those existing in electromagnetism. We have nine equations for the 14 quantities $g_{\mu\nu}$ and $\gamma_{\mu\nu}$. From new calculations it follows, apparently, that these equations give the motion of the electron. However, it is doubtful whether a place will be found here for quanta. ([227], p. 15)

gravitational and electromagnetic fields in a single formal scheme; however, as a consequence of numerous failures we have arrived at the conviction that in this way it is not possible to advance to the truth. [Ibid.]

A theory with the field equations (25) also did not yet lead to real physical progress, but it appeared less speculative and related to deep cosmological considerations.[15]

In this paper, Einstein attempted to demonstrate that Eqs. (25) were mathematically natural and transparent, by showing, in particular, that the vanishing of the antisymmetric part $A_{ik,lm}$ (which Einstein defined in a certain manner) of the Riemann–Christoffel tensor $R_{ik,lm}$ was equivalent to the vanishing of the left-hand side of these equations. He also added to the Riemann–Christoffel tensor a corresponding electromagnetic analog $E_{ik,lm}$:

$$R^*_{ik,lm} = R_{ik,lm} + \kappa E_{ik,lm}, \tag{26}$$

and showed that the vanishing of $A^*_{ik,lm}$ was equivalent to the following system of equations without the right-hand side:

$$R^*_{im} - {}^1\!/_4 \, g_{im} R^* = 0. \tag{27}$$

It is possible that in these superficially simple field equations, which unified gravitation and electromagnetism, Einstein saw, at the beginning of 1926, a possible solution to the problem. By the middle of 1926, however, he had become disenchanted with the "1/4 theory" (i.e., the theory with Eqs. (25)).

He wrote to Besso on August 11, 1926:

> The equation
> $$R_{ik} = g_{ik} f_{\alpha\beta} f^{\alpha\beta} - {}^1\!/_2 \, f_{i\alpha} f_{k\beta} g^{\alpha\beta},$$
> which I have established also gives me little satisfaction. It does not admit any electric masses free of singularities. In addition, I cannot bring myself to relate two things (like right- and left-hand sides of the equation) which do not have anything in common from the logical–mathematical point of view. ([227], p. 22; [56], p. 230, German original)

At this point in his book, Tonnelat makes the following remark:

---

[15] It is interesting to note that even in 1926, despite recognition of Friedmann's results, Einstein, as can be seen from the quoted paper, had not completely abandoned his original views about the structure of the universe (static closed model of a spacially finite world).

Einstein immediately saw the weakness of the "naive" theory, which appeared to be glued together out of a geometrical term and a second term, the validity of which derived solely from Maxwell's theory. [Ibid.]

In addition, he corrected a slip of Einstein ($\frac{1}{2} g_{ik} f_{\alpha\beta} f^{\alpha\beta} - f_{i\alpha} f_{k\beta} g^{\alpha\beta}$ in place of the right-hand side of the above equation) and noted its equivalence to Eq. (25). Thus, in the framework of the "1/4 theory," which Tonnelat calls "naive," the attempt to find nonsingular particle-like solutions failed again. Another defect was the absence of a unified geometrical foundation for the gravitational and electromagnetic fields.

Einstein gave an even more pessimistic evaluation of his efforts in the field of unified theories in a letter to Sommerfeld on August 21. Sommerfeld had invited him to give some lectures at Munich, but Einstein turned down the request, explaining his decision as follows:

To speak the truth is sometimes very difficult but necessary: I have nothing to say that would be sufficiently important for the lectures that you have in mind. Therefore, I must be silent and allow someone to speak who can say something that others do not know. I have long been plagued by the attempt to find the connection between gravitation and electromagnetism, but I am now convinced that everything done up to now in this direction has been in vain. ([68], p. 236)

Against the background of the failures of the unification program, above all the efforts of Einstein himself, the theoretical breakthrough in quantum physics made a particularly sharp contrast. Even from the correspondence of the leader of the unification program, Einstein, with Sommerfeld, Born, Schrödinger, and Besso one can recognize the momentum and depth of the breakthrough. "The most interesting theoretical achievement of recent time," wrote Einstein to Besso in a letter of December 25, 1925, "is the Heisenberg–Born–Jordan theory of quantum states. There is a truly magical calculus, in which infinite determinants (matrices) appear instead of Cartesian coordinates" ([227], p. 15). From the very beginning, however, one senses a cautious, skeptical attitude by Einstein toward the new theory. In the same letter to Besso, he characterizes the "magical calculus" as follows: "To the highest degree ingenuous and, by virtue of its complexity, insured against proof of incorrectness." Schrödinger's wave mechanics, with its continuum basis, evoked much greater sympathy from him. At the beginning of May 1926, he wrote to Besso: "Schrödinger has written two remarkable papers on the quantization rules (*Ann. d. Phys.* 1926, Vol. 79). A deep truth is uncovered" ([227], p. 19).[16]

---

[16] Planck acquainted Einstein with Schrödinger's papers. Einstein was delighted

In August 1926, when the equivalence of the wave and matrix formalisms of quantum mechanics had already been established, Einstein (in a letter to Sommerfeld on August 21) rated these two approaches as follows:

> Of the new attempts to obtain a deeper formulation of the quantum laws, I prefer that due to Schrödinger. If only it were possible to transfer the wave fields introduced there from the n-dimensional coordinate space to three- and, accordingly, four-dimensional spaces! The Heisenberg–Dirac theories force my admiration, but for me they do not have the smell of reality. ([68], p. 236)

The "material," or "scalar," fields of Schrödinger, described by wave functions, were not, as Einstein soon understood, classical fields described by continuous functions in a real three- or four-dimensional space. From this point of view, Schrödinger's theory too did not correspond to the aims of the field program, despite being much closer to it than the matrix-operator form of quantum mechanics. Einstein probably thought of the possibility of "transferring" Schrödinger's wave fields to physical space (or space-time), but this did not succeed, and by the end of 1926 the enthusiasm for Schrödinger's wave mechanics had significantly decreased. Einstein wrote to Sommerfeld on November 28, 1926:

> The conclusions from Schrödinger's theory make a great impression, but I still do not know whether there is more here than in the old quantum rules, i.e., something that corresponds to one side of what is truly happening. Have we really got closer to guessing the secret? ([68], p. 237)

The probabilistic interpretation of the wave function disappointed Einstein even more. He wrote to Born on December 4, 1926:

> Quantum mechanics is certainly imposing, but an inner voice tells me that it is not yet the real thing. The theory says a lot, but does not really bring

---

by the wave mechanics, and, in a letter to Schrödinger on April 16, 1926, he noted: "The idea of your paper indicates true genius" ([237], p. 330). Einstein initially thought, however, that Schrödinger's equation was in conflict with the condition of additivity of the quantum states of independent systems. After a week, he recognized the incorrectness of his conclusion, and in the letter of April 26 gave preference to the Schrödinger approach over the matrix form of quantum mechanics:

> I am convinced that with your formulation of the conditions of quantization you have achieved a decisive success. I am also convinced that the path chosen by Heisenberg and Born does not go in the right direction. ([237], p. 332)

us any closer to the secret of the "old one" [i.e., God]. I, at any rate, am convinced that *He* is not playing at dice. Waves in $3n$-dimensional space, whose velocity is regulated by potential energy? ([247], p. 91)[17]

We have already noted that the decisive successes of quantum theory in 1925–1926 coincided in time with a certain crisis in the unification program (see, for example, the recognition of Einstein in his letter to Sommerfeld on August 21, 1926 from which we quoted above). Neither the purely affine theories, nor the affine-metric direction, which even recently had seemed so promising to Einstein, nor the concept of overdetermination, nor the "1/4 theory" gave truly tangible physical results. They appeared particularly speculative and fruitless compared to the successes of quantum mechanics, which progressed rapidly despite appreciable difficulties of a fundamental, theoretical nature.

An interesting episode relating to this period was recalled by A.F. Ioffe, who traveled with Einstein via train to a meeting of the Solvay committee in Amsterdam in 1926. Ioffe at that time regarded the direction of unified geometrized field theories as having no promise. He subsequently recalled:

> New hypotheses, their analysis, the recognition that they were not convincing, and more and more new attempts—this filled his [Einstein's] scientific life but did not bring the expected fruits. At the time of our walks, especially at night, the question of unified field theory, like a maniacal distraction from which there was no escape, was often raised by Einstein himself, but the discussion always boiled down to an exposition of his most recent hypothesis, from which he expected success and after which he could return to the sphere of physics [i.e., the sphere of physics not associated with the development of unified field theories]. The hypothesis failed, but after a year or two a new one appeared. ([249], pp. 226–227)

On the journey to Amsterdam, Ioffe "attempted to deflect him from the dead end" and switch his interest to the sphere of quantum physics, which at this time was going through great difficulties:

> Having sketched the deep contradictions between the phenomena in the microscopic world and the disorder in the thoughts if the physicists, I expressed the conviction that Einstein, with his exceptional physical intuition, was more likely than any other to find the way out.

---

[17] Ultimately, these doubts about the truth of quantum mechanics led Einstein to regard it as an incomplete theory (incompleteness, according to Einstein, was the circumstance that the description by means of "the ψ function is incapable of describing certain qualities of an individual system, whose 'reality' none of us doubt..." ([247], p. 188). A modern discussion of this problem in a philosophical and methodological approach is contained in [248].

The reference is evidently to the complexities of the physical interpretation of quantum mechanics. Ioffe persuaded Einstein:

> One cannot fail to see the fog of mysticism that clouds the clear contours of physics; science is losing faith in its strengths and is abandoning the reality of nature itself. There is only one solution—Einstein must do his duty, and he does not have the right to hide in the abyss of the unified field.

According to Ioffe's words, Einstein "promised to make every effort to change but doubted whether he would succeed ([249], p. 228).

In fact, in 1926 Einstein published practically no papers on unified field theories but did publish several papers in which he proposed and discussed an experiment with radiation emitted by canal rays, which, in Einstein's opinion, would permit a more definite choice in favor of the wave or the corpuscular concept of light [250–253]. Subsequently, an error in Einstein's idea was demonstrated [253]. At the end of 1926, Einstein told Born: "I am working very hard at deducing the equations of motion of material points regarded as singularities, given the differential equation of general relativity" ([254], p. 91).[18] It is possible that on this more realistic but roundabout route on which Einstein had returned to the solid ground of general relativity, he hoped, without giving up the geometrized field program, to arrive at a mechanics of particles and in particular quantum properties of such mechanics.

Returning to the affine-metric theory of 1925, we emphasize that this form attracted the interest of Einstein and other investigators (for example, Schrödinger) in the 1940s and 1950s. In the opinion of Tonnelat,

> by means of hypotheses corresponding to theories of the last type [i.e., affine-metric type], Einstein succeeded in creating a unified theory possessing great generality and making it possible to obtain the most natural generalization of the theory of gravitation. ([201], p. 373 of the Russian translation)

In the review of unified field theories written in 1956 for the English edition of his encyclopedia article on relativity theory, Pauli identified two classes of theories that warranted consideration: affine-metric theories with asymmetric g_{ik} and Γ^l_{ik} and five-dimensional theories. He considered that the most serious objection to affine-metric theories was their deviation from the "principle of irreducibility": "All these theories are exposed to the objection that they are in disagreement with the *principle that only irreducible quantities should be*

---

[18] The Russian translation [247] is inaccurate here.

used in field theories" ([63], p. 226).[19] Such irreducible entities are symmetric and antisymmetric quantities (in the considered case, tensors and geometrical objects). Therefore, according to Pauli and Weyl

> *Cogent mathematical reasons* (for instance invariance postulates of a wider group of transformations) *have to be given why a decomposition of the reducible quantities used in the theory (for instance R_{ik}, g_{ik}, and Γ^l_{ik}) does not occur,* ([63], p. 226)

and this was not done sufficiently convincingly in either the 1940s or the 1950s and still less in the 1920s and 1930s.

In fact, in the cited review, Pauli noted that Einstein himself clearly understood the shortcomings of theories with asymmetric g_{ik} and Γ^l_{ik} and, in his last papers of 1955, of rather formal requirements of λ invariance and transpositional symmetry [255, 256]. The first symmetry is associated with invariance of a theory with respect to the transformations

$$\Gamma^{l'}_{ik} = \Gamma^l_{ik} + \delta^l_i \lambda_{,k},$$

where $\lambda(x)$ is an arbitrary function. The second symmetry corresponds to invariance of the transition from the quantities A_{ik} to the corresponding transposed quantities. Einstein believed that its physical significance consisted of "invariance of the field equations with respect to a change in the sign of the electric charges" [255], but neither λ symmetry, which placed a prohibition on the use of symmetric Γ^l_{ik}, nor transpositional invariance, whose connection to charge symmetry is very problematic, had, in Pauli's opinion, a physical justification.

---

[19] Compare this with the earlier statement of Weyl:

> If there is an important lesson that one can take from mathematics for the formulation of physical theories, it is that only quantities which are *non-decomposable* under their specific transformation law represent a single physical entity; such quantities are the symmetric and the skewsymmetric tensor, g and f, but not their combination—Pauli formulates this principle as follows: what God has put asunder, man shall not put together. (*Wenn man der Mathematik eine für die Aufstellung physikalischer Theorien wichtige Lehre entnehmen kann, so ist es die dass nur Grössen, die unter ihrem spezifischen Transformationsgesetz unzerlegbar sind, eine einheitliche physikalische Entität darstellen; eine solche Grösse ist der symmetrische und der schiefsymmetrische Tensor, g und f, aber nicht ihre Zusammenfassung.... Pauli formuliert dieses Prinzip so: Was Gott getrennt hat, soll der Mensch nicht zusammenfügen*). ([18], p. 431)

THE PROBLEM OF THE EQUATIONS OF MOTION AND THE FIVE-DIMENSIONAL APPROACH, 1927

In the second half of 1926 and the beginning of 1927, discoveries in the field of quantum mechanics seemed to spill from a cornucopia: Born's probabilistic interpretation of the wave function (July 1926); the development of transformation theory, which made it possible to prove rigorously the equivalence of matrix and wave mechanics (November–December 1926, Dirac, Jordan, and others); the establishment of the connection between quantum statistics and the quantum mechanics of systems of identical particles (summer and fall 1926, Heisenberg, Dirac, Fowler, and others); the development by Dirac of the quantum theory of radiation (January–February 1927), etc. By this time, Heisenberg had already come very close to the discovery of the uncertainty principle (in the fall of 1926, he and Pauli discussed preliminary formulations of it) [257].

At the same time, there was a wave of investigations into five-dimensional unified field theories, not in opposition to the quantum theoretical program but in connection with the establishment of a deep relationship between the Schrödinger equation (or rather its relativistic generalization) and a scalar wave equation in five-dimensional space. From April to May 1926 (on April 28, O. Klein's paper was received by the editors of the *Zeitschrift für Physik* [258]) to the spring of 1927 not less than 20 papers were published on this problem by O. Klein, Fock, Mandel', de Donder, de Broglie, Ehrenfest and Uhlenbeck, Gamow and Ivanenko, London, Frederiks, Wiener and Struik, and others.

As we have seen, Einstein's position in the second half of 1926 was not very definite. On the one hand, he was clearly very disappointed in the efforts so far made to create unified geometrized field theories. On the other hand, he was inspired by the wave form of quantum mechanics. By the end of 1926, however, after the equivalence of the wave and matrix mechanics had been proved and the probabilistic interpretation of the wave function had been developed, Einstein's enthusiasm for quantum mechanics faded and his doubts as to its truth became stronger and stronger. In quantum mechanics, he felt the greatest unease about its fundamentally probabilistic nature and the impossibility of ascribing to the ψ function the meaning of a classical field like the electromagnetic and gravitational fields. The ineradicable wave–corpuscle dualism inherent in quantum mechanics—to whose creation Einstein had made a significant contribution—appeared to him extremely unsatisfactory from the point of view of the unification program.

At the end of 1926, Einstein returned to an analysis of the possibilities of this program for the problem of particles and their quantum properties. It seemed to him that the nonlinearity and general covariance of the field equations potentially contained both the corpuscular aspect of matter and the quantum features of the behavior of particles; it was only necessary to know how to extract them. Both these properties were also inherent in the equations of the gravitational field alone, which meant that one could attempt to obtain corpuscular and quantum laws without recourse, for the time being, to a unification of gravitation and electromagnetism. This led Einstein, aided by Grommer, to the first important studies on the problem of deriving the equations of motion of particles of matter from the field equations of general relativity.

The solution of this problem was given only 11 to 12 years later in the papers of Einstein (with Infeld and Hoffmann) and Fock. The solution was associated with the use of various approximate methods. Subsequently, improved approximation schemes, which also had a larger domain of validity (for a larger class of moving objects), were proposed.

Although the problem of obtaining equations of motion from the field equations was developed by Einstein (together, in part, with Grommer) in the framework of general relativity alone, this does not mean that he had become completely disenchanted with the unification program and had given up further investigations in this direction. First, he believed that results obtained in general relativity could be readily extended to a unified field theory whose equations were as yet unknown. Second, success in the solution of the problem of the motion of particles in general relativity would demonstrate the promise of the unification program as well. Finally, as is clear from the first paper of Einstein and Grommer on the equations of motion, Einstein returned to the idea of interpreting particles as field singularities. Some of the earliest forms of unified geometrized field theories, above all Kaluza's five-dimensional theory, had been rejected, in particular, because they did not contain nonsingular static centrally symmetric solutions that could be identified with charged particles. Arguments of such type against unified geometrized field theories seemed to lose validity, and, having lost faith in the affine and affine-metric theories related to the theories of Weyl and Eddington, Einstein returned to Kaluza's five-dimensional scheme.

Alongside the investigations into the problem of motion, he set up Kaluza's scheme in a simpler and more rigorous form, although he did not appear to obtain essentially new results (a paper on the law of motion was presented to a meeting of the Prussian Academy of Sciences on January 6, 1927 [228], and a paper on Kaluza's theory on February 17 [259], a first part having been presented earlier on January 20 [260]). As we have seen, it was an entirely natural step for Einstein to turn to the five-dimensional approach. Of course, this may have been due to the revival of five-dimensional field theory in connection with quantum mechanics. From the middle of 1926, there had appeared in the *Zeitschrift für Physik* alone not less than ten papers on the application of the five-dimensional approach to quantum mechanics. It appears that Einstein did not pay attention to these papers, since otherwise he would not have failed to mention the new aspect of the five-dimensional approach discovered by O. Klein and Fock and associated with the interpretation of the relativistic generalization of the Schrödinger equation as a scalar wave equation in a five-dimensional space. In a note added in proof to the second part of his communication on Kaluza's theory, he did, in fact, mention the "G. Mandel' reported ... [to him] that the results presented here are not new and are contained in papers of O. Klein. ... See also the papers of V.A. Fock ... " [259].

It was the problem of the equations of motion and the five-dimensional approach that were the main subject of Einstein's thinking and investigations in 1927, when quantum mechanics, after the formulation of the principles of uncertainty and complementarity, acquired a completed form. At the Solvay Congress in October 1927, which summarized the period of the creation of quantum mechanics, the famous discussions between Bohr and Einstein on fundamental problems of quantum mechanics began.

We return, however, to Einstein and Grommer's paper "The general theory of relativity and the law of motion" (presented on January 6, 1927) [228]. Einstein's failures in the attempt to develop the program of unified geometrized field theories, in particular, the unsuccessful attempts in all forms of such theories to interpret particles as nonsingular centrally symmetric static solutions, turned his thoughts to the consideration of particles as singularities of a field. The equations of the unified field must then also contain the equations of motion of these singularities. It was then natural to assume that in the absence of electromagnetic fields neutral particles must be described by singularities of the equations of the gravitational field, and their law of motion should be derivable from these equations.

Einstein accepted this idea only in 1927, however. He explained this delay as follows:

> We already thought of the possibility long ago that the law of motion of singularities could be contained in the field equations. However, the following argument appeared irrefutable and frightened us off. The law of the gravitational field can be approximated by a linear law with great accuracy for real cases, but a linear law, like the electrodynamic law, permits arbitrarily moving singularities [i.e., the law of motion is independent of

the field equations]. It appears self-evident that from such an approximate solution one could, by the method of successive approximation, go over to a rigorous solution that differed very little from it. If this were the case, then there could be a field corresponding to the exact equations for arbitrarily specified motion of the singularities, i.e., the law of motion of the singularities would not be contained in the field equations. [228]

Another reason has already been given: Einstein was immersed in the development of unified field theories and regarded particles not as singularities of field equations but, on the contrary, as nonsingular solutions to these equations. In the introduction to their paper, Einstein and Grommer compared the relationship between the field equations and the equations of motion in electrodynamics, and also in the classical theory of gravitation, was such that there the equations were completely independent, and this, in the opinion of the authors, was due to the linearity of the field equations. To the field–particle dualism there corresponded the dualism of the equations that described theory motion.

The authors mentioned Mie's attempts to overcome this dualism in his nonlinear electrodynamics. According to his approach,

all of physical reality is described by a field free of singularities; the field describes not only "empty space" but also material particles, and the laws of this reality are completely determined by partial differential equations. [228]

The same idea was, in essence, taken up by the program of unified geometrized field theories, which was also developed by Einstein, following Weyl and Eddington: "Although this attempt has as yet been unsuccessful, it continues to be the leading program even outside the purely electrodynamic domain (Weyl, Eddington)" [ibid.].

In general relativity, there were, in the opinion of Einstein and Grommer, three approaches to the problem of the relationship between the field equations and the equations of motion. The first was completely analogous to what was adopted in electrodynamics and Newtonian theory: the equations of motion of a material point, identical with the equations of a geodesic, were postulated independently of the equations of the gravitational field (for empty space). The second approach was based on the introduction of an energy–momentum tensor of matter and on the assumption, close to Mie's concept, that the corresponding equations of the gravitational field do not contain singularities. Then the four Noether identities (or Bianchi identities) imposed a divergence condition on the matter tensor, from which, under the assumption that the matter is situated on narrow "world tubes," it follows that the axes of these "tubes" coincided with geodesics. This actually meant that the equations of motion were a consequence of the field equations. This approach, however, intimately related to the idea of Mie and the program of unified geometrized field theories, had not led to success precisely because it had not been possible, either in the framework of general relativity or on the basis of the various forms of unified geometrized field theories, to obtain particles as nonsingular solutions of the field equations.

"The suspicion that this is an altogether incorrect way to understand the particles of matter," emphasized the authors of the paper, "became after many painstaking attempts so strong that we do not wish to talk about it here" [228]. That phrase expresses disappointment in the unified theories of affine and affine-metric type as well as in the very idea of interpreting particles as nonsingular solutions to field equations.

The third approach was associated with the interpretation of particles as singularities and with the obtaining of the equations of motion of these particles from the field equations:

> Thus, we set out on the path to explain elementary particles as singular points or singular world lines. This path is also suggested by the fact that not only the equations of the pure gravitational field but also the equations augmented by the Maxwellian electromagnetic field—have a simple centrally symmetric solution with singularity. Thus, we have arrived at a third method of treating the problem, in which, besides the gravitational and electromagnetic fields, there are no other field variables—the place of which, however, is occupied by singular world lines.... It appears likely that the law of motion of the singularities is completely determined by the equations of this field and the nature of the singularities. [228]

The last conclusion was deduced from the nonlinearity of the field equations. Some doubt about the validity of this conclusion could arise from the apparent possibility of approximating a rigorous solution of the nonlinear equations by an approximate solution of the linear equations. Analysis of the axisymmetric static solutions of the gravitational equations obtained during 1918–1922 by Weyl, Levi-Civita, and Bach showed that this possibility was illusory [64]. These studies had proved the impossibility of the existence on a symmetry axis of two (or more) singular points, and this indicated the absence of static solutions with axial symmetry corresponding to two point masses at rest. This suggested that the field equations themselves precluded motions of particles that did not agree with the field equations.

In the cited paper, Einstein and Grommer showed that in the simplest case, when the equations of motion of a singularity reduced to the equation of its equilibrium in the gravitational field, these equations were completely determined by the field equations. The paper ends with a comment that further suggests a very strong stimulus for investigating the problem of the motion of particles—the successes of the quantum theory of the motion of particles. The authors summarized:

> The achieved success is . . . that for the first time is has been shown that field theory can contain a theory of the mechanical motion of discrete particles of matter. This may have significance for the theory of matter, for example, for quantum theory. [228]

During the whole of 1927, Einstein continued to investigate the problem of deriving equations of motion from field equations. At the end of November, he completed a second paper (this time without the assistance of Grommer) with the same title. It was presented to the meeting of the Berlin Academy of Sciences on December 8, 1927 [261]. In this paper, too, one senses a sharp opposition to the quantum theoretical program. Under the conditions of the rapid progress of that program, Einstein tried to prove the viability of the unification program, which had already entered a phase of steady regression. His aim was to show that the classical geometrical field program was capable of offering at least the prospect of an explanation of the discrete aspects of matter, one of which was the equations of motion of elementary particles (in a first stage, the classical equations should be obtained and then, possibly the quantum mechanical ones). Having in mind the derivation of a law of motion from the field equations, he noted:

> This result is of interest from the point of view of the general question of whether field theory conflicts with the postulates of quantum theory. The majority of physicists are now convinced that the existence of quanta rules out field theory in the usual sense. However, this conviction is based on an inadequate knowledge of the consequences of field theory. Therefore, further investigation of the conclusion of field theory with regard to the motion of singularities appears to me still justified, even though quantum mechanics, proceeding in a different way has achieved a large degree of mastery of the numerical relationships. [261]

Subsequently, using approximate methods, Einstein showed that the field equations contained the law of motion of singularities (provided the nature of the singularity is specified in the first approximation). In fact "in the case of a static and centrally symmetric singularity a geodesic law complemented with electromagnetic forces is obtained." Of course, this did not mean a complete solution of the problem, since this result was obtained up to the third order and only for point particles, understood as field singularities. A more complete and rigorous solution of the problem was given in 1937–1939 by Einstein, Infeld, and Hoffmann (1937–1938), Eddington and Clark (1938), and Fock (1938), although even now the problem cannot be regarded as exhausted. The history of the solution of this problem is a major independent theme that warrants a special investigation.[20]

We mention that in the three-author paper [231] of 1938 the true reason for the inclusion of the equations of motion in the field laws was emphasized, namely, the overdetermination of the field equations. This overdetermination was a consequence of the general covariance of the theory. It was said in this paper:

> The overdetermination is responsible for the existence of equations of motion, and the nonlinear character [of the field equations] for the existence of terms expressing the interactions of moving bodies. ([231], p. 65)

Incidentally, a somewhat more careful analysis of the question of overdetermination and its relationship with general covariance and nonlinearity of the equations of the gravitational field, and also the connection of the last two aspects with each other, is contained in Bergmann's book [12].

Fock, who at least in the 1930s was not a supporter of the unification program, emphasized that the

> general theory of relativity is above all a theory of gravitation.... We believe, therefore, that the problem of general relativity can have nothing in common with the problem of the structure of elementary particles.... In speaking of singular points of the solutions of his equations, Einstein had in mind in the first place elementary particles, and not celestial bodies. ([266], p. 234)

---

[20] We do not know of any investigations by historians of science into the problem of the equations of motion in general relativity and unified field theories. In addition to the studies on axisymmetric solutions mentioned earlier, and the fifth edition of Weyl's book, Einstein subsequently mentioned from the early papers on this problem those of Droste and De Sitter (1916), and from later ones the papers in the 1930s of Levi-Civita (1937), Mathisson (1931), his own papers written in collaboration with Rosen (1935–1936), Robertson (1938), and others [231]. A brief and clear exposition of the work of Einstein, Hoffmann, and Infeld is given in Bergmann's book [12]. The status of the problem at the beginning of the 1960s is reviewed in the book of Infeld and Plebanski [262], and also in the third chapter, written by Goldberg, of the book *Gravitation: An Introduction to Current Research*, edited by L. Witten [263]. Fock's studies are described in most detail by himself in the later reviews [264] and in the book [265]. Modern investigations in this field are discussed in the monograph of Misner, Thorne, and Wheeler [270].

In contrast, Fock solved the problem of motion for the case of finite masses, emphasizing thereby that his work was concerned in the first place with the derivation of the equations of motion of bodies of astronomical scale. From his point of view, the problems of elementary particles must be solved by means of the quantum theoretical program, not the program of unified geometrized theories.

> The incomplete and nominal recognition by Einstein of quantum mechanics undoubtedly has its basis not only in his scientific conservatism but also in the hope of explaining elementary particles as singular points of a field. ([226], p. 234)[21]

A fragment from Einstein's correspondence gives further evidence that in 1927 he placed especially large hopes on the solution of the problem of the equations of motion, hoping in this way to strengthen the position of the field program in its opposition to the quantum theoretical program. In a letter to Weyl of April 26, 1927, he wrote:

> In reply to your friendly letter I have naturally examined closely your proof of the law of motion of the electron and was very impressed by its beauty and clarity.... I value the whole subject so highly because it would be very important to know whether the field equations as such must be refuted or not owing to the factor of the quanta. One is naturally inclined to believe this and most people do believe it. But so far these do not seem to be positive proof. ([57], pp. 184–185 of the English translation)

In a paper written to mark the 200th anniversary of the death of Newton and published in 1927, Einstein expressed unease because the ideals of the Newtonian conception of nature, which he obviously also identified with the geometrized fields program, had been put under threat by the most recent achievements of quantum mechanics:

> It is not without serious grounds that many physicists assert that, in the face of these facts [from the physics of thermal radiation, spectra, and

---

[21] Fock wrote this in 1938, when the quantum theoretical program had already clearly outstripped the program of unified geometrized field theories.

This hope [i.e., of interpreting elementary particles as singularities of a unified field] we in no way share, and it seems to us that the colossal successes of quantum mechanics during the last 10 to 15 years and the complete failure of the attempts made by Einstein during the same time to explain elementary particles by means of a unified field theory have sufficiently clarified the situation in favor of quantum mechanics. ([266], p. 235)

radioactivity] not only the differential law but also the law of causality is powerless . . . the very possibility of a space-time construction in a one-to-one correspondence with physical phenomena is denied.

He recognized the successes of quantum theory, but emphasized that "those who use this method are forced to abandon localization of material particles and strictly causal laws" [267].

At the end of 1927, the Fifth Solvay Congress, which marked the triumph of quantum mechanics, took place. Einstein had critical comments on Bohr's lecture about the physical interpretation of quantum mechanics that was taking hold, the interpretation associated with the names of Born, Heisenberg, and Bohr. Having considered a thought experiment with electrons passing through a small opening and then striking a screen he attempted to find a flaw in the Bohr–Heisenberg interpretation and prove the incompleteness of the quantum-mechanical description, which, as it appeared to him, undermined the principles of the field program—continuity and classical causality associated with the possibility of a unique description of physical processes in space and time [268].

The papers of Grommer and Einstein on the problem of the law of motion in general relativity were a considerable achievement. As we have said, they led to a satisfactory solution of this problem at the end of the 1930s and generated an important direction of development of general relativity. The main stimulus for Einstein in the development of this program was the desire to prove the viability of the unification program and to show, in principle at least, its possibilities for studying the discrete aspects of matter under conditions of rapid progress of the competing program. On the background of this progress, however, the achievement of general relativity and the unificational program remained effectively unnoticed. In this connection, Fock wrote in 1939:

> Despite the great importance of this result—it is, in our opinion, one of the main foundations of the general theory of relativity—the two papers of Einstein [the first with Grommer] passed with little notice and did not receive the appropriate development. ([266], p. 233)

The paper on Kaluza's five-dimensional theory was presented on February 17, 1927 (the first part was presented on January 20) [259, 260]. Einstein did not believe that his paper contributed anything essentially new to this theory: "The differences from his [i.e., Kaluza's] arguments are purely formal" [259]. The main aim of the paper was to draw attention to the theory as a promising direction in the general unification program. After the disappointments with the affine and affine-metric type theories, it was natural to return to the five-dimensional theory. Moreover, in connection with his thinking about the problem of the motion of particles in general relativity, Einstein had changed his point of view with regard to particles. He had ceased to think of them as nonsingular centrally symmetric static solutions and instead identified them with singularities of the field equations. Then, as we have already noted, the argument against Kaluza's theory based on the absence of suitable nonsingular solutions and advanced in 1922 in the joint paper of Einstein and Grommer [269] lost its force. It is possible that the revival of the five-dimensional idea associated with the quantum-mechanical studies of O. Klein, Fock, and others was partly responsible for Einstein's revival of interest in five dimensions, although in a note added in proof to the paper he said that he was unaware of these studies, which, as Mandel' had pointed out to Einstein, also contained results of his investigation.

The content of Einstein's paper is very close to the first section of Klein's paper cited above and published half a year earlier. The main progress achieved by Klein compared with Kaluza's 1921 paper was that the unification of gravitation and electromagnetism, which had been achieved by Kaluza only in the first approximation, was now obtained as an exact result. This applied both to the field equations and the equations of motion of charged particles. In addition, Klein actually introduced a "strengthened cylinder condition" (Einstein's expression), which contained, besides the cylinder condition,

$$\partial \gamma_{\mu\nu} / \partial x^0 = 0 \qquad (\mu, \nu = 0, 1, 2, 3, 4), \tag{28}$$

where $\gamma_{\mu\nu}$ is the metric tensor of the five-dimensional Riemannian space, the requirement $\gamma_{00} = 1$ (or $\gamma_{00} = \alpha = \text{const}$). This meant that the group of general covariance was augmented by "x^0 transformations" (Einstein's terminology):

$$x^0 = \overline{x}^0 + \Psi(\overline{x}^1, \overline{x}^2, \overline{x}^3, \overline{x}^4). \tag{29}$$

The equations of the gravitational and electromagnetic fields were deduced from the unified variational principle

$$\delta \int P\sqrt{-\gamma}\, dx^0\, dx^1\, dx^2\, dx^3\, dx^4 = 0, \tag{30}$$

where P is the five-dimensional scalar curvature. Einstein in fact proceeded from the equivalent Lagrangian,

$$G = \sqrt{\gamma}\, \gamma^{\mu\nu}(\Gamma^\alpha_{\mu\nu}\Gamma^\beta_{\alpha\beta} - \Gamma^\beta_{\mu\alpha}\Gamma^\alpha_{\nu\beta}). \tag{31}$$

The identification of the equations of motion of a charged particle with the equations of a geodesic in a five-dimensional space was also made by Fock [270].

The main aim of the papers of Klein and Fock was not so much the obtaining of these results, which were of great interest from the point of view of the unification program, as the possibility of interpreting the relativistic generalization of Schrödinger's equation as a scalar wave equation in a five-dimensional space. In other words, the fifth dimension interested them above all in connection with the development of the quantum theoretical program, which in this matter was unexpectedly related to the unification program. Klein emphasized that the optical–mechanical analogy that provided the basis of wave mechanics found its complete expression precisely when a five-dimensional space was used. From his point of view, this was a more serious argument for modifying space-time notions in the spirit of Kaluza's theory than the arguments associated with realization of the unification program.

In this paper of 1926, Klein advanced the idea of replacing the cylinder condition by the condition of periodic dependence of all field variables on the fifth coordinate x^0 with period proportional to Planck's constant h.[22] There arose, thus, the idea of unifying gravitation and electromagnetism with wave mechanics on the basis of a five-dimensional geometrical scheme, although the mechanism of this unification remained unclear. "It is also an open question," noted Klein,

> whether the 14 potentials are sufficient for the description [the requirement of the unification program] or whether the inclusion of Schrödinger's method amounts to the introduction of some new field variable (*Zustandgrösse*). ([258], p. 906)

Fock[23] also regarded the establishment of a connection between five dimensions and quantum mechanics as the most important thing: "The intro-

---

[22] Klein wrote:

If we consider such solutions of this equation [i.e., scalar wave equation in a five-dimensional space], in which the fifth dimension appears as a purely harmonically varying quantity with a certain period determined by Planck's constant, then we arrive at . . . quantum-theoretical methods. ([258], p. 895)

[23] Fock did not know about Klein's paper until his was already in press. Fock noted that he had been stimulated by discussions with the Soviet theoreticians Frederiks and Mandel'. The latter, not knowing the papers of Kaluza and Klein, had independently developed a five-dimensional scheme of unification of gravitation and electromagnetism [271]. Fock learned about the five-dimensional approach from the manuscript of Mandel's paper. Concerning the role of Frederiks, Fock wrote: "The idea of this paper arose in discussion with Professor V. Frederiks, to whom I am also grateful for some valuable advice" ([270], p. 226). Frederiks, who had mastered well the unified geometrized field theories of Hilbert, Weyl, and Eddington, obviously did not know of Kaluza's paper, which he would not have failed to mention in the introduction to the

duction of a fifth coordinate parameter appears to us convenient not only for establishing Schrödinger's equation but also the equations of classical mechanics in invariant form" ([270], p. 226). Incidentally, in discussing the invariant properties of the relativistic scalar wave equation, Fock introduced for the first time the local gauge transformations

$$\psi' = e^{-i\frac{ef(x)}{\hbar c}} \psi,$$

$$\varphi' = \varphi_i + \frac{\partial f(x)}{\partial x_i}. \tag{32}$$

Subsequently, the gauge conception of the electromagnetic field arose in this way. In the 1950s this led to the gauge concept in the theory of elementary particles (see Chapter 6).

In 1927, Frederiks published an important review entitled "Schrödinger's theory and the general theory of relativity" [273], which discussed the papers of Klein, Fock, and others on the problem of five dimensions and the connection with quantum mechanics (the review was completed in the fall of 1926). It appeared to Frederiks that the results of Klein and Fock made it possible to approach quantum mechanics in the framework of the program of unified geometrized field theories in a new way: He wrote:

> We see that the five-dimensional space considered by Fock and Klein has very remarkable properties. It is true that we cannot yet speak of a "unified" field geometry on the basis of these studies, but it appears to us that for the finding of it an important and interesting step has been taken, and we would say that it is particularly interesting because it is intimately related to the quantum theory, a problem that the theory of relativity has not yet been able to approach. ([273], p. 89)

In the initial stages, Kaluza and Einstein do not appear to have been particularly concerned with the physical meaning of the fifth dimension, although from the point of view of the unification program the five-dimensional spacetime should be regarded as a generalization of the real four-dimensional spacetime of general relativity. After the papers by Klein and Fock, two points of view with regard to this question developed. The first was clearly expressed in the following words of Frederiks:

> The question of the reality or nonreality of the fifth dimension is not as important as it might appear at first glance. The invariant properties of

book *Foundations of the Theory of Relativity*, which he wrote with Friedmann [272].

a spherical surface in ordinary Euclidean space can be studied by means of the propositions of ordinary geometry, but they can also be obtained as properties of a spherical, non-Euclidean two-dimensional geometry. Just as the third coordinate of Euclidean space makes the study of a spherical surface simpler, in the study of the properties of our four-dimensional world the fifth coordinate may be merely a tool for simplifying our conclusions. ([273], p. 98)

The interpretation of the fifth dimension as a purely auxiliary construction of essentially formal computational significance was evidently associated with the fact that the five-dimensional metric

$$\gamma_{ik} = g_{ik} + \frac{e^2}{m^2 c^4} \varphi_i \varphi_k, \tag{33}$$

in contrast to its four-dimensional part, did not have a universal nature but, rather, depended on the properties of the considered particle if, for example, one was describing the motion of a charged particle with charge e and mass m. "It should not be forgotten," emphasized Frederiks in this connection, "that the measure determination (4) [i.e., (33)] is suitable only for a given particle with definite charge and mass" ([273], p. 87).

Subsequently, in the 1940s and 1950s, the Soviet physicist Rumer, emphasizing this nonuniversality of the five-dimensional Riemannian metric, wrote:

> This shows that the 5-space of 5-optics [5-optics was Rumer's name for the generalization of Kaluza–Klein theory he had developed] is not the universal space of the general theory of relativity (extended by one additional dimension) but the *configuration space* for the particle whose motion we consider. ([13], p. 27)

Recognition of the configuration nature of the five-dimensional space did not, however, mean that the search for a physical meaning of the fifth dimension was abandoned. The second point of view, which Klein, London, and others began to develop (and, in particular, Rumer later developed) was to relate it to physical quantity. The basis for this connection was provided both by the relationship of the fifth dimension to electromagnetism and quanta as well as by the cylinder condition and periodicity condition. Klein believed that the fifth dimension had a charge nature in the same sense as the three-dimensional space had a momentum nature—the unobservability of the fifth dimension resulted in charge conservation, in exactly the same way as homogeneity of three-dimensional space resulted in momentum conservation. Klein related the quantum nature of charge to cyclicity of the fifth dimension with a period of order 10^{-30} cm. As a result, the five-dimensional space was, as it were, closed with respect to the fifth coordinate. The extreme smallness of the period explained why the fifth dimension did not manifest itself in physical phenomena [274]. London attempted to relate the fifth dimension to the angle of rotation of an electron around its axis [275]. Ten years after these events, Einstein and Bergmann revived Klein's idea, assuming that the five-dimensional space was topologically closed in the fifth dimension with an extremely small period b [198]. A two-dimensional model of such a space was cylindrical surface with radius $b/2\pi$. Another 10 years later, Rumer ascribed to the fifth dimension the physical meaning of action and identified the constant b of Einstein and Bergmann with Planck's constant, generalizing thereby Klein's periodicity condition [13].

In his paper of 1927, Einstein ignored all aspects of the fifth dimension associated with quantum mechanics, probably because he hoped to obtain particles as singular solutions of unified field equations and the quantum features of their behavior as properties of these solutions. He did not obtain anything essentially new compared with the papers of Klein and Fock except, perhaps, for two comments concerning the interpretation of the "strengthened cylinder condition." First, he noted that

> if we now assume that it is not the $\gamma_{\mu\nu}$ themselves but only their ratios that have objective meaning, in other words, that in space it is not the metric $d\sigma$ [i.e., $d\sigma = \gamma_{\mu\nu} dx^\mu dx^\nu$] that is specified but the set of "light cones" ($d\sigma = 0$), then the Hamilton function will depend on only the first two of the above combinations [i.e., in fact on the ratios γ_{mn}/γ_{00}, γ_{0m}/γ_{00}, and their derivatives].... This means that the requirement $\gamma_{00} = 1$ does not lead to a restriction of the fundamental theoretical possibilities, and we arrive at a strengthened cylinder condition. [259]

Second, Einstein noted that the choice of $+1$ for γ_{00} and, thus, the choice of a space-like nature of the fifth dimension determined a positive sign of the second term in the Lagrangian of the unified field:

$$P = R + \tfrac{1}{4} f_{ik} f^{ik}. \tag{34}$$

In other words, this choice of the sign of γ_{00} was associated with the requirement "that the equations of the unified field take the usual form" [259].

At various times, Weyl, Pauli, and others advanced two objections against the five-dimensional approach to the construction of unified theory of gravitation and electromagnetism. First, there was the formal nature of the unification. The fact was that irrespective of the nature of the fields, general covariance in conjunction with gauge invariance led to a five-dimensional formalism. Second, and this objection was more serious, "There is ... no justification for the particular choice of the five-dimensional curvature scalar as integrand of the action integral from the standpoint of the restricted group of the cylindrical metric" [63, p. 230).

In connection with overcoming this difficulty, several directions aimed at generalizing Kaluza–Klein theory arose in the 1940s and 1950s. If one gave up the strengthened cylinder condition and identified γ_{55} with some scalar field (in the spirit of the original form of Kaluza's theory) there appeared a possibility to develop a five-dimensional theory with a variable gravitational constant (Dirac, Jordan, Thiry, and others), a theory that in its turn led to a number of new difficulties, in particular of operational and experimental natures [ibid.].

Another way of generalizing the five-dimensional theory was associated with the replacement of the cylinder condition by the periodicity condition (periodic dependence of all field quantities on the fifth coordinate x^0). This idea, which had been advanced by Klein in 1926, was subsequently developed by Einstein and collaborators, Rumer, and others. In this case, one could speak of the group of transformations

$$x'^i = p^i(x^0, x^k),$$
$$x'^0 = x^0 + p^0(x^0, x^k), \tag{35}$$

where p^μ are arbitrary periodic functions of x^0 with period 2π. This justified the introduction of the five-dimensional scalar curvature P as basic invariant, although the problem of the existence of other, more complicated invariants still remained open.

Interesting possibilities of relating the five-dimensional approach to quantum mechanics, for example, in the spirit of Rumer's five-optics, then developed, but the discovery of a great variety of elementary particles, especially in the 1950s, emphasized the manifest limitation of the five-dimensional theory, in which, as in other unified geometrized field theories, the main attention was devoted to the unification of only the gravitational and electromagnetic interaction.

The possibilities of the five-dimensional scheme must not, however, be regarded as exhausted. In his Nobel address, Salam, noting the "major developments in realizing Einstein's dream" called the first of them "the Kaluza–Klein miracle," ([1], p. 534). In particular, he mentioned the investigations of Julia and Cremer (1979),

> which started with an attempt to use the ideas of Kaluza and Klein to formulate extended supergravity theory in a higher (compactified) space-time—more precisely, in eleven dimensions. ([1], p. 535)[24]

---

[24] In this connection, we mention the paper [276] of Voronov and Kogan, which uses

In a modern monograph on the theory of gauge fields, there are frequent mentions of the use of the five-dimensional concept, for example, in connection with the interpretation of the mass of a gauge field and interpretation of the Lagrangian of gravitation and Yang–Mills fields as the scalar curvature of a fiber space of more than four dimensions: "Such an approach generalizes the unified Kaluza–Klein theory of gravitation and electromagnetism to non-Abelian gauge fields" ([7], p. 133). In another place, the authors write:

> The presence of the tensor field $g_{\mu\nu}$, the vector field A_μ, and the scalar field χ makes Kaluza–Klein theory attractive for generalizations to the case of gauge theories with spontaneous symmetry breaking in a Riemannian V_4 and in supersymmetric theories. ([7], p. 145)

At the end of this chapter, we shall consider a modification of the Kaluza–Klein theory undertaken at the beginning the 1930s by Einstein in collaboration with W. Mayer. The following chapter, which contains an outline of the formation and development of the gauge field concept, will show, in particular, the part played by the five-dimensional approach in the genesis of the concept of gauge symmetry.

GEOMETRY WITH ABSOLUTE PARALLELISM: 1928–1930

By 1928, various forms of affine and affine-metric schemes (including the theories of Weyl and Eddington) and the five-dimensional approach had been tried; the problem of obtaining the equations of motion of particles regarded as singularities of the field from the equations of this field had also been made the subject of a special investigation. The successes were as yet very limited, especially in comparison with the achievements of the quantum theoretical program. Despite their mathematical refinement, none of the attempts at a geometrical field synthesis of gravitation and electromagnetism undertaken up to 1928 satisfied Einstein, and they were ultimately rejected by him. The problems of the field interpretation of elementary particles and the explanation of their quantum properties (and also the quantum properties of the electromagnetic field) remained unresolved. The possibility of deriving equations of motion of particles from the equations of general relativity, demonstrated in 1927 by Einstein (in part in collaboration with Grommer), gave only some hope of an analogous solution of this problem in a future unified field theory.

---

a generalization of the five-dimensional Kaluza–Klein approach to construct a unified theory of the four fundamental interactions. We also cite the papers of D. Friedmann and P. van Nieuwenhuizen, and E. Mielke, as noted in Footnote 12a of Chapter 4.

In 1928 Einstein found a new method of unifying gravitation and electromagnetism in the framework of the program of unified geometrized field theories based on an affine geometry "that retains the concept of 'absolute' parallelism" [227]. The first exposition of the new unified theory was given in a paper presented by Planck to the Berlin Academy of Sciences (first part on June 7 [278], second part on July 14 [278], both published on July 10, 1928). The first part considered the geometrical foundations of the theory, the second the physical interpretation of the new geometry. In essence, what was proposed was a very special extension of Riemannian geometry in which the curvature tensor $R^{\nu}_{\mu\rho\sigma}$ and the corresponding form Ω^{ρ}_{ν} vanish (as in fact the homothetic curvature Ω did, too), while the torsion tensor and the corresponding form Ω^{ρ} were nonzero. In such a geometry, it is possible to compare vectors not only in absolute magnitude (as in Riemannian geometry) but also in direction, i.e., one can speak of the parallelism of vectors that are separated by a finite distance. In his first paper and in most of the following papers, Einstein used the term *Fernparallelismus* ("distant" parallelism, or in the expression of Lanczos, "teleparallelism"); the term "absolute parallelism" is also widely used [279].

In developing this geometry, Einstein took as basis the local-frame (tetrad) formalism that in the 1920s had been developed in the differential-geometrical investigations of Cartan, Weitzenböck, Levi-Civita, and others. During 1926–1927, Cartan read a course of Riemannian geometry on the basis of the orthogonal-frame approach and the method of exterior forms [280]. Even earlier, Weitzenböck had considered a Riemannian space equipped with a local-frame structure [281]. In 1926 and 1927 Eisenhart published his monographs *Riemannian Geometry* [282] and *Non-Riemannian Geometry* [283], which have since become classics. They also used the formalism of local orthogonal frames.[25]

Einstein did not use the orthogonal frame method to reformulate general relativity but to construct a new geometry designed for a unified description of the gravitational and electromagnetic fields. In his first publications, there were no references to the investigations of the differential geometers.

We shall consider in some detail Einstein's study of 1928 [277, 278]. From the beginning, he emphasized that the main stimulus for him was the search for a more natural generalization of Riemannian geometry in which there would

---

[25] As used in general relativity, this formalism subsequently became known as the tetrad formalism and became the basis of the corresponding (tetrad) formulation of general relativity [284].

be a place for a geometrical description of the electromagnetic field. Einstein wrote:

> In the geometry of Riemann, which has provided the possibility for a phys-
> ical description of the gravitational field in the general theory of relativity,
> there are absolutely no concepts that could be associated with the elec-
> tromagnetic field. Therefore, theoretical physicists, hoping to construct a
> logical theory that unifies all physical fields from a common point of view,
> have sought to find natural generalizations of Riemann's geometry con-
> taining more concepts than that geometry (*natürliche Verallgemeinerungen
> oder Ergänzungen der Riemannschen Geometrie . . . welche begriffsreichen
> sind als diese . . .*). ([277], p. 223; [285], p. 217)

He regarded as the characteristic feature of the new geometrical scheme the possibility of comparing distant vectors not only in magnitude, as in Rie-mannian geometry, but also in direction, and also the resulting possibility of speaking of a parallelism of such vectors. At the end of the first part, he compared the new geometry with the geometries of Riemann and Weyl:

> For vectors separated by a finite distance it is impossible to compare them in
> either length or direction *in Weyl's theory; in Riemann's theory* [i.e., general
> relativity] a comparison in length is possible, but not direction; *in the theory
> considered here*, comparison is possible in both length and direction. [277]

The transition from Riemannian geometry to more amorphous generalizations of it of the type of Weyl's geometry had not led to success, and Einstein attempted to go over to a more rigid structure, which, however, would be sufficiently "rich" that its geometrical characteristics would contain quantities corresponding to both the gravitational and the electromagnetic field.

Whereas for Weyl, as we have seen, one of the main stimuli was the aim of further elimination from physical geometry of elements involving action at a distance, the new geometrical scheme seemed to come into conflict with this aim, since it introduced a new element involving nonlocality, the parallelism of vectors separated by a finite distance. In other words, in a space with absolute parallelism, the change in the components of a vector undergoing parallel transport does not depend on the path of that transport.

Like the orthogonal-frame method, spaces with absolute parallelism were not a novelty for geometers. They were apparently discussed for the first time by Cartan in his long paper on the structure of continuous groups, published in 1904–1905 [286]. Cartan showed that Riemannian spaces with nonvanishing curvature tensor admit the introduction of the concept of absolute parallelism. During 1926–1927, i.e., a year or two before the appearance of Einstein's unified field theory based on geometry with absolute parallelism, this geometry was again considered in joint studies of Cartan and Schouten [287] (and also in a fundamental paper of Cartan [288], subsequently translated into Russian [289], p. 39, p. 55) on the basis of the latest achievements of differential geometry, in particular, the concept of affine connection. When Cartan learned of Einstein's unified field theory based on the idea of absolute parallelism, he published a paper in the *Mathematische Annalen* devoted to the development of the concept of absolute parallelism in differential geometry [290]. This paper appeared next to a paper by Einstein containing an exposition of the latest form of his unified theory (it was submitted to the journal on August 19, 1928 [291]). Einstein wrote:

> From Weitzenböck and Cartan, in particular, I learned that a theory of continua of the kind considered here [i.e., spaces with absolute parallelism] is not in itself new. Cartan has kindly taken upon himself the work of writing an historical review of the corresponding branch of mathematics, which complements my paper [Einstein is referring to the above paper of Cartan, which immediately followed Einstein's paper in the journal]. [291][26]

We now return to the first publication of Einstein's theory. The essence of Einstein's approach was that he equipped the space-time continuum with a field of four-dimensional pseudo-orthogonal frames (*n-Bein Feld*). With respect to such a frame, constructed at a certain point P, a vector A^ν (specified relative to a certain Gaussian system of coordinates x^ν) has components A_a. If we denote by h_ν^a the components of the orthogonal frame with respect to this system of coordinates x^ν, then the components of the vector A^ν and A_a are related by

$$A^\nu = h_a^\nu A_a. \tag{36}$$

We also have the relation

$$A_a = h_{\mu a} A^\mu, \tag{37}$$

where $h_{\mu a}$ are the normalized minors of h_a^μ, Because the continuum is locally Euclidean, taking into account (37), we obtain an expression for the metric in terms of h_a^μ or $h_{\mu a}$:

$$g_{\mu\nu} = h_{\mu a} h_{\nu a}, \tag{38}$$

---

[26] Obviously, important material on theories with absolute parallelism, especially in connection with the preceding geometrical investigations of Cartan and his reaction to Einstein's paper, is contained in the correspondence between Einstein and Cartan, of which we learned only after the present work was completed [291a].

where $h_{\mu a} h_a^\nu = \delta_\mu^\nu$ and $h_{\mu a} h_b^\mu = \delta_{ab}$, and for fixed a the quantities $h_{\mu a}$ and h_a^μ have a vector nature.

The field of orthogonal frames is determined by the 16 functions h_a^μ, whereas the Riemannian metric is determined by only 10 functions $g_{\mu\nu}$. The relation (38) makes it possible to introduce the metric, but the metric is insufficient to determine the field of orthogonal frames. Einstein showed, further, that in the presence of such a field it is possible to introduce the concept of absolute parallelism:

> ... if (A) and (B), two vectors at the points P and Q (with respect to their local orthogonal frames), have equal corresponding local coordinates (i.e., $A_a = B_a$), then they will be equal and "parallel." [277]

At the same time, the group of rotations of the four-dimensional orthogonal frames (i.e., the Lorentz group) is introduced into the geometry alongside the group of general covariance.

Absolute parallelism is incompatible with the presence of curvature. A necessary and sufficient condition for an affine space to possess the property of absolute parallelism is identical vanishing of the curvature tensor. Einstein gave a sketch of the proof of this assertion. For a more rigorous proof, see for example [292]. From the condition of absoluteness of parallel transport of the vector A_a there follows

$$dA_a = d(h_{\mu a} A^\mu) = \frac{\partial h_{\mu a}}{\partial x^\sigma} A^\mu \, dx^\sigma + h_{\mu a} \, dA^\mu = 0, \tag{39}$$

from which the law of parallel transport of the vector A^μ is obtained directly:

$$dA^\nu = -\Delta_{\mu\sigma}^\nu A^\mu \, dx^\sigma, \tag{40}$$

where

$$\Delta_{\mu\sigma}^\nu = h^{\nu a} \frac{\partial h_{\mu a}}{\partial x^\sigma} \tag{41}$$

is the affine connection. A calculation of the curvature tensor using the relation (41) leads to the result zero.

Thus, the connection $\Delta_{\mu\sigma}^\nu$ realizes an integrable parallel transport of a vector. The nonintegrable transport realized by the Riemannian connection $\Gamma_{\mu\nu}^\sigma$ generates the vector $A^\nu + \bar{d}A^\nu$, and therefore $dA^\nu - \bar{d}A^\nu = (\Gamma_{\alpha\beta}^\nu - \Delta_{\alpha\beta}^\nu)A^\alpha \, dx^\beta$ is a vector, and $\Gamma_{\alpha\beta}^\nu - \Delta_{\alpha\beta}^\nu$ is a tensor. In this geometry, a fundamental role is played by the antisymmetric part of this tensor:

$$\tfrac{1}{2}(\Delta_{\alpha\beta}^\nu - \Delta_{\beta\alpha}^\nu) = \Lambda_{\alpha\beta}^\nu, \tag{42}$$

which characterizes the extent to which the space is non-Euclidean and is the torsion tensor. Vanishing of the tensor (42) is a necessary and sufficient condition for the space to be Euclidean.

Einstein summarized,

> Moreover, since the tensor $\Delta_{\alpha\beta}^{\nu}$ is obviously the simplest from the formal point of view, the simplest characteristic properties of the considered continuum must evidently be associated with it, and not with the more complicated Riemann curvature tensor. [277]

The 16 quantities $h_{\mu a}$ make it possible to describe not only gravitation (for which the 10 quantities $g_{\mu\nu}$ are sufficient), but also electromagnetism. Instead of the Riemann and Ricci tensors and the scalar curvature, one can use for this the tensor $\Delta_{\alpha\beta}^{\nu}$ and the corresponding scalar constructions

$$g^{\mu\nu}\Lambda_{\mu\beta}^{\alpha}\Lambda_{\nu\alpha}^{\beta}, \quad g_{\mu\nu}g^{\alpha\beta}g^{\sigma\tau}\Lambda_{\alpha\beta}^{\mu}\Lambda_{\sigma\tau}^{\nu}.$$

Einstein chose the first as basic invariant and formulated a variational principle to obtain field equations in the form

$$\delta \int hg^{\mu\nu}\Lambda_{\mu\beta}^{\alpha}\Lambda_{\nu\alpha}^{\beta}\,d\tau = 0. \tag{43}$$

The new geometry appeared promising because in the first approximation the field equations that followed from the variational principle (43) led to the equations of gravitation and electromagnetism. Having come across a new, nontrivial geometry, Einstein then found that

> from such a theory one can obtain very simply and naturally, in the first approximation at least, the laws of the gravitational field and electrodynamics.... Therefore, one may suppose that this theory will displace the original form of the general theory of relativity [278].

In the first approximation ($h_{\mu a} = \delta_{\mu a} + k_{\mu a}$, where $k_{\mu a}$ is an infinitesimal of first order) the field equations associated with the variational principle (43) reduce to the following second-order equations:

$$\frac{\partial^2 k_{\beta\alpha}}{\partial x_\mu^2} - \frac{\partial^2 k_{\mu\alpha}}{\partial x_\mu \partial x_\beta} + \frac{\partial^2 k_{\alpha\mu}}{\partial x_\beta \partial x_\beta} - \frac{\partial^2 k_{\beta\mu}}{\partial x_\mu \partial x_\alpha} = 0. \tag{44}$$

Introducing in place of $k_{\mu\nu}$ the more usual quantities

$$g_{\alpha\beta} = (\delta_{\alpha a} + k_{\alpha a})(\delta_{\beta a} + k_{\beta a}),$$

or $\overline{g}_{\alpha\beta} = g_{\alpha\beta} - \delta_{\alpha\beta}$, which to first order are equal to $(k_{\alpha\beta} + k_{\beta\alpha})$, and also the electromagnetic potentials

$$\varphi_a = \Lambda^{\mu}_{\alpha\mu} = \frac{1}{2}\left(\frac{\partial k_{\alpha\mu}}{\partial x_{\mu}} - \frac{\partial k_{\mu\mu}}{\partial x_{\alpha}}\right), \tag{45}$$

Einstein obtained the field equations (44) in the form

$$\frac{1}{2}\left(-\frac{\partial^2 \overline{g}_{\alpha\beta}}{\partial x_{\mu}^2} + \frac{\partial^2 \overline{g}_{\mu\alpha}}{\partial x_{\mu}\partial x_{\beta}} + \frac{\partial^2 \overline{g}_{\mu\beta}}{\partial x_{\mu}\partial x_{\alpha}} - \frac{\partial^2 \overline{g}_{\mu\mu}}{\partial x_{\alpha}\partial x_{\beta}}\right) = \frac{\partial \varphi_{\alpha}}{\partial x_{\beta}} + \frac{\partial \varphi_{\beta}}{\partial x_{\alpha}}. \tag{46}$$

Then, in the absence of an electromagnetic field, Eqs. (44) go over in the first approximation into the gravitational equations of general relativity,

$$R_{\alpha\beta} = 0. \tag{47}$$

Similarly, simple manipulations with Eqs. (44)–(46) lead to Maxwell's equations for the vacuum:

$$\partial\varphi_{\alpha}/\partial x_{\alpha} = 0, \qquad \partial^2\varphi_{\alpha}/\partial x_{\beta}^2 = 0. \tag{48}$$

Of course, this decomposition of Eqs. (44) into Einstein's equations and Maxwell's equations occurs in the first approximation, and Eqs. (44) include more information than Eqs. (47) and (48) alone. Einstein noted in the conclusions that the use of the second invariant $g_{\mu\nu}\, g^{\alpha\sigma}\, g^{\beta\tau}\, \Lambda^{\mu}_{\alpha\beta}\Lambda^{\nu}_{\sigma\tau}$ as Lagrangian led to analogous results. Thus, the question of the choice of the Lagrangian was not completely solved.

On January 10, 1929, Planck presented to the Berlin Academy of Sciences a paper of Einstein, who at that time was not in Germany [293]. The paper dealt with a certain modification of the theory with absolute parallelism. Retaining the geometry, Einstein attempted to find field equations without using a variational principle. "Attempts to derive field equations from Hamilton's principle have not led me to a simple and unique path," wrote Einstein [293]. As before, the leading role in the derivation of the field equations was played by the torsion tensor $\Lambda^{\nu}_{\mu\sigma}$ (Einstein in fact did not give it that name).

From the condition of vanishing of the curvature tensor, one can readily obtain the identity

$$\Lambda^{\alpha}_{kl;\alpha} + \varphi_{l;k} - \varphi_{k;l} - \varphi_{\alpha}\Lambda^{\alpha}_{kl} = 0, \tag{49}$$

which, by introduction of the antisymmetric (with respect to the indices k and l) tensor density

$$B_{kl}^{\alpha} = h(\Lambda_{kl}^{\alpha} + \varphi_l \delta_k^{\alpha} - \varphi_k \delta_l^{\alpha}) \qquad (50)$$

and generalized density divergence

$$T_{\cdot|\sigma}^{\cdot\sigma} = T_{\cdot;\sigma}^{\cdot\sigma} - T_{\cdot\cdot}^{\cdot\alpha} \Lambda_{\alpha\sigma}^{\sigma}, \qquad (51)$$

can be rewritten in the form

$$(B_{kl}^{\alpha})_{|\alpha} = 0. \qquad (52)$$

After I discovered the identity (3b) [i.e., (52)], it became clear that under a natural restriction of the manifold of the form considered here an important part must be played by the tensor density $\mathfrak{B}_{kl}^{\alpha}$ [i.e., B_{kl}^{α}]. Since its divergence $\mathfrak{B}_{kl|\alpha}^{\alpha}$ is identically equal to zero, the thought immediately arises of requiring (as field equation) the the other divergence $\mathfrak{B}_{kl|l}^{\alpha}$ also vanish. [293]

This was the new, very formal approach to the construction of field equations. Ultimately, Einstein arrived at the system of field equations

$$B_{\underline{k}\,\underline{l}|l}^{\alpha} - B_{\underline{k}\,\underline{\tau}}^{\sigma} \Lambda_{\sigma\tau}^{\alpha} = 0, \qquad (53)$$

$$\left[h(\varphi_{\underline{k};\underline{\alpha}} - \varphi_{\underline{\alpha};\underline{k}})\right]_{|\alpha} = 0, \qquad (54)$$

where the underlining of the indices denotes their raising or lowering (for example, $\Lambda_{\underline{k}\,\underline{l}}^{\alpha}$ denotes the purely contravariant tensor corresponding to the tensor Λ_{kl}^{α}). In the first approximation, these equations gave the equations of gravitation and electrodynamics. Their compatibility appeared to be ensured by the fact that for the 12 variables ${}^s h_{\alpha}$ (the notation was introduced by Weitzenböck [294] and then used by Einstein; the number is explained by taking into account the requirement of general covariance: $16 - 4$) there existed the 20 equations (53)–(54) and eight identities connecting them.

Einstein concluded,

A deeper investigation of the consequences of the field equations (11) and (10a) [i.e., (53)–(54)] must show whether the Riemannian metric in conjunction with absolute parallelism gives an adequate understanding of the physical properties of space. [293]

At the end of the paper, he expressed his thanks to the mathematician Müntz and the Physics Fund of the Prussian Academy of Sciences for enabling him to have "such an assistant in the investigation as Dr. Grommer."

The beginning of 1929 was a period of high hopes for Einstein. It appeared to him that this time he was on the correct path. Just after he had completed this paper, he wrote in cheerful spirits to Besso (in a letter on January 5, 1927):

But the most beautiful thing, on which I struggled and calculated for whole days and half the nights, is now finished and compressed onto seven sides with the title "Unified field theory." It appears archaic, and dear colleagues—and you are one too—will initially stick out their tongues to the full length since the equations do not contain Planck's constant. But when the possibilities of the statistical fad (*statistische Fimmel*) have been exhausted, there will be a return to the space-time conception, and then these equations [the field equations (53), (54)] may provide a point of departure. I have actually found a geometry that possesses not only a Riemannian metric but also absolute parallelism, which hitherto we intuitively regarded as characteristic of Euclid. The simplest field equations of such a manifold lead to the known laws of electricity and gravitation. Even the equations $R_{ik} = 0$ must be put in the attic, despite the achieved successes. ([227], p. 28)

Incidentally, the program of unified geometrized field theories is here clearly opposed to the quantum theoretical program. After the creation of quantum mechanics, there arose an entire direction of theoretical investigations at the frontier of these two programs. Five-dimensional theories were particularly popular (see the previous section). The efforts in this direction were foreign to Einstein. This opposition to the quantum theoretical program was also reflected in the development of the theory with absolute parallelism. In the radically new geometry, he saw the prospect of the revenge of classical physics. ("But when the possibilities of the statistical fad have been exhausted, there will be a return to the space-time conception, and then these equations may provide a point of departure.") We shall see below that Einstein's theory with absolute parallelism was soon taken up by many theoreticians with the aim of developing relativistic quantum mechanics and establishing its connection with gravitation (papers of Wiener, Weyl, Tamm, Fock, Ivanenko, and others). Once again, Einstein did not support such efforts. From his point of view, quantum properties must be explained by investigating particle-like solutions (or singularities) of unified-field equations obtained completely independently of the propositions of quantum mechanics.

On January 26, 1929 the *Daily Chronicle* of London published an interview with Einstein on his new theory, an extract from which was reprinted in the News and Views section of *Nature* ([293], pp. 258–259; [295]). Having briefly characterized the program of unified geometrized field theories and the previous failures in the attempt to realize it, he said:

I have thought out a special construction which is differentiated from that of my relativity theory, and from other theories of four-dimensional space, through certain conditions. These conditions bring under the same mathematical equations the laws which govern the field of gravitation. ... Now, but only now, we know that the force which moves electrons in their ellipses about the nuclei of atoms is the same force which moves our earth in its annual course about the sun, and is the same force which brings to us the rays of light and heat which make life possible upon this planet. ([295], pp. 174–175)

In the two issues of February 4 and 5, 1929, the *Times* published an article by Einstein entitled "New field theory," which was shortly afterward reprinted in the journal *Observatory* [296]. In it Einstein attempted a popular and intuitive interpretation of geometry with absolute parallelism as the foundation of a unified field theory. He explained the intermediate nature of the new geometry by an example with parallel lines and a parallelogram. In both Euclidean geometry and a geometry with absolute parallelism one can speak of parallel lines. In the new geometry, however, if we attempt to intersect a pair of parallel lines with another such pair in order to obtain a parallelogram (as happens in Euclidean geometry), then we do not succeed in obtaining this figure. The straight line $Q_1 R$ parallel to the line $P_1 P_2$ (the points P_1 and Q_1 lie on the line l_1, the point P_2 on the line l_2 parallel to line l_1) does not intersect the line l_2. This property can be taken as the basis of geometry with absolute parallelism. "Einstein's world-geometry," noted Eddington, "may be briefly described as a geometry in which there are *parallels* but not *parallelograms*" ([297], p. 280). The measure of the deviation of the line $Q_1 R$ from the line l_2 is characterized by the extent to which the space possesses torsion, namely, by the torsion tensor $\Lambda_{\mu\nu}^{\sigma}$.

In March 1929, Einstein returned to the use of the variational principle to establish field equations (the paper "Unified field theory and Hamilton's principle" was presented at the meeting of the Prussian Academy of Sciences on March 21 [298]). The fact was that Müntz and Lanczos had drawn Einstein's attention to the lack of four identities needed to ensure compatibility of the field equations. Derivation of the field equations from a variational principle automatically guaranteed compatibility of these equations. The choice of the necessary Lagrangian was not simple or natural, however, and to obtain Maxwell's equations in the first approximation certain artificial constructions were needed. A more popular paper written for a collection published in honor of the 70th birthday of A. Stodola covers similar ground.

The Lagrangian of the unified field in the new form was taken to be a linear combination

$$H = AH_1 + BH_2 + CH_3,$$

where $H_1 = g^{\mu\nu}\Lambda_{\mu\beta}^{\alpha}\Lambda_{\nu\alpha}^{\beta}$, $H_2 = g^{\mu\nu}\Lambda_{\mu\alpha}^{\alpha}\Lambda_{\nu\beta}^{\beta}$, $H_3 = g^{\mu\nu}g^{\sigma\tau}g_{\lambda\xi}\Lambda_{\mu\nu}^{\lambda}\Lambda_{\sigma\tau}^{\xi}$, but there were no clear argument on how to determine the coefficients A, B, and C. "The development and physical interpretation of the theory," wrote Einstein, "is difficult because *a priori* there exist no grounds for the choice of the relations between the constants A, B, and C" [299]. Some hope of resolving this difficulty was provided by the fact that for $B = -A$, $C = 0$ one obtained equations that in the first approximation were identical to the equations of gravitation and electromagnetism and which, as calculations made in collaboration with Müntz showed, led to the Schwarzschild solution for the field of an uncharged material point.

Einstein's theory with absolute parallelism aroused great interest both in Germany and abroad. This is indicated not only by Einstein's popular articles and interviews with him in newspapers and journals (for example, in the *Times* and *Daily Chronicle*), but also by the publication, beginning in 1929, of a number of studies by eminent theoreticians and mathematicians on various problems of this theory. During a period of one and half to two years papers were published by Weitzenböck, Levi-Civita, Cartan, Eddington, Weyl, Wigner, Wiener (in part with Vallarta), Lanczos, Tamm (in part with M.A. Leontovich), Fock (in part with Ivanenko), and others that in one way or another were associated with this theory.

All these studies can be divided nominally into three groups: (1) those that concentrated the main attention on the geometrical aspects of the theory; (2) those that did not go beyond the framework of the unification program and were devoted either to the popularization and explanation of the foundations of the theory or to attempts to resolve its difficulties, or to criticism and a different interpretation of the geometrical scheme; (3) those in which attempts were made to relate the theory to quantum mechanics.

The first group included, for example, papers by Weitzenböck, Levi-Civita, and Cartan. Weitzenböck (whose paper, published in 1928 [294], was one of the first responses to Einstein's theory) investigated in more detail the basic invariant of a geometry with absolute parallelism; these were then used by Einstein. He also introduced the new notion for the tetrads ${}^{s}h_{\alpha}$. Levi-Civita, having noted that geometry with absolute parallelism and the associated geometrical methods had been developed earlier by Weitzenböck and the Italian geometer Vitali, gave a more standard and transparent construction of this geometry. He used the concept of congruence and, on its basis, introduced the concept of a "world lattice,"[27] which was equivalent to a field of tetrads [300]. He also introduced the use of the antisymmetric rotation coefficients γ_{ikl} of

---

[27] Later, Treder compared a space with absolute parallelism to a crystal lattice ([216], p. 38).

Ricci; these determine the variation of a vector specified by the orthogonal components A^a under parallel transport.

As we have already mentioned, Cartan gave an extensive historical review of earlier investigations of geometry with absolute parallelism [290]. He also drew attention to possible difficulties associated with the investigation of cosmological problems in the framework of geometry with absolute parallelism. In lectures read at the Scientific Research Institute of Mathematics and Mechanics at Moscow (June 16–20, 1930), Cartan emphasized the fundamental significance of the most recent achievements of differential geometry, in particular the symmetric classification of generalized geometries developed by him, Schouten, and others, in the development of theoretical physics; after also mentioning geometry with absolute parallelism in this connection, he noted:

> It is clear that a large number of the generalized geometries are as yet only geometrical objects worthy of interest. However, they have the double advantage of casting light on the foundations of differential geometry themselves and of forming a reservoir of geometrical schemes from which mathematics and mathematical physics can draw. For example, the Riemannian geometry with absolute parallelism that forms the basis of Einstein's most recent investigations belongs to the general scheme that we have presented. ([301], p. 62)

Apart from the latest papers of Einstein himself, the second group of papers were by Eddington, McVittie, Lanczos, Piaggio, and others. Particularly interesting was the reaction of one of the pioneers of the unification program— Eddington. He was apparently the first in Britain to respond to Einstein's new theory. His short paper appeared in *Nature* on February 23, 1929 [297]. Having characterized clearly and succinctly the foundations of geometry with absolute parallelism (it was in this paper that he noted that in this geometry there existed parallels but not parallelograms) and having described the way in which it was used to construct a unified field theory, he compared the new scheme with the affine geometry that underlay his own theory of 1921. This revealed that they were at opposite poles: In Eddington's theory (as in Weyl's theory) parallel transport of a vector was nonintegrable, the connection was symmetric, the curvature tensor $R^{\mu}_{\alpha\mu\nu}$ determined the structure of the geometry and the torsion tensor was zero; in Einstein's theory, parallel transport of a vector was integrable, the connection was asymmetric, the curvature tensor vanished, $R^{\beta}_{\alpha\mu\nu} = 0$, and the torsion tensor $\Lambda^{\nu}_{\alpha\mu}$, in contrast, determined the structure of space-time.

Adopting a division of geometry into "universal" and "natural" (the former was a nominal scheme, a kind of graph; the latter was truly physical, manifesting itself in physical measurements, and was Riemannian), Eddington was prepared to accept geometry with absolute parallelism only in the first sense, as one of a number of possible geometrical schemes, inferior, in fact, from his point of view, to the more adequate affine schemes.

> For my own part I cannot readily give up the affine picture, where gravitational and electrical quantities supplement one another as belonging, respectively, to the symmetrical and antisymmetrical features of world measurement; it is difficult to imagine a neater kind of dovetailing. Perhaps one who believes that Weyl's theory and its affine generalization afford considerable enlightenment may be excused for doubting whether the new theory offers sufficient inducement to make an exchange. ([297], p. 281)

Lanczos [302] and McVittie [303] investigated the problem of the compatibility of the field equations and their solutions in special cases (in the first approximation). Piaggio gave a simplified, popular exposition of the theory, noting the absence of connection of the new theory with experiment and the difficulty of obtaining exact solutions. He also pointed out that in this theory Einstein "may even get back to the position of Newton, who conceived absolute rotation to be a real thing" ([304], p. 879).

The third group consisted mainly of papers by younger theoreticians mainly concerned with quantum mechanics. They were mainly interested in the question of the connection between general relativity and quantum theory. In the tetrad method and the geometry of absolute parallelism they saw new possibilities for realizing a connection. What they had in mind was the relativistic quantum mechanics that Dirac had just developed and the associated notion of electron spin. One of the first papers of this kind was a note by Wiener and Vallarta in *Nature* (on March 2, 1929):

> The notion of a parallelism valid for the whole of space and of Einstein's n-uples enables us to carry over the Dirac theory into general relativity almost without alteration.... All that we need to do is to interpret Dirac's P_0, P_1, P_2, P_2 [i.e., momentum operators] not as differentiation with respect to four variables x, y, z, t defined throughout space-time, but as differentiation along the lines of the quadruple (Einstein's "4-Bein")... the quantities $^{s}h_{\lambda}$ of Einstein seem to have one foot in the macro-mechanical world formally described by Einstein's gravitational potential and characterized by the index λ, and the other foot in a Minkowskian world of micro-mechanics characterized by the index s ([305], p. 317).[28]

---

[28] See also the following paper of Wiener. [306]

Tamm wrote several papers (two of them with Leontovich) on the connection between Einstein's theory and quantum theory (the first paper was dated March 14, 1929 [307]). He attempted to translate Dirac's equation to a geometry with absolute parallelism, augmenting it in the absence of an electromagnetic field with a certain term associated with the torsion tensor $\Lambda^{\sigma}_{\mu\nu}$, namely the term

$$i\mu\psi = i\frac{h}{2\pi}\sqrt{\Lambda^{\lambda}_{\mu\lambda}\Lambda^{\nu}_{\mu\nu}}.$$

He found a basis for its introduction in the following natural assumption:

If the vectors A^s and p_s [occurring in the Dirac equation

$$(A^s p_s + Bmc)\psi = 0 \qquad (55)$$

are identical, respectively, to the Dirac matrices and momentum operators] are referred to Einstein's parallel frames, then the wave equation possesses essentially the same simple form (4) [i.e., (55)] in an arbitrary field as in the case without field. ([307], p. 185)

Since, Tamm assumed, no static spherically symmetric solution of the field equations corresponding to a charged particle had been found up to then, it was possible that

the simplest solution of Einstein's field equations corresponding to a charged particle at rest is not distinguished by spherical but by axial symmetry and thus takes into account the spin of a charged particle. ([308], p. 192)

Another feature of Einstein's theory was, in Tamm's opinion, that in it one could not associate the equations of motion of particles with geodesics (because of the absence in the equations of the geodesic term corresponding to the Lorentz force). On the other hand, it seemed to him that Dirac's equation could be naturally introduced into the theory. Therefore, Tamm assumed that

in this theory the wave-mechanical principle stands above the principle of the shortest path, so that the equations of motion of a (charged) particle should be derived by means of a limiting process from a wave equation. ([308], p. 193)

In these features, he saw indications that a theory with absolute parallelism contained important quantum properties, and therefore believed that it opened up the possibility for a natural unification with Dirac's theory. In this respect,

Tamm departed from Einstein, who regarded the introduction of Planck's constant into the basic equations of unified field theory as unacceptable.[29]

A brief note by Fock and Ivanenko in *Nature* entitled "Quantum geometry" and dated March 21, 1929 laid the foundation of important investigations of Soviet scientists that led to the correct translation of Dirac's equation to Riemannian geometry [310]. They proposed the use of Einstein's tetrad structure to construct a geometry with the fundamental linear form

$$ds = \sum_k \gamma_k \, dx_k,$$

where γ_k are the Dirac matrices, making the assumption that if gravitation and the corresponding Einstein equations are described by a quadratic differential form, then Dirac's equation must be associated with this linear form, and the Dirac matrices must be transformed by means of Einstein's h_{va}. In the opinion of the authors, this geometry must be the foundation of a synthesis of gravitation, electromagnetism, and quantum theory.

Subsequently, Fock and Ivanenko took from Einstein's theory only the tetrad method, which proved to be a very valuable tool in the study of the interaction of fermions (initially, only the electron) with the gravitational field. Ultimately, this led to one of the directions of spinor analysis, and also played a decisive role in the development of the gauge field concept (see Chapter 6).

---

[29] In his recollections of Tamm, S.M. Rytov wrote:

Having liberated himself from work on the textbook [*Fundamentals of the Theory of Electricity*], he immediately turned in his lectures to the subject that occupied him. In the spring semester of 1929, he read a course "Physical foundations of the theory of relativity" devoted mainly to the general theory of relativity and necessary for the understanding of the following course, which he began immediately after this in April 1929. The course was called "The theory of gravitation and the electromagnetic field of A. Einstein." This was about a new paper of Einstein, one of his first attempts to construct a unified field theory of the gravitational and electromagnetic fields. This theory strongly attracted Tamm. He not merely acquainted us with it but also attempted to improve it, believing that the difficulties that it encountered could be overcome by invoking quantum mechanics (Dirac's equation). In 1929 there were five publications of Tamm (two of them with Leontovich) devoted to these questions. Of course, these papers were unknown to us students, and, indeed, we could hardly have mastered them. However, in the lectures he sketched the entire situation with great clarity. ([309], p. 186)

At that time, Weyl worked in a related direction, deviating significantly, however, from Einstein's theory with absolute parallelism. His first paper "Gravitation and electron," published in the *Proceedings of the National Academy of Sciences (USA)* was completed at the beginning of 1929 (the corrections made in proof were dated March 4, 1929) [311].[30] Weyl also took up Einstein's tetrad method as a tool, but, like Fock and Ivanenko, used it in the framework of general relativity, and not in a unified theory with absolute parallelism. "Despite a certain formal agreement," wrote Weyl, "my approach differs radically in that I reject absolute parallelism and adhere to Einstein's classical theory of gravitation," ([313], p. 245).

A more important stimulus for him was the difficulties of Dirac's theory. In the 1929 papers, Weyl developed a method for relativity, and also the gauge concept of the electromagnetic field, which he also related to general relativity (see Chapter 6). Recognizing and using the tetrad approach, Weyl criticized the concept of absolute parallelism:

> There are several reasons why I do not believe in the theory of infinitely distant [i.e., absolute] parallelism. First, my mathematical intuition cannot accept *such an artificial geometry*; it is difficult for me to understand what force unifies local coordinate systems at different world points, twisted relative to each other, into some unified rigid structure.... Second, the possibility of independent rotation of the coordinate systems at different points ... is equivalent to the fulfillment of the theorem of conservation of angular momentum [in a theory with absolute parallelism, this possibility is not realized]. [Ibid.]

Weyl gave a more extended criticism of the unified theory with absolute parallelism in his Cambridge lecture "Geometrie und Physik" (May, 1930) [17]. He saw the main shortcomings of this theory in the abandonment of the level of geometrical local action that had been achieved on the transition from the special to the general theory of relativity and the absence of a physical agent that ensured the common adjustment of the local orthogonal frames; there were certain difficulties with the formulation of the conservation laws for energy and momentum and, especially, angular momentum; finally,

---

[30] With certain modifications and additions, this paper was soon published in two other journals [312, 313]. The most extensive exposition is in [313], which we cite in what follows (see also Chapter 6).

there was the presence in such a geometry of two types of straightest (or geodesics) lines, and this did not have any physical basis.[31]

Weyl's critical attitude to Einstein's theory did not mean that he gave preference to his own unified field theory of 1918–1920 or any other unified geometrized field theory. After the development of quantum mechanics, Weyl abandoned the geometrized field program. In the Cambridge lecture, he said:

> In my opinion, the entire situation has been completely changed during the last four or five years by the discovery of the matter field [he is referring to wave mechanics]. All these geometrical pirouettes (*Luftsprünge*) were premature, and we return to the solid ground of physical facts. ([17], p. 56)

We return to the summer of 1929. At that time, Soviet physicists were preparing the First All-Union Conference on Theoretical Physics at Khar'kov. The problems of unified geometrized field theories, in particular Einstein's theory with absolute parallelism, were then together with quantum mechanics at the center of attention of the Soviet theoreticians. Fock, Ivanenko, Mandel', Frederiks, Gamow, and Grommer were preparing to (and did) participate in the conference. They had been actively concerned with problems between

---

31

> For more than two years, Einstein has doggedly followed a new track... besides a Riemannian metric, he takes as a fundamental structure distant parallelism of vectors. He assumes that the local frames are related to each other in such a way that they can all only simultaneously be subjected to one and the same rotation.... Einstein has broken with the infinitesimal point of view. This has had the consequence that everything good achieved by the transition from the special to the general theory of relativity has been lost. As yet, the losses do not offset the slight gains. For example, one cannot understand how the energy–momentum conservation law is obtained in this theory. From a speculative point of view, I feel that the geometry taken *a priori* as the basis of the theory is unnatural; I cannot imagine what force can freeze the local frames in their twisted position relative to each other. A further strong argument against the theory is the conservation law for angular momentum. It... is equivalent to the requirement of an invariance which presupposes that the local frames at different world points are independent of each other and freely rotatable. Further, in Einstein's geometry there are two types of straight lines, or geodesics, associated either with the infinitesimal parallel transport of Levi-Civita or with distant parallelism. In nature there are no indications of this doubling of the property of inertia. ([17], p. 56)

them and quantum theory.[32] Einstein was also invited, but could not travel because he was in "a poor state of health and very occupied" ([279], p. 70). In a letter to the chairman of the organizing committee of the conference, I.V. Obreimov, on July 11, Einstein already characterized his theory in less optimistic terms:

> I believe that the hypothesis of distant parallelism warrants serious attention, since it may make it possible to understand the physical structure of space. However, I am now seriously convinced that my previous interpretation of the electromagnetic field is not correct. The development of the theory progresses slowly, since I have not yet been able to overcome a number of mathematical difficulties . . . there are still so many necessary points missing that I have not been able to arrive at a definite conclusion with regard to either the choice of the field equations or the success altogether of this version. [Ibid.]

Commenting on this letter almost half a century after the events just described, one of the participants, D.D. Ivanenko, noted:

> Despite the failure to achieve the desired aim, however, this work of Einstein [his complete cycle of investigations of unified field theory with absolute parallelism] played an important positive role, indirectly helping to establish the theory of spinors in Riemannian geometry [especially in the cited works of Weyl, Fock and Ivanenko, and later papers of Schrödinger, Infeld, and van der Waerden, and others [314]], the discussion of the problem of frames of reference [see, for example, the monograph of Rodichev [284] and Vladimirov [315]], the development of the theory of gauge fields [see Chapter 6], and the solution of a number of concrete problems in general relativity. In fact, this work was the first time in which gravitation was described, not by means of the 10 metric coefficients $g_{\mu\nu}$ but by the "square root" of them, or the 16 tetrad coefficients h_μ^a ($g_{\mu\nu} = h_\mu^a h_\nu^a$): they relate the axes of an orthogonal frame in the flat tangent space ("fiber") ($a = 1, 2, 3, 4$) with the axes of the Gauss–Riemann curvilinear system of coordinates in space-time (in the "base") ($\mu = 1, 2, 3, 4$). Tetrads proved to very helpful in the Riemann–Einstein geometry of general relativity itself. . . . ([279], p. 71)

Moreover, as Ivanenko emphasized, the geometry of absolute parallelism was one of the first examples of the use of spaces with torsion, and even today these have not lost their importance for the analysis of possible extensions of general relativity.

---

[32] Some leading Soviet and foreign theoreticians who had actively worked in the field of quantum physics also participated: Frenkel', Landau, Bursian, Ambartsumyan, Jordan, Heitler, and others.

When mathematicians took a lively interest in the new unified theory, Einstein decided to write an exposition of it for the journal *Mathemische Annalen* (the paper was received on August 19, 1929 [291]). It was published with the historical survey of early geometrical investigations of spaces with absolute parallelism written by Cartan [290], to whom Einstein expressed his special gratitude. An even more extensive and clearer exposition of the theory, also intended in the first place for mathematicians, was given by Einstein in his lecture of November 8, 1929 at the Institute A. Poincaré in Paris and then prepared for publication in the *Annales* of that institute [316].[33]

In these papers, Einstein adhered to the previously developed point of view with regard to the method of obtaining field equations, avoiding the use of the variational principle. As a result, the question of the compatibility of the field equations became particularly important, and he devoted great attention to it. If the variational principle is not used, the process of obtaining field equations becomes more difficult and is more arbitrary. Einstein was guided by the following considerations:

> My point of departure was the identities satisfied by the quantities $\Lambda^{\alpha}_{\mu\nu}$. In the general case, the discovery of certain identical relations can be a great help in the choice of field equations, suggesting a possible nature of the required equations. An investigation of the equations must, therefore, proceed the choice of the system of equations. However, *a priori* we do not know the quantities between which relations should be sought. [316]

Einstein argued further that the following choice of field equations was suggested:

$$\Lambda^{\alpha}_{\mu\nu} = 0. \tag{56}$$

These equations were, however, as restrictive as the vanishing of the Riemann–Christoffel tensor in general relativity; they would indicate that the "twisted"

---

[33] On November 8, 1929, Einstein reported to Solovine: "Today, at 5:30, I read a lecture on my new theory at the Institute A. Poincaré" ([227], p. 35). A few months later, he recalled with pleasure the days spent in Paris, as he wrote to Solovine on March 4, 1930:

> My field theory has good successes. Cartan is working strongly in this field. I am myself working with a mathematician (W. Mayer from Vienna), a remarkable man who would already have long ago obtained a Professor's chair had he not been a Jew. I often remember the beautiful days spent in Paris.... [317]

Einstein collaborated with Mayer on several investigations, both on the theory with absolute parallelism and the five-dimensional approach (see the beginning of the following section).

space degenerated into Euclidean space. Then, on the basis of the identities for $\Lambda^{\alpha}_{\mu\nu}$,

$$G^{\mu\alpha}_{j;\alpha} - F^{\mu\alpha}_{j;\alpha} - \Lambda^{\alpha}_{\mu\nu}F_{\nu\alpha} = 0, \qquad (57)$$

where

$$G^{\mu\alpha} \equiv \Lambda^{\alpha}_{\mu\nu;\nu} - \Lambda^{\sigma}_{\mu\tau}\Lambda^{\alpha}_{\sigma\tau}, \qquad F^{\mu\nu} \equiv \Lambda^{\alpha}_{\mu\nu;\alpha}$$

Einstein successively discussed the following possible field equations:

$$\Lambda^{\alpha}_{\mu\nu;\sigma} = 0, \qquad (58)$$

then

$$\Lambda^{\alpha}_{\mu\nu;\alpha} = 0, \qquad (59)$$

or

$$\Lambda^{\alpha}_{\mu\nu;\nu} = 0, \qquad (60)$$

or the system of equations (59) and (60).

The main criterion in choosing the most suitable form was compatibility of the investigated systems of equations, namely, the condition that "the excess of the number of equations over the number of identical relations must be equal to the number of variables minus n," i.e., $n^2 - n$. Thus, from this point of view the system (59)–(60) was unsatisfactory, since the number of corresponding equations was too large: six equations (59) and 16 equations (60) ($n^2 - n = 16 - 4 = 12$, and therefore there must exist 10 further identities, which cannot be found). "This makes clear," emphasized Einstein, "how the condition of compatibility enables us to reduce effectively the arbitrariness in the choice of the field equations" [316].

The identity (57) suggested to him the choice of field equations in the form

$$G^{\mu\alpha} = 0, \qquad (61)$$
$$F^{\mu\alpha} = 0. \qquad (62)$$

This system is close to the system (59)–(60), but makes it possible, as can be shown, to find the necessary number of identical relations that ensures its compatibility.[34]

Having then obtained Poisson's equations (for gravitation) and Maxwell's equations (for the electromagnetic field) as a first approximation of Eqs. (61)–(62), Einstein himself made some critical comments about the most recent form of his theory:

---

[34] By means of the identity

$$\Lambda^{\alpha}_{\mu\nu;\alpha} - \left(\frac{\partial\varphi_{\mu}}{\partial x^{\nu}} - \frac{\partial\varphi_{\nu}}{\partial x^{\mu}}\right) \equiv 0, \qquad (\Lambda^{\alpha}_{\mu\alpha} \equiv \varphi_{\mu}),$$

In the present state of the theory, it is not possible to judge whether the interpretation of the quantities representing the field is correct. In reality, a field is determined in the first place by its ponderomotive effect on particles. However, the law of this effect is not yet known. To discover it, it would be necessary to integrate the field equations, and it has not yet been possible to do this.

This had the consequence, emphasized Einstein, that the

results hitherto obtained do not make possible experimental verification of predictions of the theory.... The first difficulty, with the overcoming of which the development of the theory must begin, is to find integrals that are free of singularities and satisfy the differential field equations and can give a correct solution to the problem of particles and their motions. Only after this will comparison with experiment become possible. [316]

At the same time, the nontrivial geometrical structure of the theory (the "equipping" of the point manifold with a set of orthogonal frames, and the presence of absolute parallelism and torsion) and the possibility of obtaining a consistent system of equations of the unified field that, in the first approximation, gave the classical equations of gravitation and electrodynamics made it possible to hope for success. Einstein was also impressed by a further feature of the theory: "An attraction for me of the theory presented here is its unity and *large (but allowed) degree of overdetermination of the field variables* [our italics] [291]. We recall that it was precisely with overdetermination of the field equations that Einstein hoped to obtain particle-like solutions possessing quantum features.

In December, Einstein used a suggestion of Cartan to simplify the proof of compatibility of the field equations (61)–(62). From January 1930, the mathematician W. Mayer began to assist Einstein, writing several papers on unified geometrized field theories with him. In February 1930 they obtained

---

the system (62) is reduced to the four equations $F_\mu = 0$; here

$$F_\mu = \varphi_\mu - \frac{\partial \ln \psi}{\partial x^\mu},$$

where ψ is a new scalar variable. As a result, we obtain the 16 equations (61) and the four equations (62), totalling 20, or 17 variables (the 17th is the scalar ψ), among which four are arbitrary. Therefore, the number of identities must be $20 - (17 - 4) = 7$. Four relations are known; they are the identities (57). It remains to find three more, which, as Einstein showed, do indeed exist. The physical, or even geometrical, meaning of these identities was not clear, and here he had hope of assistance from the mathematicians, above all Cartan, Weitzenböck, and Eisenhart [316].

as a result of very complicated calculations two rigorous static solutions of the field equations (61)–(62), corresponding to the field of an electrically charged sphere with nonzero mass and to the field of an arbitrary number of neutral material points at rest. The latter had no analog in general relativity. An Einstein noted, "nothing in nature corresponded to it" [318]. In his opinion, however, this could not serve as an argument against the theory, since, in contrast to general relativity, a law of motion of singularities did not follow from the field equations. The obtained solutions were also not an argument in support of the physical applicability of the theory; such an argument would be the finding of nonsingular static spherically symmetric solutions that could be interpreted as electrons and photons. The question of the equations of motion still remained unanswered. The unsuccessful attempts to solve field equations of the type (61) and (62) and the serious shortcomings of the theory, noted by Einstein himself, by Weyl, and by Eddington, created more and more doubts in Einstein about the correctness and promise of the concept of absolute parallelism. No later than the fall of 1930 he began to work with Mayer on the development of a new form of unified geometrized field theory associated with five dimensions, without abandoning the theory with absolute parallelism.

The first communication associated with this latest return to five dimensions was published in the journal *Science* and was dated October 30, 1930 [319]. The theory of absolute parallelism, on the creation and development of which so much effort had been expended, now qualified as an "incorrect way" [319]. It is true that the investigation of the mathematical aspects of the theory with absolute parallelism continued in the second half of 1930 and even in 1931 [320]. The final paper devoted to the concept of absolute parallelism was apparently an academic report written together with Mayer and presented at a meeting of the physics and mathematics section of the Prussian Academy of Sciences on April 23, 1931 [321]. The authors motivated the system of field equations by requiring them to be consistent with a certain differential identity of the type of relation (57), but much more complicated. This vector identity and the system of equations contained terms that depended on $\Lambda^{\alpha}_{\mu\nu}$, $\varphi_{\alpha;\mu}$ and their derivatives up to the second order with arbitrary coefficients that were determined from the conditions of consistency. As a result, the authors obtained four types of systems of equations, which coincided in part with the systems investigated in the earlier studies.

Thus, the unification of gravitation and electromagnetism on the basis of a geometry with absolute parallelism encountered a number of serious difficulties: the absence of a physical justification for absolute parallelism; the nonuniqueness of the system of field equations; the lack of clarity of the equations of motion; the absence of exact solutions, in particular particle-like ones; the problems of constructing a sensible cosmological scheme, etc. The ever more refined investigations of the problem of compatibility of the system of field equations made by Einstein and Mayer also failed to give any significant results, and from the end of 1930 Einstein gradually lost interest in the concept of absolute parallelism.

Nevertheless, as we have seen, the development of this direction was not without fruit. This was one of the first unified field theories based on a geometry with nonvanishing torsion. Einstein's theory was based on the tetrad formalism, which had not previously been used in general relativity and unified field theories. This formalism was then used by Weyl, and also by Fock and Ivanenko to describe the interaction of fermions with the gravitational field (extension of the theory of spinors to Riemannian geometry). Subsequently, the tetrad formalism was also used as a tool in the solution of classical problems of general relativity (for example, the energy problem). In the following chapter, we shall see how the papers of Einstein, Weyl, Fock and Ivanenko, and others were among the important stages in the formation of the gauge field concept.

An echo of the unified field theory with absolute parallelism can be found in the papers of Einstein and Mayer on the theory of spinors written during 1932–1942 [322–325]. Einstein's unified theory had a strong influence on the theory of spinors in Riemannian space (Weyl, Fock and Ivanenko, Schrödinger, van der Waerden and Infeld, Bargmann, and others) [314] and thus on the solution of the problem of introducing Dirac's equations into general relativity, and these papers, in their turn, stimulated Einstein's interest in spinors, Dirac's equation, and thus in relativistic quantum mechanics. This time, and apparently for the first time, Einstein, while maintaining the position of the unification program, became interested in extending it to quantum theory. Beginning with Weyl's theory (above all the attempts to relate it to quantum mechanics made by Schrödinger, London, and others) almost every unified geometrized field theory was regarded as a way to overcome the gulf between general relativity and quantum mechanics (and accordingly between the quantum theoretical and field geometrical program). In fact, the development of unified field theories often did have a stimulating, heuristic effect on the development of quantum theory, although sometimes not at once and not directly.

Weyl's theory played a certain part in the genesis of wave mechanics and the formation of the gauge field concept. It was in the development of affine field theories that the idea of charge symmetry, which is so important in quantum field theory, was first formulated. The five-dimensional approach led to the establishment of a relativistic wave equation for spinless particles and was an important stage in the development of the concept of gauge symmetry. The theory with absolute parallelism directly influenced both the development of the gauge field concept and the solution to the problem of introducing Dirac's equation into general relativity.

As a rule, Einstein underestimated these naturally arising bridges between the competing programs, although he never ceased to think of a radical reduction of quantum mechanics to the principles of the geometrized field program, striving to obtain the quantum features in the behavior of particles as corresponding properties of nonsingular solutions of the equations of a unified field. In the case of the theory with distant parallelism, Einstein himself became seriously interested in quantum extensions of it. The result of this interest was four papers on spinors and semivectors published in 1932–1934 with Mayer. They were largely mathematical in nature.

In the second paper of this series the authors sketched a theory whose Lagrangian contained not only the scalar curvature and a Maxwellian part but also a Dirac part. Of course, this was not a unification in the spirit of unified geometrized field theories, although it did indicate the possibility of a generally relativistic treatment of particles described by the Dirac equation. The authors emphasized that although they used the Dirac equation, the unification had a classical, nonquantum nature:

> To such a field theory, one cannot apply Born's probability interpretation of the ψ field. Therefore, the question of whether such a theory admits a consistent interpretation of the atomistic structure of matter as yet remains open. [323]

In the final paper of the spinor cycle published in 1934, after his transfer to Princeton, a geometry with absolute parallelism again appeared [325]. Einstein and Mayer abandoned the frame formalism for the description of semivectors (and spinors) and showed that if only Gaussian coordinate systems are used, semivectors behave as vectors for which covariant differentiation is defined not only by the Riemannian metric, but also by quantities that are characteristic for a geometry with absolute parallelism. As we have seen, Einstein's theory of 1928–1930 took as its basis, not curvature, but torsion, which is characterized by a third-rank tensor. The vanishing of the curvature (including the segmental, or homothetic, curvature, which is absent in Riemannian geometry but is nonvanishing in Weyl's geometry), which makes it possible to introduce absolute parallelism, is an important restriction in this theory, making it easier to find field equations. From the beginning of the 1920s, affine geometries with nonvanishing torsion had been considered by Cartan, Schouten, and Weitzenböck. Not long before Einstein's theory, spaces with torsion had been used to construct unified field theories by the French mathematician H. Eyraud (1926) [326] and Einstein's future coauthor L. Infeld (1928) [327–329]. These theories were not sufficiently constructive, however, since they did not contain restrictive conditions of the type of absolute parallelism, and as a result the arbitrariness in the choice of the field equations was too large.

Eyraud used a geometry with nonvanishing curvatures and torsion. In the Cartan classification, this corresponded to the most general case: $\Omega_\mu^\nu \neq 0$, $\Omega \neq 0$, $\Omega^\mu \neq 0$. Infeld assumed that in the presence of Riemannian curvature and torsion the length of transported vectors did not change ($\Omega = 0$), but this condition too did not lead to a theory more realistic than the previous unified theories [22]. Nevertheless, it should not be forgotten that the first unified field theories based on an affine geometry with nonvanishing torsion were developed before Einstein's theory with absolute parallelism, although his direction was not much refined (see, however, the series of papers of the Italian theoretician P. Straneo [22]).

A NEW FIVE-DIMENSIONAL APPROACH (EINSTEIN AND MAYER, 1931–1932)

In conclusion, we consider the five-dimensional theory of Einstein and Mayer, the last unified theory of the Berlin period, for which Einstein did not in fact have great hopes. In the lecture "The present status of the theory of relativity," read at the Physics Institute of the University of Vienna on October 14, 1931, Einstein gave a very pessimistic estimation of the many years of attempts to create a unified field theory:

> Attempts to find unified laws of matter and to generate field theory and quantum theory have not ceased. The task has been to find a structure of space that satisfies the conditions advanced by both theories. The result has been a graveyard of buried hopes. [330]

Having mentioned the failure with the concept of absolute parallelism, Einstein gave a brief sketch of the next modification of the five-dimensional approach,

> in which electromagnetic phenomena find their place, and the architectonic unity is not lost.... I and Mayer assume that the fifth dimension should not be manifested. It is used only mathematically to construct components whose use gives equations for electromagnetic phenomena.... [330]

This theory too did not appear to lead to success:

However, the hope has not been realized. I assume that if this law [i.e., the equations of the unified field] could be found, then a theory that would apply to quanta and matter would be obtained. But this is not so. The constructed theory appears to founder on the problem of matter and quanta. Between the two ideas there is still a chasm. [330]

At the beginning of the 1930s, before the Einstein–Mayer theory, there arose a new wave of investigations into five-dimensional theories associated with an interpretation of space-time in terms of projective geometry. Between 1930 and 1933 more than ten papers were published on the "projective theory of relativity," in other words, on a projective treatment of the five-dimensional Kaluza–Klein theory. The first paper of the projective direction appears to have been a paper by Princeton mathematicians Veblen and Hoffmann, which was received by the editors of *The Physical Review* in June 1930 and was published in the September issue of the journal [332]. There followed other papers by these authors, as well as papers by Schouten and van Dantzig, Pauli, and others. The formalism of homogeneous five-dimensional coordinates made it possible to formulate the Kaluza–Klein scheme in a simple and mathematically elegant manner. Particularly transparent in this respect was the paper of Pauli [333], which was subsequently reproduced almost in its entirely in the first edition of *The Classical Theory of Fields* of Landau and Lifshitz (§100).[35]

At first, it appeared that the projective scheme was more general than the Kaluza–Klein scheme and that systematic development of this direction could lead to solution of the problem of a unified field theory. Then, at the beginning of the 1940s, Bergmann proved the equivalence of the two formulations [12].

Einstein and Mayer presented their five-dimensional theory in two papers presented at meetings of the Prussian Academy of Science on October 22, 1931 [335] and April 14, 1932 [336]. They assumed that real space-time has four dimensions, but that physical quantities are described by vectors and tensors whose indices take values from 1 to 5, although the components are functions of only four coordinates. In effect, they used the frame formalism developed by Einstein in the theory with absolute parallelism. The only difference was that at each point of the four-dimensional Riemannian space they considered not a tetrad but a five-dimensional linear vector space ("pentad").

---

[35] Landau and Lifshitz did not regard the projective five-dimensional method of unified formulation of the equations of gravitation and electrodynamics as a nontrivial unified physical field theory. They emphasized that "this ... is in no way a reduction of the electromagnetic field to the metric properties of space-time as occurs for the gravitational field" and the "the metric of real space and time ... remains as before ([334], p. 266).

Thus, a five-dimensional tensor analysis was used without the introduction of a five-dimensional space. As a result, all the results of the Kaluza–Klein theory were reproduced, and it appeared to be possible to avoid the difficulties associated with the unobservability of the five-dimensional continuum, the lack of justification for the cylinder condition, and the absence of a physical interpretation of the g_{55} component of the metric tensor. In a letter to Besso on October 30, 1931, Einstein characterized this approach as follows:

> All the subtlety is introducing 5-vectors a^σ in a four-dimensional space, the vectors being associated with the space by some linear mechanism. Let a^s be the 4-vector corresponding to a^σ; then there exists the following dependence:
>
> $$a^s = \gamma_\sigma^s a^\sigma.$$
>
> Therefore, in our theory the only meaningful equations are those that are satisfied independently of the relations introduced by means of the γ_σ. Further, we define an infinitesimal displacement of the 5-vectors a^σ in the four-dimensional space, and also a corresponding 5-curvature, from which we obtain the field equation. ([227], pp. 5–6)

Einstein did not have very great hopes for the new form of the five-dimensional approach. He spoke of this in the Vienna lecture cited above, a week before the meeting of the Prussian Academy of Sciences at which the first part of the joint paper with Mayer was presented. In Einstein's opinion, the theory lacked promise because it "founders on the problem of matter and quanta." In the same letter to Besso, Einstein again emphasized the inability of the new theory to solve the maximum problem of the unification program:

> All that has succeeded in our investigation is the unification of gravitation and electricity, the equations of the latter being identical to those of Maxwell for empty space ... no physical progress is thereby achieved, except that it becomes clear Maxwell's equations are not only a first approximation but are rationally justified just as well as the equations of gravitation for empty space. Neither charge density nor mass density exists, the great enterprise collapses, and we already have the quantum problem, which no one as yet succeeded in solving from the standpoint of field theory (just as no one has yet succeeded in constructing relativity theory on the basis of quantum mechanics). ([227], p. 5)[36]

---

[36] The paper of Einstein and Mayer (part 1) ended just as pessimistically:

> The theory presented here leads in a unified manner to the equations of the gravitational and electromagnetic fields. However, it gives nothing for the understanding of the nature of corpuscles or for the understanding of the results established in quantum mechanics. [335]

Six months later, in the middle of April 1932, Einstein and Mayer presented the second part of the study, in which they discussed "whether the considered... spatial structure admits a generalization leading to equations of the electromagnetic field with nonzero density of electric charge" [336]. Using some results from the first part, they showed that "there is an entirely natural generalization of such kind permitting one to obtain a combined system of field equations." The paper did not go beyond a discussion of purely mathematical aspects associated with obtaining generalizations of the system of field equations and proof of their compatibility. "The question of the applicability of this system of equations to reality is not considered here," emphasized Einstein and Mayer at the end of the introduction to the paper [ibid.].

AN OUTLINE OF EINSTEIN'S FURTHER ATTEMPTS (PRINCETON, 1933–1955)

In the second half of 1932, called a "fantastic year in physics" by Weisskopf because of the fireworks of outstanding discoveries made in this year in the physics of nuclei and elementary particles ([337], p. 7), Einstein developed the theory of spinors with Mayer. In a letter of October 21, 1932, Einstein reported to Besso:

> Together with my Dr. Mayer, I have been working on the theory of spinors. We are already able to explain the mathematical dependences. To an understanding of the physical side of the matter there is still a long way to go, much further than is now believed. As before, I am particularly convinced that the attempt to create an essentially statistical theory is doomed to failure. ([227], p. 7)

Einstein and Mayer presented their paper "Semivectors and spinors" to the Prussian Academy of Sciences on November 10, 1932 [322]. This was the last paper that Einstein presented to a meeting of the Prussian Academy of Sciences.

After Dirac had created the relativistic quantum theory of the electron (1928), the question of the mathematical nature and physical significance of spinors became very topical.[37] During 1932–1933, very important investigations on the theory of spinors, made by Schrödinger (February 1932),

---

[37] According to the evidence of van der Waerden, the author of several of the first investigations into spinors, the name itself was introduced by Ehrenfest, who was also one of the main initiators of the systematic development of spinor analysis ([314], p. 273 of the Russian translation).

Bargmann (July 1932) and Infeld and van der Waerden (January, 1933) were published in the *Sitzungsberichte* of the Prussian Academy of Sciences [338–340]. This spinor study of Einstein and Mayer did not appear to have a direct bearing on the program of unified geometrized field theories, although as we have seen, the problems of the theory of Dirac and of spinors were intimately related to Einstein's unified theory with absolute parallelism.

In the following paper on the theory of spinors (presented to the meeting of the Amsterdam Academy of Sciences on May 27, 1933),[38] Einstein and Mayer touched on the problem of deriving the basic field equations from a unified variational principle. Besides the gravitational and electromagnetic parts of the Lagrangian, they also considered a "semivector" (or "spinor") component, which should lead to the Dirac equation. It appeared to the authors that in their approach "an explanation is given for the first time why there exist two electrically charged elementary particles with different masses and electric charges of opposite signs" [323].

From the beginning of October 1933 Einstein was established at Princeton. During 1933–1934, he continued to work with Mayer on the development of the spinor subject [325]. He then returned to the problem of particles and equations of motion in general relativity—the interpretation of particles as "bridges" connecting two different "sheets" of space (1935–1936, in collaboration with Rosen) [341], the classical derivation of equations of motion from field equations (1938–1940, in collaboration with Infeld and Hoffmann) [231], etc. It was only in 1938 that he again turned his attention to unified geometrized field theories, and again to the five-dimensional form of them, which together with Bergmann, and then also with Bargmann, he continued to study until 1943. The proof of the nonexistence of regular stationary solutions of the five-dimensional field theory by Bargmann, and also by Einstein and Pauli [343], brought this period of Einstein's investigations in the unification program to a close and was the main motive for his abandoning further work in this direction.

Disappointed with the five-dimensional approach, Einstein devoted the last decade of his life to studying different possibilities of four-dimensional

---

[38] At the beginning of 1933, Einstein returned from the United States to Europe. After having given lectures at Oxford and Cambridge, he visited the Belgian spa Le Coq not far from Ostend. Earlier, in the United Stated, Einstein had issued a statement, published in the *New York World Telegram* on March 11, 1933, in which he spoke of his wish not to return to Nazi Germany. His letters to the Prussian and Bavarian Academy of Sciences (April 5, 12, 21), in which he announced his resignation from those organizations, followed.

generalization of the metric tensor g_{ik}: replacement of a Riemannian g_{ik} by a bivector field $g_{\alpha\beta}^{ik}$ that depended on the coordinates of two points (1944, in collaboration with Bargmann) [344]; transition to a complex Hermitian-symmetric generalization of g_{ik} ($g_{ik} = s_{ik} + ia_{ik}$) (1945–1948, in part in collaboration with E. Strauss); the use of an asymmetric g_{ik} (1950–1955, in part in collaboration with Kaufmann). The first two ideas were quite quickly rejected, but Einstein worked on the last idea, the "relativistic theory of an asymmetric field," until his final days.

Attempts to reconcile this theory with the principle of irreducibility, by the introduction of the requirements of "λ invariance" and transpositional symmetry, did not lead to any physical progress.

The problems of singularities and boundary conditions remained obscure. Einstein was inclined toward the thought that it was necessary to eliminate singularities and postulate boundary conditions. In this way, he still hoped to relate the theory more intimately to quantum structure and cosmology. He became more and more doubtful, however, about the possibility of extending the program of unified geometrized field theories to quantum phenomena. In Appendix II to the fifth edition of *The Meaning of Relativity*, his last published paper,[39] Einstein wrote:

> One can give good reasons why reality cannot at all be represented by a continuous field. From the quantum phenomena it appears to follow with certainty that a finite system of finite energy can be completely described by a finite set of numbers (quantum numbers). This does not seem to be in accordance with a continuum theory, and must lead to an attempt to find a purely algebraic theory for the description of reality. But nobody knows how to obtain the basis of such a theory. ([345], pp. 157–158)

It was just at this time, however, that Yang and Mills (whose paper, which was to become a classic, appeared in 1954 [346]) discovered a new possibility to develop the field program, which was realized during the following 20 to 25 years through efforts of many tens of investigators. It was in this way that the modern gauge treatment of the interactions of elementary particles arose, including the theory of electroweak interactions, quantum chromodynamics, and the theory of supergravity. We note in this connection that gauge theories admit an elegant geometrical formulation in the language of the theory of fiber bundles.

---

[39] This book appeared in 1955, the year of Einstein's death.

SUMMARY

Einstein's efforts over decades to achieve a geometrical field synthesis of physics, like his subsequent investigations in this direction (from the time of his transfer to Princeton until his death) did not lead to decisive success.[40] In the 1920s, especially before the creation of quantum mechanics, the majority of physicists followed the work on unified geometrized field theories with hope and interest. Quantum mechanics opened up rich new possibilities for theoretical physics. On the background of the rapidly progressing quantum theoretical program, the limited mathematical successes of the unification program, which in addition failed to reach an experimental–empirical level, led to a steady loss of interest in the approach. The beginning of the 1930s (the breakthrough year of 1932 and the following years) marked a brilliant turning point in the field of fundamental physics, associated on the one hand with the discoveries in nuclear physics and the finding of new elementary particles (neutron, positron, and more), and on the other with the development of a quantum field theory of the fundamental physical interactions (quantum electrodynamics and the first versions of the theories of strong and weak interactions). The investigation of unified geometrized field theories at the beginning of the 1930s mainly attracted the attention of specialists in differential geometry. Even Einstein interrupted his work on unified field theories before the end of the 1930s. It was only in 1938 that he revived his work on the unification program. The leading theoreticians were agitated by the problems of quantum field theory and elementary particles, and in the 1940s and 1950s the unification program appeared manifestly anachronistic in the eyes of the majority.

Returning to the main decade we have considered, we list the stages in the development of Einstein's investigations of unified field theories: 1923–1925, affine field theories and the concept of overdetermination of the field equations (advanced with the aim of solving the maximum problem of the unification program); 1925–1926, affine-metric theories and the subsequent crisis of the field geometrical program compared to the triumph of the quantum theoretical program; 1927, the return to a five-dimensional theory in the spirit of Kaluza's theory and development of the possibility of deriving the equations of motion from the field equations; 1928–1931, theories with teleparallelism; 1931–1932, a new form of five-dimensional theory (Einstein–Mayer theory). These

---

[40] In the book A. Pais, *Subtle is the Lord: The Science and the Life of Albert Einstein*, Oxford University Press, 1982 (Ch. 17), it is possible to find some useful supplementary details of Einstein's efforts and of subsequent work on unified field theories.

numerous refined differential-geometrical schemes proved to be physically ineffective. "A graveyard of buried hopes"—such was the summary of the decade of development of unified geometrized field theories and especially the efforts of Einstein himself, in his own estimation.

Of course, from 1923 Einstein was recognized as the leader of the unification program (he took over the baton from Weyl), and the above sequence of unified theories to a large degree reflects the main direction in the search for a solution to the problem of a field geometrical synthesis of physics. Since there was no real physical progress, however, the number of mathematical schemes became even greater, and moreover, because of the weak connection between the theoretical constructions and experimentation, the majority of these schemes could not be definitively refuted, and many forms of unified geometrized field theories were developed simultaneously. Unified theories rejected by Einstein continued to be developed by others. At the beginning of the 1930s the theories at the center of attention were those with teleparallelism and five dimensions, in particular projective schemes, but affine theories and different forms of theories with torsion (the theories of Straneo, Infeld, and others) continued to be developed.

The arsenal of differential-geometrical constructions discovered by the studies of Cartan, Schouten, Eisenhart, and others was not assimilated by symmetric testing of the different possibilities but, as a rule, on the basis of intuitive considerations of a physical and geometrical nature. In addition, the choice of a particular scheme was sometimes related to investigations in the framework of the quantum theoretical program. Thus, the burst of interest in five-dimensional theories in 1926–1927 was due to the discovery of quantum mechanics and, in particular, the interpretation of the quantum relativistic scalar wave equation as a d'Alembert equation in a five-dimensional Riemannian space.

"Mathematics is good and beautiful," Einstein wrote in a letter to Weyl, having in mind the geometrical structures of the unified theories, but the abundance of theoretical constructions and the difficulties of adapting them to reality made the problem of choosing a suitable scheme extremely complicated. For this the program criteria were insufficient. Besides the very vague and subjective criteria of simplicity and aesthetic appeal, Einstein used, as one of the main criteria of the viability of a unified geometrized field theory, the existence of nonsingular centrally symmetric static solutions of the field equations. After the proof of the existence of such solutions, one would be able, according to Einstein, to obtain a complete solution to the maximum problem of the unification program by obtaining in one way or another (for example, by overdetermination of the system of field equations) a description of the quantum properties of particles and the description of the quantum properties of particles and the electomagnetic field. In almost all the theoretical schemes used by Einstein, however, nonsingular particle-like solutions did not exist, and this was one of the main reasons for abandoning these schemes.

We mention some further aspects of Einstein's investigations in the unification problem. Whereas at the beginning of the program, and the beginning of the 1920s, Einstein devoted great attention to the truly physical aspects of a theory and the search for new experimental applications (recall his criticism of Weyl's theory and his persistent searches for a key experimental fact in the development of affine theories), the experimental–empirical aspects subsequently played an even smaller role in his developments. Another characteristic feature was the clear opposition between the geometrical field program and the quantum theoretical program. Einstein did not believe it was possible to introduce quantum representations and concepts into unified field theories in any form. Everything quantum must, according to Einstein, be obtained automatically, as a prize for the successful choice of the geometrical construction. Therefore, he did not notice, or clearly underestimated, some very interesting studies that brought to light various nontrivial connections between unified theories and the concepts and formalism of quantum theory. For example, he attributed no significance to the various quantum-theoretical extensions of five-dimensional schemes and theories with absolute parallelism. One further feature of Einstein's studies of unified field theories was the almost periodic return to certain previously rejected versions (admittedly, as a rule, with certain modifications). For example, the five-dimensional approach attracted his attention in 1921–1922, 1927, 1931–1932, and, after the considered decade, in 1938–1943.

In the example of Einstein's studies in the unification program, we find a significant influence of the geometrical field program on investigations in both the theory of gravitation and quantum theory, although as we have noted, Einstein often did not pay particular attention to this side of the problem. The rigorous proof of charge symmetry of any generally covariant field theory (in the framework of studies into affine and affine-metric theories); the development of the variational method with independent variation of the metric and the affine connection (the Palatini method—in affine-metric theories); the development of the idea of overdetermination of the field equations and the possibility of deducing equations of motion of matter (particles) from field equations (in connection with the discussion of the maximum problem of the program of unified geometrized field theories); the development of the tetrad and spinor formalism in general relativity and the use of spaces with torsion (in connection with the theory with absolute parallelism)—this is a far from complete list of the results that were obtained either by Einstein himself in his unification studies or, in part, by other investigators on the basis of these studies and that played a significant part in the development of both the relativistic theory of gravitation and quantum theory. In the following chapter, we shall consider in more detail the heuristic influences of the unification program on the emergence and development of quantum theory (both quantum mechanics and quantum field theory).

CHAPTER 6

The Role of Unified Geometrized Field Theories in the Genesis and Development of Quantum Theory

In Chapter 4 we emphasized that at the beginning of the 1920s, before the creation of quantum mechanics, quantum theory had progressed to the level of a global research program, or at least to the prototype of such a program. Despite the existence of two directions in the development of quantum theory (the theory of radiation and the theory of the atom) and a certain opposition between them, by the middle of the 1920s the characteristic features of the quantum theoretical program, which contrasted sharply with those of the program of unified geometrized field theories, became steadily clearer: discreteness, the probabilistic nature of causality, the "material" (nongeometrical!) nature of the fundamental entities, etc. (see Chapter 4).[1] As is well known, quantum mechanics arose precisely through the development of the quantum theoretical program, and both directions were important in its genesis. In Chapter 5, we saw how the unification program gradually receded into the background after the emergence of quantum mechanics. The creation of the foundation of relativistic quantum mechanics and quantum electrodynamics and the discovery of a number of new elementary particles and new types of interaction already

---

[1] It should be said that many adherents to the first direction of the quantum theoretical program did not fully share these aims. They regarded as inadmissible the abandonment of the principles of a continuum space-time description, causality, and conservation laws; they insisted on the use of differential equations as the main tool for describing physical phenomena and were convinced of the reality of light quanta. Developing the quantum theory of radiation, some of these theoreticians remained at a deeper level in the framework of the field program.

at the beginning of the 1930s revealed even more clearly the difficulties and lingering fruitlessness of the geometrical field program.

Nevertheless, investigations of unified geometrized field theories continued. The most persistent adherent to and leader of the program continued to be Einstein, although from time to time other major theoreticians returned to the geometrized field concepts (Schrödinger, Pauli, Jordan, Bergmann, Klein, and others). Of course, compared with the many outstanding achievements of the quantum field program, the direction of unified geometrized field theories appeared to be cut off from the living tree of physics and almost fruitless. At the beginning of the 1920s, however, especially after the recent triumph of general relativity, this direction appeared theoretically very deep and promising. In addition, the scientific authority of Einstein, Hilbert, Weyl, and Eddington was not of little significance. At that time, many physicists regarded the geometrization program highly and considered general relativity or Weyl's theory as the ideal of fundamental physical theory. Therefore, although unified theories by themselves did not lead to important physical progress, in the 1920s they could influence the genesis and development of quantum theory.

In this chapter we shall show that there was indeed such an influence. We shall see that the unified geometrized field theories played a significant part in the genesis of quantum mechanics and the creation of some of the most important concepts of quantum theory. In fact, in the earlier chapters we already touched on this. Thus, we may recall that the discovery of the relativistic scalar wave equation for spinless particles (Klein–Gordon–Fock equation) was associated with the five-dimensional Kaluza–Klein theory (see Chapter 5). Another important example was Einstein's discovery of antimatter ([347], p. 9), or rather charge symmetry (see Chapter 5). We shall consider the importance of Weyl's theory in the genesis of Schrödinger's wave mechanics and the history of the development of the concept of gauge symmetry and the gauge conception of the electromagnetic field, which played a key role in the development of modern quantum field theory and also in the present-day revival of the concept of a unified field theory. Thus, the opinion widely held from the 1940s to the 1970s, that the investigations of unified geometrized field theories had been completely fruitless and had no prospects, all things considered, is incorrect. The efforts of Einstein and the other supporters of the geometrical field concept were not entirely in vain. Moreover, there was more to it than simply that the unification program stimulated various branches of modern differential geometry or had a significance like that played in science by negative results.

The fact is that although the unification program was in opposition to the quantum theoretical program, it did have a heuristic influence on the latter.

Moreover, when the quantum theoretical program was extended to field processes (above all to electrodynamics), the fundamental role of some of the concepts that had arisen and been developed in the framework of the unification program was clarified. Some of these concepts became important tools in the field synthesis of physics in the present stage, especially the gauge field concept.

WEYL'S THEORY AND WAVE MECHANICS

In 1921 Schrödinger became a Professor at the University of Zurich, where Weyl had worked since 1913 [348]. They became great friends. At Zurich, wrote Schrödinger, "I enjoyed the society, friendship, and assistance of Hermann Weyl, Peter Debye, and others..." [349]. Shortly before coming to Zurich, Schrödinger had made some important investigations in the general theory of relativity, one of which played an important part in the development of the energy problem in the theory (1918) [350]. In the same year he wrote two manuscripts: *Tensor Analytical Mechanics* (118 pages) and *Hertz's Mechanics and Einstein's Theory of Gravitation* (31 pages) [351]. Thus, general relativity, its problems, and methods were close to and understood by Schrödinger. From 1921, he also began to work actively on problems of quantum theory, doing so simultaneously in the two directions mentioned above (quantum statistical and atomic spectroscopic). With respect to the two programs, during this period (up to the end of 1924) in a certain sense he occupied an intermediate position and did not associate himself with any of the main schools or programs. To this period one could indeed apply the remark that he himself made about the style of his scientific activity: "In my scientific studies, as in life generally, I have never belonged to any general line or followed a guiding program designed for long periods" [349]. It was precisely these circumstances that led to Schrödinger's investigations in which he attempted to reformulate Bohr's well-known quantum conditions in the language of Weyl's unified field theory (1922) [352].

Relating the Weyl vector φ_i to the electromagnetic potential by means of the equations (see Chapter 3)

$$\varphi_0 = \gamma^{-1} eV, \quad \varphi_1 = -\gamma^{-1}\frac{e}{c}A_x, \quad \ldots, \quad \varphi_3 = -\gamma^{-1}\frac{e}{c}A_z, \qquad (1)$$

where V is the scalar potential, (A_x, A_y, A_z) is the vector potential, and γ is a certain constant coefficient of proportionality having the dimensions of action, Schrödinger represented Weyl's exponential factor that occurs in Eq. (12) of

Chapter 3 as follows:

$$\exp\left\{-\frac{e}{\gamma}\int\left(V\,dt - \frac{1}{c}A_x\,dx - \frac{1}{c}A_y\,dy - \frac{1}{c}A_z\,dz\right)\right\}. \tag{2}$$

Schrödinger considered the argument of this exponential for the simplest problems of quantum theory (unperturbed Kepler motion, Zeeman and Stark effects, etc.) and arrived at the following reformulation of the quantum conditions (for spatially closed periodic systems):

> The ... remarkable, as it seems to me, property of the quantum trajectories is that the "true" quantum conditions, i.e., the conditions that lead to determination of the energy and, thus, the spectrum, consist of the requirement that the argument of the scale factor (5) [i.e., (2) in our numbering] must be an integral multiple of $\gamma^{-1}h$ ... for all approximate periods of the system. ([352], p. 14)

Bohr's quantum condition for the unperturbed Kepler motion was

$$2\tau\overline{T} = nh, \tag{3}$$

where τ is the period of the orbital motion, and $\overline{T}$ is the time-averaged kinetic energy. In accordance with the virial theorem,

$$\overline{T} = \tfrac{1}{2}e\overline{V},$$

where V is the electric potential of the nucleus. Therefore, we have

$$e\overline{V}\tau = e\int_0^\tau V\,dt = nh.$$

Since the field of the nucleus in the given case is purely electrostatic, $A_x = A_y = A_z = 0$, and the argument of the exponential factor is

$$-\frac{e}{\gamma}\int_0^\tau V\,dt,$$

we obtain the relation

$$-\frac{e}{\gamma}\int_0^\tau V\,dt = n\gamma^{-1}h, \tag{4}$$

which is identical to Schrödinger's quantum condition.

Similar results are obtained for the other problems of atomic spectroscopy (Zeeman and Stark effects, relativistic precession, etc.). Discussing the possible physical meaning of the obtained result, Schrödinger noted that if the electron were to carry with it in its orbit some "interval" that is transported unchanged in the motion, then, calculated from some arbitrary point of the orbit, the measure (*Masszahl*) of this interval would always arrive multiplied by an almost exactly integral power of exp $\frac{h}{\gamma}$ whenever the electron returned with good accuracy to the original position and, simultaneously, to the initial state of motion.... It is hard to believe that this result is merely a fortuitous mathematical consequence of the quantum conditions and does not have a deep physical significance. ([352], p. 22)

Schrödinger also thought it possible that the "remarkable property" of the orbital electron that he had established indicated an "adjustment" of it in the electromagnetic field (as a magnetic needle is "aligned" in a magnetic field). In conclusion, he discussed the possible physical significance of the constant γ, which had the dimensions of action. One of the possibilities was to set $\gamma \simeq h$. This led to the thought

of whether γ might not have the purely imaginary value $\gamma = \frac{h}{2\pi\sqrt{-1}}$, for then the universal factor (22) [i.e., $\exp(h\gamma^{-1})$] would be equal to 1, and the measure of the transported interval would be reproduced after every quasiperiod. ([352], p. 23)

The expression for the "measure of the transported interval" would then take the form

$$l = l_0 \exp\left\{-\frac{2\pi i}{h} \int \left(V\,dt - \frac{1}{c}A_x\,dx - \frac{1}{c}A_y\,dy - \frac{1}{c}A_z\,dz\right)\right\}, \quad (5)$$

which strongly resembles the expression for the wave function introduced by Schrödinger more than three years later.[2]

In the autobiographical sketch written by Schrödinger after he had received the Nobel Prize, listing his most important papers, he mentioned this paper of 1922 and emphasized that

the establishment that ... the scale factor of Weyl for closed quantum orbits is an integral power of some universal constant, which, possibly, is equal to unity, appears important as an anticipation, unconscious of course ... of L. de Broglie's theory.... [349]

---

[2] A similar result (in particular, the introduction of imaginary values of the potentials for the unperturbed Kepler motion) was also obtained two years before this by Fokker, who did not publish his result, merely communicating it to Weyl. After being acquainted with Schrödinger's paper, Weyl wrote to him in Arosa about this, where, as it happens, Fokker too obtained his result ([352], p. 162 of the Russian text).

Indeed, the reduction of the quantum conditions to a certain resonance property of the Weyl measure, which is related in its structure to de Broglie's phase wave, can be regarded as an undoubted anticipation of de Broglie's resonance interpretation of the Bohr–Sommerfeld quantum conditions. It is true that in the series of famous papers by Schrödinger published in the *Annalen der Physik* during 1926 (the four communications "Quantization as an eigenvalue problem") [353] we do not find any mention that the 1922 paper had any significant influence on the genesis of Schrödinger's ideas, and there is not even a reference to it. There are also no indications that it influenced de Broglie, on whose work Schrödinger based his own.

The first to draw attention to Schrödinger's paper "On a remarkable property . . . ," and who regarded it as a prototype and analog of the wave conception of microscopic particles, was London, who subsequently became known through his fundamental studies in low-temperature physics. On December 10, 1926, having become acquainted with this paper, he wrote an excited letter to Schrödinger, brief extracts from which we now give:

> . . . you showed that on the discrete actual orbits the gauge unit (*Eicheinheit*) (with $\gamma = 2\pi i / h$) reproduces itself in the case of a spatially closed path; moreover, you noted at the same time that on the nth orbit the unit of length swells and contracts n times exactly as in the case of the standing wave that describes the position of the charge. . . . You even held in your hands the resonance nature of the quantum postulate long before de Broglie. ([354], p. 304)

A week later, at a meeting of the Württemberg branch of the German Physical Society at Stuttgart, London gave a paper in which he attempted to identify the Weyl measure with the wave function and on this basis remove Einstein's well-known objections to Weyl's theory. The paper was then published in the *Naturwissenschaften* (brief communication) [275] and the *Zeitschrift für Physik* (in February 1927) [355]. As in the quoted letter, London held the Schrödinger paper of 1922 in high esteem, repeating and developing the main thoughts of his letter:

> Quantum theory permits only a discrete series of states of motion. One may suppose that these distinguished motions make possible only such transport of the scale that as a result of its return to the original point the phase [wave function] passes through an integer number of cycles. . . . In essence, this recalls the resonance property of de Broglie waves, namely, the property by means of which de Broglie so fruitfully reformulated for the first time the old quantum conditions of Sommerfeld and Einstein. . . . Thus, if we add [to the system of postulates of Weyl's theory] the requirement of uniqueness of length, as a widely recognized experimental fact, then this necessarily leads

to a system of discrete states of motion of a "classical" quantum theory and their de Broglie waves. I would not like to miss the welcome opportunity of pointing out that this resonance property of the Weyl measure, which we encounter here as a characteristic theorem of wave mechanics, was advanced by Schrödinger already in 1922 as a "remarkable property of quantum trajectories" and demonstrated in a number of examples, although the physical significance of this was at that time not yet fully recognized .... Thus, Schrödinger already had in his hands the characteristic wave-mechanical periodicities which he later again encountered on an entirely different basis. ([355], p. 380–381)

Obviously, Schrödinger agreed with London's estimation of his 1922 paper, since he did not object to it in either letters or in press (after publication of the Stuttgart paper). In his paper, London also did not change his position from the December 10 letter. Finally, in the autobiographical sketch (1933), Schrödinger himself regarded his 1922 paper as an "anticipation, unconscious of course ... of L. de Broglie's theory" [349].

What was the true significance of this anticipation in the genesis of Schrödinger's wave mechanics? In addition to some other facts (the absence of prejudice of Schrödinger against the ideas of de Broglie, in contrast to the physicists of the Göttingen, Copenhagen, and Munich schools; the adherence of Schrödinger and de Broglie to similar program aims; the recognition of the reality of light quanta; the need for a space-time continuum and causal description of the microscopic world; the use for this purpose of differential equations, etc.; the identical interests of Schrödinger and de Broglie—quantum statistics and atomic spectroscopy, the optico-mechanical analogy and variational approach, the use of the relativistic approach to problems of atomic physics, etc.), the anticipation appears to have been decisive in determining why it was precisely Schrödinger who developed de Broglie's wave conception [354]. The 1922 paper can be seen as a bridge between de Broglie's resonance interpretation of quantum conditions and Schrödinger's approach.

Indeed, as Schrödinger noted,

in the first place, the stimulus to these deliberations was provided by the deep ideas of Louis de Broglie and reflection on the spatial distribution of his "phase wave," of which he showed that an integer number, measured along the orbit, always corresponds to a period or quasiperiod of the electron. ([353], pp. 372–3)

The central problem, the attempt at the solution of which led Schrödinger to wave mechanics, was the problem of the wave reformulation of the quantum conditions for the hydrogen atom. This is indicated by a letter of Schrödinger to Landé (on November 16, 1925), in which, in particular, he emphasized that

"I have been deeply involved with Louis de Broglie's ingenious thesis. . . . I have vainly attempted to make myself a picture of the phase wave of an electron in an elliptical orbit" ([354], p. 313). These attempts indicate that Schrödinger attempted, in the language of de Broglie waves, to obtain a transparent (as in his 1922 paper) and physically sensible wave formulation of the quantum conditions. Thus, the 1922 paper, and with it Weyl's unified theory, served as a point of departure for the development by Schrödinger of de Broglie's wave ideas and played a very important part in the genesis of Schrödinger's wave mechanics.

It appears that this cross fertilization is a rather general feature of the development of scientific knowledge under conditions when competing research programs exist. Programs that appear to be in regression, have come to a dead end, or are premature can, in the process of competition with a program that ultimately comes out on top, significantly influence the development of the latter.

THE GAUGE FIELD CONCEPT AND UNIFIED FIELD THEORIES (FORMULATION OF THE PROBLEM)

The main achievements of the quantum theory of fields and the interactions of elementary particles during the last 10–15 years have been largely associated with the gauge concept, the development of which was initiated by the paper of Yang and Mills in 1954 [346]. As stated in the foreword to the book on the quantum theory of gauge fields by Slavnov and Faddeev,

> Gauge fields appear in the majority of modern models of not only the strong but also the weak and electromagnetic interactions. An extremely attractive prospect of the unification of all interactions in a single universal interaction is opened up. ([356], p. 5)

The gravitational field also has a gauge structure. Thus, during the last decade the idea of a unified field theory of the strong, weak, and electromagnetic interactions ("grand unification") has acquired real features, and there have also appeared approaches that take into account gravitation (supergravity)— all of this on the basis of the gauge concept. Einstein's cherished idea of a geometrical field synthesis of physics has been reborn at a new level, and in this rebirth the gauge concept plays a decisive part.

It is true that in gauge theories geometrization is understood somewhat differently than in the classical theories of the program of unified geometrized field theories, but nevertheless the geometry of fiber bundles makes it possible to include gauge fields in a unified geometrical structure [7]. The fact is that the gauge field can be regarded as a connection in a principal fiber bundle whose base is formed by the four-dimensional space-time, while the fiber is the internal symmetry group.

As we have already said, the gauge philosophy, which has led to such significant successes, derives from the 1954 paper written by Yang and Mills. Incidentally, there is a different, no less popular designation for gauge fields: Yang–Mills fields. Proceeding from the requirement of local isotopic invariance, they introduced a new field responsible for the interaction of nucleons. In accordance with Noether's theorem, the law of conservation of isospin is associated with invariance of the theory with respect to rotations in the isotopic space. Physically, this means that the proton and neutron cannot be distinguished if electromagnetic interactions are ignored. Yang and Mills noted:

> As usually conceived, however, this arbitrariness [i.e., this indistinguishability] is subject to the following limitation: once one chooses what to call a proton, what a neutron, at one space-time point, one is then not free to make any choices at other space-time points. ([346], p. 192)

In other words, the internal symmetry is usually characterized by transformations that act in the same way at all points of space-time, i.e., is global. "It seems," they noted, "that it is not consistent with the localized field concept that underlies the usual physical theories." The systematic implementation of this concept, which combines the principles of symmetry and local interaction, leads in the case considered to requiring

> all interactions to be invariant under *independent* rotations of the isotopic spin at all space-time points, so that the relative orientation of the isotopic spin at two space-time points becomes a physically meaningless quantity. [Ibid.]

To ensure this condition, it is necessity to introduce a new vector field satisfying quite definite equations. In 1960, Sakurai introduced introduced vector fields associated with the conservation laws of baryon number and hypercharge in a similar manner. It was in this way that the basic ideas of the modern gauge theory of elementary particles arose later [358].

In a discussion at the section "Einstein and the physics of the future" held in the framework of the jubilee symposium at Princeton in 1979 for the centenary of Einstein's birth, Ne'eman and Pais recalled the Kammerlingh–Onnes Conference of 1953, at which Pauli advanced the idea of applying the principle of local gauge invariance to the concept of isospin ([359], pp. 505–506). He advanced this thought in connection with a discussion of a paper in which Pais had considered problems whose subsequent solution led to the concept of "strangeness." Pais also noted that after this Pauli developed a corresponding formalism, which initially appeared to him different from the method of Yang and Mills but was then recognized as equivalent to it, as Pauli wrote in a letter to Pais. Thus, although Pauli published nothing about this, the noted documentary evidence (the discussion on Pais's paper at the end of 1953 conference and Pauli's question, which advanced the idea of localizing the isospin group, were then published in the journal *Physica*) indicate how close he was to developing a generalized gauge concept.[3] He did formally develop the idea he had put forward, but apparently only after the publication of the paper by Yang and Mills. It will be shown below that Pauli was one of the main figures in the history of the development of the idea of gauge symmetry in electrodynamics and the gauge concept of the electromagnetic field. Therefore, his anticipation in a certain form of the Yang–Mills concept is not so surprising.

It also appears that Wigner was close to the gauge concept at the end of the 1940s and the beginning of the 1950s, when he thought that nuclear forces could be associated with the localization of a gauge symmetry corresponding to the law of conservation of baryon charge. He compared the entire situation with the law of conservation of electric charge, which corresponds to a local gauge symmetry and the electromagnetic interaction. In 1960, Sakurai wrote that the gauge ideas

> arose under the influence of the ideas of many authors—Wigner, Yang and Mills, Lee and Yang, Schwinger, Gell-Mann, Fujii, and some others. All these authors, as I was, were concerned with the question of the possible connection between the conservation laws for internal qualities such as isospin and baryon number, on the one hand, and the interaction dynamics of elementary particles, on the other. ([360], p. 105)

The first person named in this list is Wigner (there was, in particular, his 1949 paper [361]). Yang and Mills appear after him.

We return to the paper of Yang and Mills. How did they get the idea of local invariance and the idea that localization of a global symmetry would necessarily lead to a certain vector field? It is not difficult to answer this question: the example of the electromagnetic field stood before their eyes. This is how they themselves wrote about about this:

> ...an entirely similar situation arises with respect to the ordinary gauge invariance of a charged field which is described by a complex wave function

---

[3] My attention was drawn to this material by I.Yu. Kobzarev, for which I am most grateful to him.

ψ. A change of gauge means a change of phase factor $\psi \to \psi'$, $\psi' = (\exp i x)\psi$, a change that is devoid of any physical consequence. Since ψ may depend on x, y, z, and t, the relative phase factor of ψ at two different space-time points is therefore completely arbitrary. In other words, the arbitrariness in choosing the phase factor is local in character. ([346], p. 192)

The transition from α to $\alpha(x, y, z, t)$ destroys the invariance of the theory and, as is well known,

... in electrodynamics it is necessary to counteract the variation of α with x, y, z, and t by introducing the electromagnetic field A_μ which changes under a gauge transformation as

$$A'_\mu = A_\mu + \frac{1}{e}\frac{\partial \alpha}{\partial x_\mu}. \quad \text{[Ibid.]}$$

The authors write of the connection between local gauge invariance and the electromagnetic field as a thing fairly well known and refer in this connection to Pauli's famous review of 1941 [362].

Thus, already in the 1930s and 1940s the connection between the electromagnetic field and local gauge symmetry was fairly well known. The local gauge transformations

$$\begin{cases} \psi' = e^{i\alpha(x)}\psi, \\ A'_\mu = A_\mu + \dfrac{1}{e}\dfrac{\partial \alpha(x)}{\partial x_\mu} \end{cases} \tag{6}$$

could be discovered only after the creation of quantum mechanics. Therefore the gauge treatment of the electromagnetic field could appear only at the end of the 1920s and the beginning of the 1930s.

As we shall see, a study of the history of the discovery of gauge symmetry returns us to the unified geometrized field theories and the attempts to relate them to quantum mechanics. The concept of gauge symmetry goes back to the problem of the correspondence between the potentials and field strengths in electrodynamics. Therefore, before we turn directly to the genesis of the gauge field concept in the context of unified geometrized field theories, we make an excursion into the history of this problem, which is associated with the names of Faraday and Maxwell.

POTENTIALS AND FIELD STRENGTHS IN ELECTRODYNAMICS

In classical electrodynamics the main physical characteristics of the electromagnetic field were long assumed to be the strengths of the electric and magnetic fields; the potentials were regarded as merely auxiliary quantities. This point of view is represented in all the well-known courses of electrodynamics, from the classical works of Lorentz and Abraham to the modern texts of Landau and Lifshitz, Jackson, and others. Let us go back to the sources, however; it turns out that both Maxwell, and in a certain sense Faraday, did not regard the potential as a merely auxiliary, secondary quantity. Sometimes they even regarded it as a more fundamental quantity than the field strengths. It is true that at that time the electrostatic (scalar) potential and the vector potential did not have the same status. Whereas the scalar potential was a well-known quantity, and the theory of it had been developed not only in electrostatics and magnetostatics but also in celestial mechanics, the concept of the vector potential had a difficult path of development.

A precursor of the vector potential was the concept of the electrotonic state, which first appeared in the works of Faraday.[3a] The evolution of this concept in Faraday's work was described briefly and clearly by Maxwell in the third chapter of the fourth part of the *Treatise on Electricity and Magnetism* (1873):

> The conception of such a quantity, on the changes of which, and not on its absolute magnitude, the induction current depends, occurred to Faraday at an early stage of his *Researches* (*Exp. Res.*, Series i. 60). He observed that the secondary circuit, when at rest in an electromagnetic field which remains of constant intensity, does not [show] any electrical effect, whereas, if the same state of the field had been suddenly produced, there would have been a current.... He therefore recognized in the secondary circuit, when in the electromagnetic field, a "peculiar electrical condition of matter," to which he gave the name of the Electrotonic State. He afterwards found that he could dispense with this idea by means of considerations founded on the lines of magnetic force (*Ibid.*, Series ii. 242), but even in his latest *Researches* (*Ibid.*, 3269), he says, "Again and again the idea of an *electrotonic* state (*Ibid.*, 60, 1114, 1661, 1729, 1733) has been forced on my mind.
>
> The whole history of this idea in the mind of Faraday, as [shown] in his published *Researches*, is well worthy of study. By a course of experiments, guided by intense application of thought, but without the aid of mathematical calculations, he was led to recognize the existence of something which we now know to be a mathematical quantity, and which may even be called

---

[3a] It is possible to suppose that Faraday's concept of the electrotonic state had romantic origins and was connected with the romantic opposition between the observable phenomena and hidden, essential states which give rise to these phenomena when the mentioned hidden states change suddenly and rapidly (V.P. Vizgin, *The Romantic Origins of Faraday's Electrotonic State Concept*, unpublished manuscript).

the fundamental quantity in the theory of electromagnetism. But as he was led up to this conception by a purely experimental path, he ascribed to it a physical existence and supposed it to be a peculiar condition of matter, though he was ready to abandon this theory as soon as he could explain the phenomena by any more familiar forms of thought.

Other investigators were long afterwards led up to the same idea by a purely mathematical path, but, so far as I know, none of them recognized, in the refined mathematical idea of potential of two circuits, Faraday's bold hypothesis of an electrotonic state....

The scientific value of Faraday's conception of an electrotonic state consists in its directing the mind to lay hold of a certain quantity, on the changes of which the actual phenomena depend.

We note, in passing, that Maxwell is revealed here as a true historian of science. This extract gives a brilliant sketch of the "whole history of this idea [i.e., the idea of an electrotonic state] in the mind of Faraday" ([363], pp. 187–188).

Thus, the idea that the electrotonic state was a fundamental characteristic of the electromagnetic field was introduced by Faraday not later than 1831 and was originally published in the first series of *Experimental Researches* (dated November 1831). On November 24, Faraday read the first series at the Royal Society [364]. The entire third section of the first series is devoted to the electrotonic state. Faraday emphasizes the unobservability of this state: "This peculiar condition shows no known electrical effects whilst it continues" ([364], p. 16, §60). He also noted the circumstance that the observed phenomena arise only when this state is changed: "This peculiar state appears to be a state of tension, and may be considered as *equivalent* to a current of electricity ... produced either when the condition is induced or destroyed" ([364], p. 19, §71). It is interesting that not only did the contemporaries of Faraday and Maxwell poorly understand the concept of the electrotonic state, but also later investigators—both physicists and historians of science. For example, T.P. Kravets, an important Russian physicist and historian of electrodynamics, discussing the content of the first series of Faraday's *Researches*, wrote in 1947:

There then follows an unsuccessful section of a peculiar "electrotonic state," into which matter is supposed to pass when a current flow through in (60–80); with this state, Faraday attempts to explain all the phenomena observed by him. We shall not dwell on this hypothesis, since Faraday himself decisively abandoned it already in the second series of his investigations (231). This is the only section in the studies of electromagnetic induction that has not become a classic and has not found its way into all textbooks. ([365], p. 747)

In a footnote Kravets commented: "This hypothesis had a tight grip on Faraday, and he broke with it only reluctantly (242). It again took hold of his later (1661, 1729, 1733, possibly also 1114)."[4]

It is difficult to suppose that Kravets did not recognize in the electrotonic state the concept of the vector potential. This, as we shall see, is clearly revealed in an examination of the main works of Maxwell, who had in contrast, as we have already noted, a very high estimation of this concept of Faraday's. It was simply that for Kravets, as for the overwhelming majority of physicists of the twentieth century, it was the field strengths, and not the potentials, that had deep physical meaning. In fact Faraday, as Maxwell noted, felt certain doubts about the concept of the electrotonic state. The investigations of the second series led him to the conclusion that the basic characteristic of the field must be the lines of force (field strengths). Therefore, in a note on the third section of the first series quoted above (on its publication), he said: "... later investigations ... of the laws governing these phenomena, induce me to think that the latter can be fully explained without admitting the electro-tonic state" ([364], p. 16).

As both Maxwell and Kravets noted, Faraday soon returned to the concept, dear to his heart, of the electrotonic state. Thus, in the eighth series (April 1834) he attempted to apply it to explain electrolytic phenomena (955), and in the thirteenth series (February 1838) to the analysis of electric phenomena in nonconductors (1658–1661, and also 1729–1733). At the same time he was attempting to find a way to detect the electrotonic state experimentally and attempted to generalize it, introducing in his late investigations the concept of a "magnetotonic state" (1845). In his final investigations, published in the third volume (January–February 1855), he returned to the idea of the electrotonic state as a key theoretical concept of electrodynamics:

> Again and again the idea of an *electro-tonic* state (60, 1114, 1661, 1729, 1733) has been forced on my mind; such a state would coincide and become identified with that which would then constitute the physical lines of magnetic force. ([366], pp. 420–421, §3269)

Faraday succeeded neither in detecting the electrotonic state experimentally nor in developing sufficiently clear theoretical ideas on the basis of this concept (the concept of lines of field strengths proved at that time more viable), but intuitively he felt its depth and fruitfulness and constantly returned to it.

---

[4] The numbers given here are the numbers of the points in the corresponding sections of Faraday's *Researches*.

For Maxwell, Faraday's concept of the electrotonic state was to a large degree the original one in the development of the theory of the electromagnetic field. The first mention of this concept, indicating its key role in the genesis of Maxwell's ideas, we find in a letter from Maxwell to W. Thomson on September 13, 1855:

> I intend to apply to these facts [Maxwell is referring to facts about the lines of force] Faraday's notion of an *electrotonic* state. I have worked a good deal of mathematical material out of this vein and I believe that I have got hold of several truths in the electrotonic state. One thing at least it succeeds in, it reduces to one principle not only the attraction of currents and the induction of currents but also the attraction of electrified bodies without any new assumption. ([367], p. 211)

In his first paper on field theory, "On Faraday's lines of force" (1855–1856), Maxwell did indeed start with the concept of the electrotonic state, which thus proved to be crucial in the development of the mathematical formulation of Faraday's ideas. Commenting on his prophetic statements about the importance of this concept, Maxwell noted

> The conjecture of a philosopher so familiar with nature may sometimes be more pregnant with truth than the best established experimental law discovered by empirical inquirers, and though not bound to admit it as a physical truth, we may accept it as a new idea by which our mathematical conceptions may be rendered clearer. ([368], p. 187)

It is interesting that Maxwell subsequently hoped to give a "mechanical analogy" of this state. The second part of the paper was in fact called "On Faraday's 'electro-tonic' state." In it Maxwell showed that one con introduce a triplet of quantities, i.e., a vector in modern terminology $(\alpha_0, \beta_0, \gamma_0)$, which he called the components of the electrotonic strength, such that by means of it one can, at each point of space at which the triplet is specified, determine the values of the strengths of the electric and magnetic fields:

$$
\begin{cases}
\alpha_2 = -\dfrac{1}{4\pi}\dfrac{d\alpha_0}{dt}, & \beta_2 = -\dfrac{1}{4\pi}\dfrac{d\beta_0}{dt}, & \gamma_2 = -\dfrac{1}{4\pi}\dfrac{d\gamma_0}{dt}, \\[2ex]
\alpha_1 = \dfrac{d\beta_0}{dz} - \dfrac{d\gamma_0}{dy} + \dfrac{dV}{dx}, & \beta_1 = \dfrac{d\gamma_0}{dx} - \dfrac{d\alpha_0}{dz} + \dfrac{dV}{dy}, \\[2ex]
& \gamma_1 = \dfrac{d\alpha_0}{dy} - \dfrac{d\beta_0}{dx} + \dfrac{dV}{dz}.
\end{cases}
\tag{7}
$$

Here, $(\alpha_1, \beta_1, \gamma_1)$ are the magnetic field strengths, $(\alpha_2, \beta_2, \gamma_2)$ are the electric field strengths, and V is the scalar magnetic potential. In modern vector notation (if **A** is the vector potential, V is the scalar potential, and **E** and **H** are the field strengths), the relations (7) take the form

$$\mathbf{E} = -\frac{1}{4\pi}\frac{d\mathbf{A}}{dt}, \quad \mathbf{H} = \operatorname{rot}\mathbf{A} + \operatorname{grad} V. \tag{7a}$$

An important part in Maxwell's arguments was played by theorems and formulas of vector analysis that are well known today and are associated with the names of Gauss, Green, Ostrogradsky, Stokes, and W. Thomson. Maxwell directly quoted their studies, especially the fundamental study of Thomson on the mathematical theory of magnetism (1851), which, essentially, contained Maxwell's Theorem 5, according to which the relation $\operatorname{div}\mathbf{H} = 0$ has the consequence that there exists a vector **A** such that $\mathbf{H} = \operatorname{curl}\mathbf{A}$. We note here that Thomson in no way related the vector **A** to Faraday's electrotonic state. Thus, one of the most important starting points of Maxwell's theory and its mathematical formulation was Faraday's concept of the electrotonic state.

In his next paper "On physical lines of force," dated 1861–1862, Maxwell attempted to give the promised "mechanical analogy" of the electrotonic state, emphasizing that he uses "mechanical analogies in order to assist the imagination without in any way identifying them with the causes of the phenomena" ([369], p. 109).* We shall not reproduce here the well-known Fig. 2 of this paper, which shows a mechanism intended to give a mechanical interpretation to Eqs. (7) together with the gauge condition

$$\frac{d\alpha_0}{dx} + \frac{d\beta_0}{dy} + \frac{d\gamma_0}{dz} = 0, \tag{8}$$

and, at the same time, interpreting the concept of the electrotonic state. As a result, Maxwell concluded that the "mechanical analogy" of the electrotonic state was the

> *impulse* [i.e., the momentum] which would act on the axle of a wheel in a machine if the actual velocity were suddenly given to the driving wheel, the machine being previously at rest. ([398], p. 478; [369], p. 146)

(In the same paper, Maxwell considered a hydrodynamic analogy.)

This enabled him to call the electotonic state the "electromagnetic momentum" in the paper "A dynamical theory of the electromagnetic field."

---

* *Translator's note*: This is my translation of the Russian; I have not been able to locate a matching passage in Maxwell's original paper.

What I have called electromagnetic momentum is the same quantity which is called by Faraday the electrotonic of the circuit, every change of which involves the action of an electromotive force, just as change of momentum involves the action of mechanical force. ([398], p. 542; [370], p. 273)

We note that the 20 basic quantities that characterize electromagnetic processes and occur in the 20 basic equations of electrodynamics include the three components of the electrotonic state (the "electronic tension," "electromagnetic momentum," and the future vector potential). Moreover, for the free field the "electromagnetic momentum" appears as the more fundamental quantity, since the field strengths can be expressed in terms of it and in terms of the scalar "electric potential." The circumstance that the "electromagnetic momentum" was determined only up to the gradient of an arbitrary function of the spatial coordinates was known to Maxwell from the time of his first work on electrodynamics. This gauge nonuniqueness of the electrotonic state, and the clear understanding that this quantity itself was not observable and that observable phenomena depended only on its changes, did not prevent Maxwell from according a fundamental physical character to this state.

Actually, three years later, in 1868, Maxwell systematized the basic equations of his theory without using potentials [371]. This was pointed out by Bork, who also noted that Maxwell's aim in giving up the use of potentials is not clear from the text of this paper. Some of Maxwell's letters to Tait, above all dated January 23, 1871, show that at least in 1870 his thoughts again returned to the potentials [367]. One could suppose that the temporary abandonment of them was due to the two noted shortcomings of these quantities (unobservability and nonuniqueness), doubts about the expediency of mechanical models, and the desire to give the most economic phenomenological scheme of electrodynamics. Maxwell's attachment to the potentials was too great, however; the part that they had played in the development of the foundations of the theory was too important, and they were too deeply embedded as fundamental quantities in the conceptual structure of the theory, revealing also its deep connection with the structure of classical mechanics, for the abandonment of them to be final. There was one last circumstance that may have stimulated Maxwell's return to the potentials, which is fully reflected in the *Treatise on Electricity and Magnetism*. This was the possibility of applying a new mathematical method, the vector and quaternion formalism, which took a particularly transparent form when the potentials were used. In the quoted letter to Tait, the quaternion representation of the potentials is probably encountered for the first time, moreover in a "four-dimensional" form:

$$T = P + iL + jM + kN, \tag{9}$$

where P is the scalar magnetic* potential, and (L, M, N) is the electrotonic state, called here for the first time the vector potential.

We have already quoted Maxwell from the fourth part of the *Treatise* on Faraday's concept of the electrotonic state, which he esteemed very highly, saying that it "may even be called the fundamental quantity in the theory of electromagnetism" ([363], p. 197). He first called it the vector potential in Section 405, entitled "The vector-potential of magnetic induction." As Bork notes, this name was chosen by analogy with the scalar potential (whose derivatives with respect to the spatial coordinates are equal to the components of the electric field), since linear combinations of the derivatives of the components of the vector potential with respect to the spatial coordinates gave the components of the magnetic field strength. In the *Treatise*, the vector potential occupies first place for Maxwell among the "basic vectors" of the theory, and among the scalars the electric potential occupies first place. We note also that in the chapter devoted to the electromagnetic theory of light Maxwell used the potentials to write down wave equations [ibid.].

In all the monographs and textbooks on electrodynamics written at the beginning of this century and containing Maxwell's field theory we find the system of Maxwell's equations in the Hertz–Heaviside form, in which only the field strengths are present, while the potentials are absent. Indeed, it was Hertz and Heaviside who in the first place cast doubt, in the middle of the 1880s and the beginning of the 1890s, on the fundamental nature of the potentials and showed that the system of basic equations of electrodynamics could be expressed solely in terms of the field strengths, without recourse to the potentials. We may suppose that the elimination of the potentials from the basic equations of the electromagnetic field by Hertz and Heaviside was related to their experimental (in the case of the first) and engineering (in the case of the second) interests, since the potentials, in contrast to the field strengths, are not directly measurable quantities ([372], pp. 221–223). Heaviside emphasized that in the general case knowledge of the potentials was not sufficient to determine the state of the field uniquely. He called them "highly artificial quantities" possessing an artificial nature, and assumed that only the field strengths "have physical significance in really defining the state of the medium" ([373], p. 226).

Although Hertz, like Heaviside, attempted to eliminate the potentials from physical considerations and use them only as a computational tool, he was not,

---

* *Translator's note*: The word "magnetic" does not appear in the letter of Tait quoted here ([367], p. 216); rather, P is said to be the potential of a scalar quantity m.

like Heaviside (and also Fitzgerald, who also avoided the use of potentials, and Larmor, who assumed that they "do not represent actual physical quantities" ([374]; [367], p. 222), so categoric in regarding them as quantities devoid of any physical meaning. Admittedly, he assumed that the potentials could have physical meaning only in the consideration of electromagnetic processes in a medium ("and not in a vacuum") ([373], pp. 227–230).

At the end of the 1890s and the beginning of the 1900s this point of view became almost standard. In the widely studied course of electrodynamics by Gray (1898), we can read: "The use of the vector-potential is sometimes convenient as an analytical expedient. But it is not a physical quantity which can be observed experimentally ... " ([367], p. 222). There are similar comments in Lorentz's *The Theory of Electrons* (1909), which emphasized that it was Hertz and Heaviside who gave the modern formulation of Maxwell's theory in its "clearer and more condensed form" ([375], p. 2). "Since, however, the second members of the formulae are somewhat complicated," Lorentz wrote in this book, referring to the wave equations expressed in terms of **d** and **h**,

> we prefer not directly to determine **d** and **h**, but to calculate in the first place certain auxiliary functions on which the electric and magnetic forces may be made to depend, and which are called potentials. ([375], p. 19)

The potentials are mentioned in much the same context in Abraham's famous course of electrodynamics (first edition 1904) [376]. Thus, the potentials lost the nature of fundamental physical quantities and acquired the status of auxiliary characteristics of the field, used only to simplify the mathematical calculations in the solution of the field equations.

In the middle of the 1960s Wigner, expressing the widely accepted point of view going back to Hertz and Heaviside, likened the potentials to "ghosts":

> ... the potentials cannot be measurable, and, in fact, only such quantities can be measurable which are invariant under the transformations which are arbitrary in the potential [namely, gauge transformations of the second kind]. This invariance is, of course, an artificial one, similar to that which we could obtain by introducing into our equations the location of a ghost. The equations then must be invariant with respect to changes of the coordinate of the ghost. One does not see, in fact, what good the introduction of the coordinate of the ghost does. ([377], p. 22)

Nevertheless, the attitude to the potentials was not completely uniform. At the beginning of the 1910s, Mie devised a nonlinear electrodynamics having the aim of explaining the existence of charged particles on the basis of the field concept. In Mie's theory, not only the field strengths but also the potentials had real meaning, and the theory was gauge noninvariant. This led, as Pauli subsequently noted, to a "melting away" (*rasplyvanie* = deliquescense)* of charged particles in a constant external potential field ([63], p. 192). Despite this difficulty, Mie's theory attracted the interest of theoreticians. In 1915, Hilbert, having included Mie's concept in the framework of the general theory of relativity, sketched a first unified physical theory based on the general theory of relativity (see Chapter 2).

In 1959, there appeared a paper of Aharonov and Bohm which stimulated much discussion among physicists. They quite unambiguously emphasized that in the framework of quantum theory, i.e., when allowance is made for the quantum laws of the motion of charged particles in an external electromagnetic field, the potentials and field strengths again change place. They proposed a simple interference experiment with electrons that would make it possible to reveal the direct influence of the field potential on an interference pattern. Aharonov and Bohm showed that "in quantum theory, an electron... can be influenced by the potentials even it all the field regions are excluded from it" ([378], p. 490). Although such effects, as the authors noted, depend on a gauge-invariant integral characteristic, the conception of a localized field requires recognition of the influence on the electron of precisely the potentials.

In 1960–1962, R.G. Chambers, G. Möllenstedt, W. Bayh, and others detected the Aharonov–Bohm effect experimentally. Referring the reader to the *The Feynman Lectures on Physics*, where the Aharonov–Bohm experiment and the consequences that flow from it are analyzed with great depth and clarity, we give merely the concluding words of the corresponding paragraph:

> ... the vector potential **A** (together with the scalar potential φ that goes with it) appears to give the most direct description of the physics. This becomes more and more apparent the more deeply we go into the quantum theory. In the general theory of quantum electrodynamics, one takes the vector and scalar potentials as the fundamental quantities in a set of equations that replace the Maxwell equations: **E** and **B** are slowly disappearing from the modern expression of physical laws; they are being replaced by **A** and φ. ([379], Vol. 2, Sec. 15–16)

The "materialization of ghosts" in quantum theory is explained by the fact that in that theory the concept of force loses its significance, while the concepts of energy and momentum come to the fore.[5] The phases of the wave functions,

---

* *Translator's note*: What Pauli actually wrote is: "*A material particle will therefore not be able to exist in a constant external potential field.*"

[5] A more detailed discussion of the physical side of this effect and an analysis of the corresponding experiments can be found in the papers of Feinberg [380] and Erlichson [381]. L.H. Ryder gives a more modern interpretation of Aharonov–Bohm

which determine the motion of particles, depend on the wavelengths and frequencies directly related to the energies and the momenta. The change in phase associated with the passage around a certain path l in a magnetic field with potential $\mathbf{A}$ is determined by the relation

$$\frac{e}{h} \int \mathbf{A} \, d\mathbf{s}. \tag{10}$$

The calculation of the phase shift in the Aharonov–Bohm experiment leads to an analog of this integral that is taken around a closed contour and reduces, by Stokes theorem, to the flux of the magnetic field strength through the surface bounded by the contour.[6] Thus, the result is still gauge-invariant.

The fundamental role of potentials in quantum theory was suggested by the very creation of quantum mechanics, since they occur explicitly in the Schrödinger equation for a particle moving in a field. It is true that the Hamiltonian formulation of the classical electron theory also led to expressions for the momentum and Hamiltonian containing the potential

$$P_k = p_k + \frac{e}{c} A_k, \qquad H = \sum_k \frac{1}{2m} \left(p_k - \frac{e}{c} A_k \right)^2 + e\varphi. \tag{11}$$

In this case it was one of many possible formulations, whereas in quantum mechanics the potentials acquired a fundamental nature in the operator generalization of these expressions. In fact, a Schrödinger equation with the above expression for the Hamilton operator had already been obtained in Schrödinger's first classic papers on wave mechanics. Despite this, the classical understanding of the potentials and gauge invariance of the second kind remained predominant, as we have seen, right up to the beginning of the 1960s.

The reevaluation of the role of the potentials that occurred in the 1960s and 1970s was due not only to Aharonov–Bohm-type experiments but also to the confirmation of the gauge field concept, in which the connection nature

---

effect as a phenomenon caused by the nontrivial topology of the vacuum and the gauge nature of the electrodynamics (L.H. Ryder, *Quantum Field Theory*, Cambridge University Press, 1985 (Ch. 3, §3.4)).

[6] In a more general situation, an analogous role is played by an integral

$$\frac{e}{h} \int A_\mu \, dx_\mu$$

along a four-dimensional path.

of gauge fields and, thus, the fundamental physical nature of potentials were fully demonstrated [7, 356, 358]. The interpretation of the electromagnetic potential as a connection first appeared, however, in the framework of the first unified geometrized field theories, namely in Weyl's theory, in which the prototype of local gauge transformations also appeared for the first time.

GAUGE TRANSFORMATIONS IN WEYL'S THEORY

Originally, Weyl gave the name "scale transformations" to the transformations (see Eq. (7) of Chapter 3)

$$
\begin{cases}
g'_{ik} = \lambda(x)g_{ik}, \\
\varphi'_i = \varphi_i - \dfrac{1}{\lambda(x)} \dfrac{\partial \lambda(x)}{\partial x_i},
\end{cases}
\tag{12}
$$

which, together with the arbitrary continuous transformations of the general theory of relativity, are the basis of Weyl's theory, and he called the corresponding invariance "scale invariance" (*Maßstab-Invarianz*). If the notation $e^{\alpha(x)} = \lambda(x)$ is introduced, then it becomes obvious that the potentials satisfy gauge transformations of the second kind, which thus acquired a definite geometrical meaning in Weyl's theory, since the arbitrariness in the choice of the potential was associated with the arbitrariness in the choice of the sale for measuring length. Under a gauge transformation, the square of the length of a certain interval had to be multiplied by the scale coefficient $e^{\alpha(x)}$.

In fact, in one of the 1919 papers Weyl replaced the term "scale invariance" by the term *Eich-Invarianz* [164]. In the English translation (namely, in the translation *Space, Time, Matter* of the fourth edition of *Raum-Zeit-Materie* in 1921, translated by G.L. Brose), the term "calibration" was used [128]. The term that has now become established in the English language literature, "gauge," apparently first appeared in Weyl's paper of 1929, which was translated into English by G.P. Robertson [311]. Before Weyl, gauge transformations of the second kind had no special name.[7] The gauge transformations (12) of Weyl's theory, written in the form

$$
g'_{ik} = e^{\alpha(x)} g_{ik}, \qquad \varphi'_i = \varphi_i + \dfrac{\partial \alpha(x)}{\partial x_i},
\tag{13}
$$

resemble the local gauge transformations of the wave functions and potentials (6). A difference is that the role of the wave function in Weyl's theory

---

[7] On the history of the introduction of the term "gauge," see also [8, 382].

is played by the metric tensor g_{ik} and the argument of the exponential is a real, not imaginary, function. As we have seen, Weyl believed that the corresponding geometry agreed better with the concept of local interaction. In the construction of this geometry, the idea of parallel transport was brought into foreground, as was the connection nature of the gravitational and electromagnetic fields.

In Weyl's theory, the fundamental nature of gauge transformations was recognized for the first time, and the gauge concept of the electromagnetic field was anticipated, admittedly in an incorrect form. The prototype of the gauge concept of the gravitational field had already been developed in the general theory of relativity, for in it there was realized for the first time the principle of local invariance, which led to the need to introduce the gravitational field. The transition from a global space-time symmetry (Lorentz group) to the corresponding local symmetry led to a curved space-time, which could be interpreted through the connection as a gravitational field. Einstein, it is true, put his emphasis on the metric and not the connection aspects of geometry and gravitation. It was only after the studies of Levi-Civita, Weyl, and Eddington that the fundamental nature of the connection concept began to be clarified (see Chapters 3 to 5).

The connection nature of a field, specifically, the electromagnetic field, was brought into clearer relief in Weyl's theory. Localization of the similarity transformation and change of the length vectors in accordance with such transformations under parallel transport of vectors were directly related to a four-vector possessing the properties of the electromagnetic potential. As we have seen, ten years later it had been established that the electromagnetic field is indeed associated with a local gauge symmetry and has a gauge nature, but the role of the exponential transformation of the metric was played by a corresponding transformation of the electron wave functions, moreover with an imaginary exponent. The connection nature of the electromagnetic field was also confirmed but shown to apply not to the parallel transport of intervals but to the parallel transport of electron wave functions, which are spinors, in Riemannian space. Using the language of fiber bundles, Manin noted in this connection:

Hermann Weyl was the first to propose that the electromagnetic field is a connection; however, in pre-quantum physics he could not identify the bundle on which this connection determines parallel transport and decided that the field changes the lengths of intervals transported along different paths in space-time. ([383], p. 26)

LOCAL GAUGE TRANSFORMATIONS
AND QUANTUM MECHANICS

The local gauge transformations of wave functions and their connection with the gauge transformations of the potentials, and also the invariance of the wave equations of quantum mechanics with allowance for the electromagnetic interaction with respect to the combination of these transformations, were established immediately after the creation of quantum mechanics in 1926. On this occasion, as well, the development did not occur without the assistance of a form of unified geometrized field theory, namely the five-dimensional Kaluza–Klein theory (see Chapters 4 and 5).

Having become acquainted with the idea of the five-dimensional approach from the manuscript of Mandel's paper [271], Fock found it possible to interpret the scalar wave equation in a five-dimensional Riemannian space as a relativistic generalization of Schrödinger's equation [270]. Subsequently, it became clear that the electron possesses spin, and therefore Fock's equation describes not an electron but a spinless particle. At approximately the same time, this equation was obtained by O. Klein (also on the basis of the five-dimensional approach), Gordon, Schrödinger, and others (Klein–Gordon–Fock equation). Fock also showed that the trajectory of a charged particle can be regarded as a geodesic in a five-dimensional Riemannian space.

In Fock's paper, received by the editors of the *Zeitschrift für Physik* on July 30, 1926, local gauge transformations first appeared. Having noted that the five-dimensional wave equation must be invariant with respect to the generalized gauge transformations

$$A'_\mu = A_\mu + \frac{\partial f}{\partial x_\mu}, \qquad u' = u - f, \qquad (14)$$

where f is an arbitrary function of the space-time coordinates and u is the fifth coordinate, Fock showed that, for the corresponding four-dimensional wave equation, invariance with respect to local gauge transformations remained valid only if the wave function was multiplied by the exponential factor $e^{\frac{2\pi i}{hc} f}$. "The function ψ_1 and $\overline{\psi}_i$, related to the potentials $\overline{A}$ and $\overline{A} = A - \text{grad } f$, would differ only by the factor $e^{\frac{2\pi i}{hc} f}$, equal in absolute magnitude to $1 \ldots$" wrote Fock (here ψ_1 is the quantum-mechanical wave function) ([270], p. 228). Naturally, these conclusions could also be extended to the nonrelativistic Schrödinger equation, although Fock believed that the physical value of the fifth dimension had a direct connection with the gauge of the potentials.[8]

Although Klein, proceeding from Kaluza's theory, obtained results close to those of Fock, and his paper appeared somewhat earlier, it sis not contain results relating to gauge invariance [258]. Both Klein and Fock introduced the condition of periodic dependence of the wave function on the fifth coordinate. In introducing this condition, Fock referred to the 1922 paper of Schrödinger considered above [352]. Fock assumed that the cyclicity of the "Weyl measure" of length on quantum orbits, having a direct relation to the properties of the electromagnetic potential, was related to the condition of periodicity of the wave function with respect to the fifth coordinate. On the basis of Schrödinger's paper, London showed, at the end of 1926 and the beginning of 1927, that if in the expression for the Weyl measure of length,

$$l = l_0 e^{\alpha \int \varphi_i \, dx_i},$$

one takes α, following Schrödinger, to have the value $\frac{2\pi i}{h} \frac{e}{c}$, then this quantity is, in its properties, identical to the de Broglie wave function (in its five-dimensional generalization) [355]. London demonstrated, in essence, the same for the wave function that occurs in quantum mechanics in Schrödinger's formulation. This automatically transferred the transformation properties of the symmetry of Weyl's theory to quantum mechanics; in other words, it led to a transition of the local gauge transformations of Weyl's theory into the quantum-mechanical local gauge transformations that had already been established by Fock.

We note that London knew of Fock's paper, for he referred to it in his comment published in the *Naturwissenschaften*, but only in connection with the five-dimensional relativistic generalizations of Schrödinger's equation [275]. Both he and Fock regarded as their main results not the establishment of gauge symmetry but the discovery of a connection between quantum theory and unified geometrized field theories.

Thus in 1926 Fock and London established the local gauge invariance of the electromagnetic interaction. Common features of their papers were (1) the desire to relate quantum mechanics to unified geometrized field theories, and

---

[8] Fock wrote:

The value of the additional coordinate parameter [i.e., the fifth dimension] is evidently precisely that it is responsible for invariance of the equations with respect to the addition of an arbitrary gradient to the four-potential. ([270], p. 228)

(2) the secondary nature of the demonstration of gauge symmetry. No less interesting are the differences in their papers. Fock gave preference to the five-dimensional unified field theory, London to Weyl's theory. Fock noted purely formally the need to augment the gauge transformations of the potentials by local gauge transformations of the wave functions, i.e., he pointed out the combined gauge invariance of the equations of quantum mechanics with electromagnetic interaction as a mathematical property of these equations. It seemed to London that he could understand the physical meaning of the wave function, which he interpreted as the Weyl measure of length. At the same time, the gauge symmetry of the wave functions acquired the same geometrical significance as the gauge symmetry in Weyl's theory. London's approach was, one can say, more physical than Fock's but also more speculative.

Incidentally, London assumed that after his paper the objections of Einstein and Pauli to Weyl's geometry and his theory would weaken, at least in some cases:

> Thus, Einstein's objections lose their force when one takes into account only states of motion possible from the quantum-theoretical point of view, and also the resonance properties associated with them. ([275], p. 187)

Because the incorrectness of Weyl's theory is now recognized, at least in its usual interpretation, London's basic construction lost its significance, although it was precisely in this way that he arrived at the gauge symmetry of wave functions. In contrast, Fock's conclusions and the scheme of his arguments relating to gauge symmetry still kept their significance when divorced from the question of the connection between quantum mechanics and unified field theories.

FROM GAUGE SYMMETRY TO GAUGE FIELDS

After the creation of quantum mechanics, Weyl's became interested in that field. His distinguished investigations into the theory of the representations of groups and their application to quantum mechanics were summarized in the book *Gruppentheorie und Quantenmechanik* (*The Theory of Groups and Quantum Mechanics*), the first edition of which was published in 1928 [384]. The idea of Gauge symmetry was formulated in this book, following the papers of London and Fock, with all clarity. Unlike London, Weyl did not attempt to rehabilitate his unified theory of 1918–1923; at the same time, he went somewhat further than Fock in the physical interpretation of this symmetry. First, he emphasized that the requirement of gauge invariance was related to the unobservability of wave functions and the fact that it is their squares, i.e., $\psi^*\psi$, that have direct physical significance. Second, he believed that the new "principle of gauge invariance," in contrast to his geometrized field theory, "does not tie together electricity and gravitation, but rather *electricity and matter*." By "matter" he understood charged microscopic particles whose motion was described by Schrödinger or Dirac equations. Weyl's formulation of the principle of gauge invariance was

> The field equations for the potentials φ and φ [he also called the functions ψ a potential in view of its unobservability] of the material and electromagnetic waves are invariant under the simultaneous replacement of
>
> $$\psi \text{ on } e^{i\lambda}\psi \quad \text{and} \quad \varphi_\alpha \text{ on } \varphi_\alpha - \frac{h}{e}\frac{\partial\lambda}{\partial x_\alpha};$$
>
> here, λ is an arbitrary function of the space-time coordinates. ([384], p. 100)

Comparing these transformations with the gauge transformations of his unified theory, Weyl noted the solid experimental basis of the former; the association of the exponential fact with wave functions, and not the metric (as in the 1918 theory); the naturalness of introducing atomic units for the potential; and also the remarkable circumstance that the argument of the exponential was purely imaginary and not real (as in the 1918 theory). Weyl said in 1930:

> Already at the time when I advanced my old theory, I had the feeling that the gauge factor must have the form $e^{i\lambda}$. However, at that time I was naturally unable to find any geometrical justification for such a choice. The studies of Schrödinger and London have confirmed this idea in the process of an ever clearer recognition of the connection between the problem and quantum theory. ([17], p. 57

The transition from the gauge symmetry of 1918 to the gauge symmetry of 1926–1928 led directly to the conclusion of a gauge nature of the electromagnetic field, the outline of which, as we have seen, was already suggested in 1918. This conclusion was drawn in full measure by Weyl himself in his famous paper of 1929 [313]. A shortened version of this paper appeared in English (submitted in February–March 1929) [311]. Although Weyl did abandon his own unified field theory, and indeed the geometrical direction altogether, the problem of the synthesis of gravitation, electromagnetism, and quantum mechanics still interested him.

In 1928, Dirac constructed a relativistic quantum mechanics in which the electron was described by a four-component wave function, subsequently called a spinor. Dirac's theory automatically took into account the spin properties of the electron and gave the correct results for the hydrogen atom. A difficulty of the theory was that it allowed states with negative energy, which could not be given a clear physical meaning. Weyl believed that by going over to the two-component form of the theory one would be able to eliminate the "extra" levels, which he interpreted as the energy levels of a "positive electron." The theory would then lose charge symmetry and symmetry with respect to spatial reflection. This, however, required removal of the term with the rest mass from the Dirac equation. Since mass is inseperably coupled with gravitation, Weyl assumed that mass could be taken into account correctly at a later stage, i.e., when the theory was made consistent with the general theory of relativity. Weyl believed that consideration of general relativity was also unavoidable because gauge symmetry, which was so close to his heart, had, in his own words, a generally relativistic nature:

> Since the gauge invariance includes the arbitrary function λ, it has the nature of "general relativity" and, naturally, can be understood only in the framework of the general theory of relativity. ([313], p. 245)

Here, the paths of development of the gauge concept again crossed those of the unified field theories. This time the connection was to Einstein's theory with absolute parallelism (see Chapter 5). Although this theory too, as we recall, did not lead to real advances in the solution of the problem of unifying gravitation and electromagnetism, it did play an important part in the development of the tetrad formalism, which proved to be very suitable for generalizing Dirac's equation to Riemannian spaces, i.e., for correct allowance for gravitation in Dirac's theory. This formalism was used by Weyl, and also, independently, by Fock and Ivanenko, about whose studies we shall say something a little later. In Weyl's concept, both the gravitational and the electromagnetic fields were originally related to the requirements of local symmetry, entirely in the spirit of the gauge concept. Two-component spinor equations of the Dirac type for electrons and protons were assumed to be primary. The requirements of local validity of the special theory of relativity and general covariance must introduce gravitation, which in its turn would make it possible to solve the problem of the particle masses. The requirement of local gauge invariance, consistent with the gauge symmetry of the electromagnetic potentials, must introduce the electromagnetic field determined by these potentials.

> In my opinion, the origin and necessity (*die Ursprung und Notwendigkeit*) of the electromagnetic field are determined as follows. The components [of the wave function] ψ_1 and ψ_2 are in reality not uniquely fixed by the choice of the coordinate system but only to the extent that they can still be multiplied by a certain arbitrary "gauge factor" $e^{i\lambda}$ with absolute value 1.... In the special theory of relativity, this gauge factor must be assumed to be constant because in this case we have a unique system of coordinates not associated with a particular point. The situation is different in the general theory of relativity: each point has its own coordinate system, and therefore

its own arbitrary gauge factor; as a result, in this theory the rigid connection between the coordinate systems at different points is eliminated, and the gauge factor necessarily becomes an arbitrary function of position. ([313], p. 245)

Thus, in the spirit of his unified field theory of 1918, Weyl related local gauge invariance to the need to introduce at each point of space-time a local Lorentzian coordinate system. This signified the transition to an arbitrary curved Riemannian space-time. "Only by this loosening (*Lockerung*) of space," wrote Weyl, "can one understand the gauge invariance that exists in reality" ([313], p. 246). At this point, there is a difference between Weyl, on the one hand, and Yang and Mills, on the other, on the justification of local gauge transformations. Weyl associated it with the transition to a Riemannian space and the general theory of relativity of relativity, and Yang and Mills "with the localized field concept that underlies the usual physical theories" ([346], p. 192). In Weyl's case, the mechanism of "generating" the electromagnetic field corresponded exactly to the scheme of introducing a gauge field: the transition to local gauge invariance leads to loss of invariance of the original Lagrangian (or the variational principle) for the free spinor field, and then, in order to recover invariance, a certain vector field, which can be interpreted as the electromagnetic field, is introduced. In the scheme presented by Weyl, the electromagnetic field was not assumed given; rather, as we have seen, Weyl spoke of the "origin and necessity" of this field as a result of localization of the gauge symmetry.

In the same paper Weyl solved the problem of the form of the Dirac equation in Riemannian space. This problem was also investigated in 1929 in a series of papers by Fock and Ivanenko [264, 385–388]. Weyl related the transition to Riemannian geometry with the construction of a certain generalization of Dirac's theory to a system of an electron plus a proton. In contrast, Fock (initially in collaboration with Ivanenko) attempted to give a correct formulation of Dirac's theory in the framework of the general theory of relativity. Without going into details, we mention that for both Weyl and the Soviet physicists the point of departure was the tetrad formalism developed a year earlier by Einstein in the framework of his unified field theory with absolute parallelism. Using the concept of parallel transport of a spinor, they found an expression for the covariant derivative of the spinor. On the transition to flat space, this did not transform simply into the ordinary derivative but into the expression

$$\frac{\partial}{\partial x_\mu} - ieA_\mu, \tag{15}$$

which from the beginning appears in quantum theory and takes into account
the presence of the electromagnetic field. Thus it was found that the electromagnetic potentials are manifested as a connection in the parallel transport
of a spinor. The connection nature of the electromagnetic field was very
clearly manifested as a result. In addition, a simple prescription was found
and geometrically justified for restoring invariance of the Lagrangian of the
free spinor field on the localization of the gauge transformations with constant
phase, namely replacement of the ordinary derivatives of the wave functions
by so-called compensating derivatives, i.e., the substitution

$$\frac{\partial}{\partial x_\mu} \rightarrow \frac{\partial}{\partial x_\mu} - ieA_\mu.$$

We may note that a term equivalent to "compensation" was also used in the
case considered in the paper by Yang and Mills: "To preserve invariance one
notices that in electrodynamics it is necessary to counteract the variation of
α with x, y, z, and t by introducing the electromagnetic field $A_\mu \ldots$" ([346],
p. 192).

When Fock became acquainted with Weyl's 1929 papers, he emphasized
their similarity to his own papers (written in part with Ivanenko) with regard to the concept of parallel transport of a spinor and the introduction of
the "compensating derivative," but criticized the physical side of Weyl's idea.
Nevertheless, it should be remembered that Weyl's physical argument, now
recognized as incorrect, led him to wave equations for two-component spinors
that are not invariant with respect to spatial reflections and describe the motion
of massless fermions. In 1956–1957, these equations were used by the same
Yang and T.D. Lee to describe processes that involve neutrinos and do not
conserve parity, and they were then used in Feynman and Gell-Mann's four-
fermion theory of weak interaction. In his famous 1932 survey of quantum
mechanics, Pauli did not neglect to mention Weyl's nontrivial two-component
equations, but having noted their noninvariance with respect to spatial reflection, remarked that "as a consequence of this, they cannot be applied to
physical objects" ([389], p. 254 of the Russian translation).

In May 1930, Weyl gave his lecture "Geometry and physics" at Cambridge;
it was subsequently published in the *Naturwissenschaften* [17]. He spoke
very strongly against the program of unified geometrized field theories with
its neglect of quantum theory, although it was precisely in the framework of
this program, as we have seen, that the various aspects of gauge symmetry
and the concept of gauge fields had been discovered and developed:

> In my opinion, the entire situation has been completely changed during the
> last four or five years by the discovery of the matter field [i.e., the quantum

mechanics of microscopic particles described by wave functions ψ]. All these geometrical pirouettes (*Luftsprünge*) were premature, and we return to the solid ground of physical facts. ([17], p. 58)

Now that the gauge nature of the electromagnetic field had been clarified, one should not seek a unified geometrical description of gravitation and electromagnetism, continued Weyl;

the electromagnetic field follows the ship of matter and not gravitation as wake ... if we are to speak of geometrization, then ... we must proceed from geometrization of the matter field; if this can be done, the electromagnetic field will be obtained as a bonus. [Ibid.]

Thus, the new principle of gauge invariance, compared to the gauge symmetry of Weyl's old theory, was the strongest argument for him against unified geometrized field theories and an important argument for the fundamental nature of the quantum theoretical approach.

Weyl also understood that the gauge nature of the electromagnetic field was associated with the dual interpretation of the electric charge as a conserved quantity and as a field source. Already in his unified theory of 1918 he showed that the law of charge conservation followed from gauge symmetry. He also obtained this conclusion for the new gauge symmetry depending on an arbitrary function of space-time. From the point of view of Noether's theorems on invariant variational problems, however, conservation laws are associated with finite-parameter continuous groups. In particular, the charge conservation law is a consequence of invariance of the action under a gauge transformation of the wave functions with constant phase. This fact was clearly formulated in Heisenberg and Pauli's paper "On the quantum theory of wave fields" published in 1930 [390]. In this paper, they demonstrated the important heuristic role of local gauge symmetry in the construction of a correct quantum field theory. In a number of places in this paper, we find assertions of the following type: "Nevertheless, it appears justified to assume that only quantities invariant with respect to (13) [i.e., a local gauge transformation] have physical meaning," or "Since directly measurable physical quantities are already gradient invariant [i.e., gauge invariant], etc." [390]. Therefore, Weyl was completely justified in saying in his Cambridge lecture that

for the further development of quantum theory, this new gauge invariance ... evidently has an even greater importance, as will be shown in a paper of Heisenberg and Pauli on the quantization of field equations. ([17], p. 58)

The further development of quantum field theory confirmed this early evaluation of the importance of gauge transformations.

At the end of the 1970s, Yang spoke of three arguments for the gauge conception [8]. We have already discussed two of them (in the framework of electrodynamics). They were the requirement of a local nature of gauge transformations and the associated dual interpretation of charge as a conserved quantity and a field source. The third argument was based on the notion of nonuniqueness or nonintegrability of the phase of wave functions. This means that in the presence of an electromagnetic field the gauge factor connecting wave functions at certain points A and B,

$$\exp\left[-i(e/hc)\int_A^B A_\mu \, dx_\mu\right],$$

depends on the path of integration; in other words, the total change of phase in a passage around a closed contour is not equal to zero. These arguments were first developed by Dirac in his classic monopole paper (1931) [391]. As was already known from Weyl's papers and those of Fock and Ivenenko, Dirac emphasized:

> ... we must have the wave function ψ always satisfying the same wave equation, whether there is a field or not, and the whole effect of the field when there is one is in making the phase non-integrable.... This gives a physical meaning to our non-integrability of phase

and relates it to the gauge conception. Dirac further noted:

> The connection between non-integrability of phase and the electromagnetic field ... is not new, being essentially just Weyl's Principle of Gauge Invariance in its modern form [Dirac is referring to Weyl's 1929 paper published in the *Zeitschrift für Physik* that we considered above].... It is also contained in the work of Ivanenko and Fock, who consider a more general kind of non-integrability based on a general theory of parallel displacement of half-vectors [i.e., spinors]. ([391], pp. 65–66)

By means of a remarkably simple argument using this aspect of local gauge invariance, Dirac arrived at his ideas about a monopole. In the middle of the 1970s, Yang and Wu wrote that "electromagnetism is the gauge invariant manifestation of a non-integrable phase factor" ([8], p. 95).

Thus, by the beginning of the 1930s, the principle of local gauge invariance and the associated gauge concept of the electromagnetic field had entered organically into the arsenal of theoretical tools of quantum theory, and their "illegitimate" origin on the ground of the unification program was soon forgotten. The contribution of this program, remember, (and above all the unified theories of Weyl and Kaluza and Klein, together with Einstein's theory with absolute parallelism) was decisive in the development of the gauge ideas.

ON THE PATH TO YANG–MILLS THEORY

As we have seen, the principle of local gauge symmetry, and to a certain extent the gauge concept of the electromagnetic field, were the center of attention for theoreticians at the end of the 1920s and the beginning of the 1930s (Weyl, 1928–1931; Fock and Ivanenko, 1929; Heisenberg and Pauli, 1929–1930; Dirac, 1931, etc.). The first monograph containing a clearly formulated principle of local gauge invariance was the first edition of Weyl's book *Gruppentheorie und Quantenmechanik* (1928) [384]. In the second edition, published in 1931 in German and English, the above papers of Weyl and Fock of 1929 were already taken into account [392]. In 1933, Pauli's famous article on quantum mechanics, including relativistic theory and the foundations of quantum electrodynamics, was published in Vol. 24 of the *Handbuch der Physik* [389]. In this remarkable monograph review, which even today has scientific value, Pauli considered the principle of local gauge invariance in several parts of the article (see pp. 52–53, 264–266, 317–318, and elsewhere in the Russian edition of 1947). In fact, he even gave a short historical review of the principle, referring to the papers by Fock, London, and Weyl. At the same time, Pauli did not especially concentrate attention on the gauge concept of the electromagnetic field, although it was undoubtedly contained in implicit form in his article.

From this time on, the development of gauge ideas took place almost entirely in the framework of quantum electrodynamics and the quantum field theory of elementary particles. The year in which Pauli wrote his review, 1932, was a turning point in the development of physics. One could say that it was in this year that the physics of elementary particles became the physics of three fundamental interactions—electromagnetic, weak, and strong. Theoretical studies by Fermi, Dirac, and Fock, and the discovery of the positron by Anderson, completed the creation of quantum electrodynamics. Chadwick's discovery of the neutron and the subsequent introduction of isospin by Heisenberg created the basis for the construction of a theory of strong interactions; Fermi developed the first theory of weak interactions.

After quantum mechanics, this was the second and decisive blow to the program of unified geometrized field theories, which had aimed in the first place to find a geometrical unification of only the electromagnetic and gravitational fields. This first period of development of the physics of elementary particles and their interactions, which was also associated with subsequent papers by Heisenberg, Wigner, Yukawa, Pauli, Dirac, Fermi, Tamm, Weisskopf, Proca, Kemmer, Landau, and others, was reviewed in a further paper by Pauli published in the *Reviews of Modern Physics* in 1941; it too became a classic [362]. In this paper, which is distinguished by a unified approach to the theoretical description of elementary particles based on the quantum field concept, the variational principle, and the principles of invariance and conservation laws, ample attention is devoted to gauge invariance. It was here that Pauli introduced the concepts of gauge transformations of the first and second kind, clearly demonstrated once more the connection between the law of charge conservation and global gauge transformations of the first kind, and emphasized the possible existence of other charge conservation laws and associated gauge symmetries. The review also clearly emphasized, in the spirit of the gauge concept of the electromagnetic field, that this field could be introduced by going over the "compensating" derivative,

$$D_k = \frac{\partial}{\partial x_k} - ie\varphi_k,$$

and it was noted that this was equivalent to the requirement of local gauge invariance. This 30-page review by Pauli, like his previous review on quantum mechanics, was the point of departure for all those who at the end of the 1940s and beginning of the 1950s began to develop the generalized concept of gauge fields (Yang, Mills, Lee, Utiyama, Sakurai, Schwinger, and others).

We are grateful to professor L.B. Okun for pointing out a fundamental paper by O. Klein (1938), in which he gives the first non-Abelian gauge field theory (O. Klein, "On the Theory of Charged Fields," in *New Theories in Physics* a conference organized in collaboration with the International Union of Physicists, Polish Intellectual Cooperation Committee, Warsaw, May 30–June 3, 1938). Klein's theory was developed in the frame of the five-dimensional approach and unified gravitational, electromagnetic, and nuclear interactions. This theory was forgotten until it was pointed out by C. Jarlsckog in 1981 (*Physica Scripta*, 1981, Vol. 24, pp. 867–872).

In 1943 the first textbook on the quantum theory of elementary particles and quantum field theory was published—the book by Wentzel [393]. With regard to gauge symmetry, Wentzel's textbook contained nothing more than the earlier review by Pauli; most attention was devoted to the formal aspects of the requirement of gauge symmetry; the gauge nature of the electromagnetic field remained in the background.

At the end of the 1940s and beginning of the 1950s, the requirement of gauge invariance was often applied as an important formal device in a number of papers on quantum electrodynamics, including those of Feynman, Tomonaga, and Schwinger, who were subsequently awarded the Nobel Prize. In the monographs and textbooks at the beginning of the 1950s the gauge concept of the electromagnetic field was not put in the foreground and was not described sufficiently clearly, although Pauli's reviews discussed above were regarded as constant references for all theoreticians working in the field of elementary particle physics and quantum field theory. A year before Yang and Mill's paper, Schwinger, in the second paper of his series entitled "The theory of quantized field" (1953) [394], wrote of the gauge nature of the electromagnetic field as a generally known interpretation. In those years, Schwinger's name was well known to all those concerned with problems of quantum field theory and elementary particles. In 1956, these papers were translated into Russian and published as a separate book. Schwinger wrote:

> The postulate of general [i.e., local] gauge invariance motivates the introduction of the electromagnetic field. If all fields and sources are subjected to the general gauge transformation
>
> $$'\chi = \exp\left(-i\lambda(x)\varepsilon\right)\chi,$$
>
> the Lagrange function we have been considering alters in the following manner:
>
> $$'\mathcal{L} = \mathcal{L} + j_\mu \partial_\mu \lambda.$$
>
> The addition of the electromagnetic field Lagrange function... provides a compensating [note, thus, that the characteristic term "compensation" did not occur for the first time, in the context considered in the paper by Yang and Mills but in the paper by Schwinger] quantity through the associated gauge transformation $'A_\mu = A_\mu - \partial_\mu \lambda$. ([394], Part II, p. 724)

It is noteworthy that Schwinger called Pauli's approach in his 1933 review the "closest in spirit to our procedure." It does not appear to have been by chance that Schwinger was one of the first to become involved in the development of gauge field theory and made an important contribution to the solution of the problem of massive gauge fields. In this connection, a passage in Sakurai's first paper on the theory of gauge field warrants consideration. Having in mind the generalized gauge concept, he wrote:

> This is a very profound idea—perhaps the most profound idea in theoretical physics since the invention of the Dirac theory. It essentially states that, if we have a conservation law of some internal attribute, there must necessarily exist a vector-type interaction corresponding to it in order that the conservation law in question be consistent with the concept of localized fields.

Sakurai then adds: "To borrow Schwinger's words, internal attributes should have 'dynamical manifestations'" ([357], p. 9).[9]

---

[9] Another pioneer of this concept at the end of the 1940s and beginning of the 1950s was, as Sakurai emphasized, Wigner.

The fundamental laws of electrodynamics, both classical and quantum, were essentially formulated without use of the gauge concept of the electromagnetic field. In any case:

> It is supererogatory to observe that the photon was not discovered by requiring local gauge invariance. Rather, gauge transformations were discovered as a useful property of Maxwell's equations,

as Abers and Lee wrote in their review of gauge theories [395]. It is true that the gauge treatment of the electromagnetic field developed in the 1920s and 1930s by Weyl Fock and then supported by Pauli and Dirac made it possible to deepen significantly the connections with other fields and elementary particles. The main importance of this treatment, however, as became clear in the middle of the 1950s and at the start of the 1960s, was the possibility of generalizing it to the various internal symmetries of the elementary particles. This time the gauge concept did work, and on its basis both the unified theory of the electromagnetic and weak interactions of Weinberg and Salam and quantum chromodynamics, which is the modern theory of the strong interactions, were developed.

Thus, there was a kind of "inversion" of the situation, namely, having been constructed on the basis of the gauge interpretation of the already existing theory of the electromagnetic field, in which this interpretation did not play an important part, the generalized concept of gauge fields proved to be fruitful beyond expectation. This kind of inversion is a very characteristic feature in the development of scientific knowledge. This is why, in particular, it is helpful and important to seek different equivalent reformulations of fundamental theories.

Why did the generalized gauge conception arise so late, i.e., 25–30 years after its formulation in electrodynamics? The main reason is probably the fact that it was only at the beginning of the 1950s that sufficiently extensive empirical material on the interactions of elementary particles had been accumulated and partly systematized in terms of conservation laws and associated internal symmetries.[10] Moreover, in the 1930s and 1940s, despite the successes of

---

[10] Yang himself wrote in 1977:

> With the discovery of many new particles after World War II, physicists explored various couplings between the "elementary particles" ... the desire to find a *principle to choose among the many possibilities* [of theoretical description of the resulting situation] was one of the motivations for an attempt to generalize Weyl's gauge principle for electromagnetism. ([8], p. 86)

the invariance principles and group-theoretical methods, physicists initially attempted to solve many of the problems of the incipient meson dynamics purely dynamically. A negative role may also have been played by the skeptical attitude to the Lagrangian formalism in quantum field theory that arose at the beginning and middle of the 1950s.

It is interesting that the discovery of gauge symmetry and the gauge nature of the electromagnetic field, which initially appeared formal and physically somewhat unimportant aspects of electrodynamics, showed that many fundamental physical theories and concepts were related in one way or another: the general theory of relativity; unified geometrized field theories (Weyl's theory, the five-dimensional approach, the theory with absolute parallelism, etc.); quantum mechanics (both nonrelativistic and relativistic); quantum field theory; Noether's theorems, which related conservation laws to invariance principles; and also important methodological ideas and principles of physics—the principles of observability, locality, symmetry, etc. Many apparently unrelated theoretical developments came together in the gauge concept, as at focus. A quarter of a century later, this concept, suitably generalized, gradually came to the forefront of the development of fundamental physics, and in the period from the 1960s to the 1980s became the basis for the construction of a unified quantum field theory of the interaction of elementary particles. Thus, having arisen at the intersection of the main lines of development of physics, the gauge concept has repaid its debt to physics and become the main tool in the further synthesis of physical knowledge.

Moreover, the adequate mathematical formalism of the theory of gauge fields is one of the most beautiful and powerful of mathematical theories— the theory of fiber bundles [7]. The fields that characterize the particles, or sources (for example, electrons), are described by sections of a fiber bundle. The gauge fields (photons, for example) are described by the connection of the fiber bundle. The internal symmetry whose localization "generates" the gauge field (for example, a gauge symmetry of the first kind) is the symmetry group of the fiber. The base of the fiber bundle is four-dimensional space-time. The "compensating" derivative, by means of which the transition to the local symmetry and the interaction of the original field with the gauge field is taken into account in the Lagrangian of the field and in the field equations, is identified with the covariant derivative in the fiber bundle. A special case of fiber bundles is provided by spaces with connections (for example, spaces with affine, Euclidean, conformal, projective, and other connections). The development of the geometry of these spaces was stimulated by the development of the general theory of relativity and the unified geometrized field theories. A decisive contribution to the development of these geometries in the 1920s and 1930s was made by Weyl, Cartan, Schouten, and others. It is very interesting that the concept of a fiber *space*, equivalent to the constructions with Yang–Mills fields, was developed at the beginning of the 1950s by the mathematician C. Ehresmann, only a few years before the publication of Yang and Mill's paper.

SUMMARY

The prolonged regression of the program of unified geometrized field theories against the background of the brilliant successes of the quantum theoretical program, which was already apparent in 1926 (when the foundations of quantum mechanics were laid), became completely obvious from the beginning of the 1930s (when nuclear physics and the physics of elementary particles commenced). A widespread belief was subsequently generated that the entire unification program and, in particular, Einstein's own efforts for more than 30 years in this direction, had been of no value, without prospects, and led to a dead end.

The program aim of the two competing global strategies of fundamental physics in the 1920s and 1930s appeared to be diametrically opposed, although the quantum theoretical program, the outlines of which were somewhat obscure before 1925–1926, acquired really clear contours only after the construction of quantum mechanics. There is no doubt that the presence of competing programs must positively influence each of them—under conditions of competition, each program mobilizes its resources to the maximum. From this point of view, the very existence of the unification program was an important factor in the development of quantum theory.

There were other, more constructive aspects of the interaction of these programs, in particular the influence of the geometrized field program on investigations in the domain of quantum physics. Within the broad unification program, there was a very extensive development, admittedly somewhat separated from the lifeblood of experimental physics, of the abstract theoretical and mathematical aspects of the field concept (at the classical level), space-time, relativity, invariance, etc. The quantum theoretical program too, particularly when its main direction shifted to the theory of quantum fields, did not reject the use (at times implicit) of the ideas and methods that have been developed within the unification program.

As a result, the progressive program (in our case, the quantum theoretical program) gained, as a rule. Schrödinger's attempts to interpret quantum conditions in the language of Weyl's theory led to his prototype of a wave function in 1922, and in all probability this influenced the genesis of wave mechanics in Schrödinger's papers in 1926. Although one cannot detect an influence of Schrödinger's ideas of 1922 on the investigations of de Broglie, there is every reason to believe that the successful acceptance and further development of de Broglie's work by Schrödiner were largely related to the 1922 paper. Thus, the unification program had a heuristic influence on the formation of quantum mechanics. Of course, the effect of this influence would have been more impressive if Schrödiner paper had explicitly influenced de Broglie and if Schrödiner himself, already in 1922–1923, had on the basis of the "resonance property" of the quantum orbits established by him, discovered de Broglie waves, and on their basis wave mechanics.

Clear evidence of the fruitful influence of the unification program on the development of quantum theory includes the following: the discovery of the scalar relativistic wave equation for spinless particles (the Klein–Gordon, or Klein–Fock, equation) through the study of the important connection between quantum mechanics and five-dimensional Kaluza–Klein theory; the proof of charge invariance of any field theory of elementary particles in the development of the affine form of unified field theory; and the development of the ideas of gauge symmetry, and also the gauge nature of physical fields (in direct connection with the investigations on the unified theories of Weyl, Kaluza and Klein, Einstein, etc.).

The final example is more than an illustration and has independent importance. The fact is that the modern period of development of quantum field theory, and simultaneously the modern approach to the unification of the four fundamental physical interactions, i.e., the modern approach to Einstein's problem of a unified field theory, are essentially "gauge" approaches. As we have seen, the concept of gauge symmetry and the gauge field concept (for the example of the gravitational field, and especially the electromagnetic field) were largely developed in the framework of the problem of unified geometrized field theories and through the establishment of the connections between unified field theories and quantum mechanics. From the middle of the 1950s, the gauge field concept began to play an ever greater role in the theory of elementary particles and soon established itself as the main theoretical means for unifying the fundamental physical interactions.

Concluding Remarks

Having considered the conclusions drawn in each chapter, and comparing them with what was written in the Introduction, readers will themselves be able to judge the extent to which the author has accomplished his task. There is no point in completely summarizing these conclusions. Nevertheless, as a general summary, we shall make some concluding remarks.

The unified geometrized field theories were connected successfully to the unified electromagnetic field theories, for it was in the framework of the electromagnetic field program that the "field ideal of unity" was formulated. The general theory of relativity, which adopted this ideal, added to it the idea of geometrization of physical interaction (the field). Attempting to realize the "field ideal of unity" of physics, Hilbert tried to unify the concept of an electromagnetic-field structure of matter (Mie) with the idea of field geometrization (Einstein). This method of unification, though—developed in an axiomatic framework and based on Noether's second theorem on invariant variational problems—was extremely formal and physically incorrect.

The theory that provided the basis for all subsequent attempts to realize the geometrical field synthesis of physics was created (in 1918) by Weyl, a graduate of Göttingen and a student of Hilbert. This was indeed a unification of the gravitational and electromagnetic fields on a unified geometrical basis, entirely in the spirit of the general theory of relativity, whose main creator was for several years a serious opponent of the Göttingen theories. Despite the serious physical difficulties that faced Weyl's theory, at the beginning of the 1920s it was regarded as an extremely bold and promising direction.

The program nature of Weyl's theory was fully manifested only in 1921, when several unified theories constructed after the manner of Weyl's arose. The authors of these theories—Eddington, Kaluza, Einstein—directly relied on Weyl's construction, which was for them a paradigm. At this time Einstein himself joined the new program of unified geometrized field theories. This very theoretical and highly mathematical program, which aimed not only at a unified description of gravitation and electromagnetism (minimum problem) but also at the recovery from the field equations of particle-like solutions possessing quantum properties (maximum problem), was opposed by the quantum theoretical program, which was theoretically less clear and developed, at least at the beginning of the 1920s, but much more intimately related to extensive experimental material in the fields of atomic, electron, and quantum physics. The quantum program was based essentially on the quantum notions of Planck, Bohr, and Einstein; it questioned the causality of the primary physical structures, classical causality, and the idea of geometrization of the basic physical entities. Between the programs there existed complex interrelationships that by no means always reduced to a competitive struggle.

From 1923, the role of leader of the unification program was taken by Einstein, who concentrated his efforts more and more on the investigation of unified geometrized field theories, in contrast to Weyl, Pauli, and Eddington, who gradually moved away from it or actively developed more concrete problems of physics. Therefore, we have related the subsequent history of the unification program with the work of Einstein. As a result, aiming to capture the basic directions in the development of the geometrized field program, we have deliberately not made an exhaustive survey of unified field theories, which were advanced in great numbers in the 1920s and 1930s. Following Einstein, we have taken the pulse of this program, its flights of inspiration and disappointments throughout an entire heroic decade. In the course of this decade quantum mechanics took shape and established itself firmly (after which the quantum theoretical program acquired clear outlines), the foundations of quantum electrodynamics were laid, and the first serious developments in the physics of strong and weak interactions appeared. Einstein and others rejected, one after the other, no fewer than ten unified field theories: the various forms of affine, affine-metric, and five-dimensional theories, theories with torsion and absolute parallelism, etc. If the minimum problem was, with important reservations, solved after a fashion in some of these schemes, essentially no progress at all was made on the maximum problem. For this, Einstein assumed, the field equations of the theory had to contain nonsingular spherically symmetric static solutions that could be identified with particles, for example the electron. The absence of such solutions was always a main reason for abandoning a particular theory, at least for Einstein, who consistently strove to solve the maximum problem. The unification program had its first crisis soon after the creation of quantum mechanics. The beginning of the physics of strong and weak interactions (1932) was simultaneously associated with the onset of a new stage of steady regression of Einstein's program. From that time on, investigations of unified geometrized field theories became peripheral.

The widely held view that the unification program, and in particular the efforts of Einstein to realize it during more than three decades, had been completely fruitless requires important corrections. The very existence of programs that are competing with a progressive program, say the quantum theoretical program, is a valuable stimulating factor for it. But this was not all. As we have seen, many important theoretical aspects and mathematical methods of key concepts of modern physical theory—fields, space-time, relativity, symmetry—were developed and refined, admittedly at the classical level, in the framework of unified field theories. Therefore when, following the development of quantum mechanics, the relativistic theory of elementary particles and quantum field theory came to fore, the methods, structure, and concepts created in unified field theories, or at least some of them, were taken up as tools by the quantum theoretical program. As a rule, the insufficiently formulated but theoretically more flexible quantum program gained from its contacts with the unification program. In this respect, the attempts to relate quantum theory to unified geometrized field theories had great heuristic value. We recall the part played by Weyl's theory in the genesis of Schrödinger's wave mechanics and the establishment of the relativistic scalar wave equation in connection with five-dimensional theories.

There are many typical examples of the fruitful heuristic influence of the program of unified geometrized field theories on the relativistic theory of gravitation and quantum field theory: Einstein's proof of the charge symmetry of any unified theory of the electromagnetic and gravitational fields (in the framework of studies of affine and affine-metric theories), the development of the problem of deriving the equations of motion from field equations, and the development of a more flexible variational formalism (the Palatini method) are just a few.

We have seen that there exists an organic connection between modern unified field theories based on the gauge concept and the development of unified geometrized field theories in the 1920s and 1930s. The important thing is not so much the common aim as the fact that the ideas of gauge symmetry and the actual gauge field concept were developed initially to a large degree in the framework of investigations of unified geometrized field theories. One could imagine that the experience gained from the study of nonlinear generally covariant field equations by the investigators of unified geometrized field theories using the refined geometrical methods and the gauge philosophy will also be helpful in the future.

It is now clear that the original aim of the unification program, to reduce reality to one of the possible generalized geometries of space-time, could not lead to success because it did not take into account the quantum structure of this reality. At the same time, the geometrical field approach, together with the idea of unification of the four basic interactions, takes its revenge in the modern gauge concept of quantum field theory. As Manin put it figuratively, even now "geometry acts as a preservative for perishable physics" and "the geometrical ideas of Riemann, Einstein, Hermann Weyl, and Elie Cartan are still at work in fundamental physics" ([396], pp. 15–16).

Our work by no means exhausts the problem of studying the history of unified field theories. It should be extended in the following directions: a more complete account of the theories of the 1920s and 1930s (especially the mass of theories with nonzero torsion), the study (or at least, historical survey) of the theories from the 1930s to the 1950s, a more complete analysis of the scientific communicative aspects, and also a study of the influence of investigations into unified theories on the development of differential geometry.

Bibliography

[1] A. Salam, "Gauge unification of fundamental forces," *Rev. Mod. Phys.*, *52*, 525–538 (1980) (Lectures given December 8, 1979, at presentation of 1979 Nobel Prizes in Physics, Russial translation in: *Na Puti k Eginoi Teorii Polya, Znanie*, Moscow (1980), pp. 3–35).

[2] W. Pauli, *The Theory of Relativity*, Pergamon Press, Oxford (1958), "Supplementary Notes," pp. 207–232 (Russian translation in: *Teoreticheskaya Fizika XX Veka (Theoretical Physics in the 20th Century)* (ed. Ya. A. Smorodinskii), Izd-vo Inos. Lit., Moscow (1962), pp. 419–431).

[3] A.Z. Petrov, "Some considerations on 'unified' field theories," in: *Gravitatsiya i teoriya otnositel'nosti (Gravitation and the Theory of Relativity)*, Izd-vo Kazan. Un-ta, Kazan' (1965), No. 2, pp. 3–8.

[4] D.D. Ivanenko (ed.), *Nelineinaya kvantovaya teoriya polya (Nonlinear Quantum Field Theory*, Russian translations), Izd-vo Inostr. Lit., Moscow (1959).

[5] W. Heisenberg, *Einführung in die einheitliche Feldtheorie der Elementarteilchen*, Hirzel, Stuttgart (1967) (English translation: *Introduction to the Unified Field Theory of Elementary Particles*, Wiley (1966); Russian translation published by Mir, Moscow (1968)).

[6] W. Heisenberg, "Comments on Einstein's sketch of a unified field theory," Russian translation in *Einshtein i Razvitie Fiziko-Matematicheskoi Mysli (Einstein and the Development of Physical and Mathematical Thought)* (ed. A.T. Grigor'yan), Izd-vo Akad. Nauk SSSR, Moscow (1962), pp. 63–68.

Translator's note: The bibliography in the original Russian lists a great many articles and books in their Russian translations; the originals are seldom given. In nearly all cases, I have managed to locate the originals (mostly in German and English) and give them first, followed by brief details of the Russian translations. To avoid repeated complete references to the four-volume Russian collection of Einstein's scientific works (from which over 70 papers are cited—I fortunately possess a copy of the Russian translations), I have added a further reference [397] giving the details of this Soviet publication. In the bibliography that follows, I give the original references for Einstein's papers, followed by the details in [397]. See also my note in the introductory material at the front of the book.

[7] N.P. Konopleva and V.N. Popov, *Kalibrovochnye Polya (Gauge Fields)*, 2nd ed., revised and reworked, Atomizdat, Moscow (1980), 239 pp.

[8] C.N. Yang, "Magnetic monopoles, fiber bundles and gauge fields," *Ann. Acad. Sci.*, New York, *294*, 86–97 (1977).

[9] Ya.B. Zel'dovich, "Albert Einstein, his time and creation," *Priroda*, No. 3, 5–7 (1979).

[10] W. Pauli, "Relativitätstheorie," in: *Encyklopädie der mathematischen Wissenschaften*, Vol. V19, Teubner, Leipzig (1921) (Russian translation publ. by Gostekhteorizdat, Moscow-Leningrad (1947). For English translation, see Ref. 63).

[11] A.S. Eddington, *The Mathematical Theory of Relativity*, Cambridge University Press, Cambridge (1923) (Russian translation published by ONTI, Moscow–Leningrad (1934).)

[12] P.G. Bergmann, *Introduction to the Theroy of Relativity*, Prentice-Hall, Englewood Cliffs (1942) (Russian translation published by Izd-vo Inostr. Lit., Moscow (1947).)

[13] Yu.B. Rumer, *Issledovaniya po 5-optike (Investigations into 5-Optics)*, Gostekhteorizdat, Moscow (1956), 152 pp.

[14] A.Z. Petrov, "Geometry and physical space–time," in: *Algebra, Topologiya, Geometriya. 1966 (Algebra, Topology, Geometry. 1966). Itogi nauki (Reviews of Science)*, VINITI, Moscow (1968), pp. 221–265.

[15] G. Beck, "Allgemeine Relativitätstheorie," in" *Handbuch der Physik*, Vol. 4, Springer, Berlin (1929), pp. 381–408.

[16] A. Landë, "Optik, Mechanik und Wellenmechanik," in: *Handbuch der Physik*, Vol. 20, Springer, Berlin, pp. 418–425.

[17] H. Weyl, "Geometrie und Physik," *Naturwissenschaften*, *19*, 49–58 (1931).

[18] H. Weyl, "50 Jahre Relativitätstheorie," in: H Weyl, *Gesammelte Abhandlungen*, Vol. 4, Springer, Berlin (1968), p. 421–431.

[19] E. Schrödinger, *Space-Time Structure*, Cambridge University Press, Cambridge (1959), 119 pp.

[20] P. Jordan, *Schwerkraft und Weltall*, Vieweg, Braunschweig (1955).

[21] A. Lichnerowicz, *Théories relativistes de la gravitation et de l'electromagnétisme*, Masson et Cie, Paris (1955).

[22] M.A. Tonnelat, *Les théories unitaires de l'electromagnétisme et de la gravitation*, Gauthier-Villars, Paris (1959), 522 pp.

[23] E. Schmutzer, *Relativistische Physik*, Teubner, Leipzig (1968), 974 pp.

[24] See, for example: L. Pyenson, "History of physics," in: *Encyclopedia of Physics* (eds. R. G. Lerner, G. L. Trigg, etc.), Addison-Wesley (1981), pp. 404–414.

[25] I. Lakatos, "History of science and its rational reconstructions," in: R. C. Buck and R. S. Cohen (eds.): *P.S.A. 1970 Boston Studies in the Philosophy of Science*, Vol. 8, 91–135, Reidel, Dordrecht (Russian translation publ. in Ref. 26, pp. 203–269).

[26] B.S. Gryaznov and V.N. Sadovskii, "Problems of the structure and development of science in the boston Studies in the Phyilosophy of Science," in: *Struktura i Razvitie Nauki (Structure and Development of Science)*, Progress, Moscow (1978), pp. 5–39.

[27] P.P. Gaidenko, *Évolyutsiya Ponyatiya Nauki. Stanovlenie i Razvitie Pervykh Na-uchnykh Programm (Evolution of the Concept of Science. The Establishment and Development of the First Scientific Programs)*, Nauka, Moscow (1980), 568 pp.

[28] N.I. Rodnyi, *Ocherki po Istorii i Metodologii Estestvoznaniya (Essays on the History and Methodology of Natural Science)*, Nauka, Moscow (1975), 424 pp.

[29] I.S. Alekseev, "Unity of the physical picture of the world as a methodological principle," in: *Metodologicheskie Printsipy Fiziki. Istoriya i Sovremennost' (Methodological Principles of Physics. History and the Present Day)* (eds. B. M. Kedrov and N. F. Ovchinnikov), Nauka, Moscow (1975), pp. 128–203.

[30] V.S. Stepin, *Stanovlenie Nauchoi Teorii (The Genesis of Scientific Theory)*, BGU, Minsk (1976), 320 pp.

[31] M.J. Klein, "Mechanical explanation at the end of the nineteenth century," *Centaurus, 17*, 58–82 (1972).

[32] S. Goldberg, "Max Planck's philosophy of nature and his elaboration of special theory of relativity," *Hist. Stud. Phys. Sci., 7*, 125–160 (1976).

[33] T. Hirosige, "The ether problem, the mechanistic world view and the origins of the theory of relativity," *Hist. Stud. Phys. Sci., 7*, 3–82 (1976).

[34] R.H.A. McCormmach, "Lorentz and the electromagnetic view of nature," *Isis, 61*, 459–497 (1970).

[35] B.G. Doran, "Origins and consolidation of field theory in 19th century Britain: From the mechanical to the electromagnetic view of nature," *Hist. Stud. Phys. Sci., 6*, 133-260 (1975).

[36] *Metodologicheskie Printsipy Fiziki. Istoriya Sovremennost' (Methodological Principles of Physics. History and the Present Day)* (eds. B.M. Kedrov and N.F. Ovchinnokov), Nauka, Moscow (1975), 512 pp.

[37] V.P. Vizgin, "Methodological principles and scientific research programs," in: *Metodologicheskie problemy Istoriko-Nauchnykh Issledovanii (Methodological Problems of Investigations in the History of Science)* (ed. I.S. Timofeev), Nauka, Moscow (1982), pp. 172–197.

[38] V.P. Vizgin, *Relyativistskaya Teoriya Tyagoteniya: Istoki i Formirovanie (The Relativistic Theory of Gravitation: Sources and Development, 1900–1915)*, Nauka, Moscow (1981), 352 pp.

[39] T. Hirosige, "Electrodynamics before the theory of relativity, 1890–1905," *Jpn. Stud. Hist. Sci.*, No. 5, 1–49 (1966).

[40] J. Illy, "Revolutions in a revolution," *Stud, Hist. Philos. Sci., 12*, 175–210 (1981).

[41] M. Jammer, *Concepts of Mass in Classical and Modern Physics*, Harvard University Press, Cambridge, Mass. (1961) (Russian translation published by Progress, Moscow (1967)).

[42] P.S. Kudryavtsev, *Istoriya Fiziki (The History of Physics)*, Vol. 2, Uchpedgiz, Moscow (1956), 488 pp.

[43] H.A. Lorentz, *Ergebnisse und Probleme der Elektronentheorie*, Berlin (1905) (Russian translation published by Obrazovanie, St. Peterburg (1910)).

[44] O.J. Lodge, "On electrons," *J. Inst. Elect. Eng., 32* 45–115 (1903) (Russian translation as book published by Izd. O.N. Popovoi, St. Petersburg (1904)).

[45] M.K.E.L. Planck, *Acht Vorlesungen über theoretische Physik,* Hirzel, Leipzig (1910) (English translation: *Eight Lectures on Theoretical Physics,* Columbia University Press, New York (1915). Russian translation published by Obrazovanie, St. Petersburg (1911)).

[46] N.R. Campbell, "The aether," *Philos. Mag., 19,* 181–191 (1910) (Russain translation published in: *Novye Idei v Fizike: Éfir i Materiya,* No. 2, St. Petersburg (1913), pp. 107–124).

[47] D.A. Gol'dgammer (Goldhammer), "New ideas in modern physics," *Fizicheskoe Obozrenie, 12,,* 1–35 (1911).

[48] G.A.F.W.L. Mie, *Moleküle, Atome, Weltäther,* Teubner, Leipzig (1904) (Russian translation published by Izd. P. P. Soikina, St. Petersburg (1913)).

[49] Quoted from: I.Yu. Kobzarev, "Einstein, Planck, and atomic theory," *Priroda,* No. 3, 8–26 (1979).

[50] A. Einstein, "Zu Theorie der Lichterzeugung und Lichtabsorption," *Ann. Phys. 20,* 199–206 (1906) (Russian translation: [397], Vol. 3, pp. 128–133 (1966)).

[51] R. McCormmach, "Einstein, Lorentz and the electron theory," *Hist. Stud. Phys. Sci., 2,* 41–87 (1970).

[52] A. Einstein, "Zum gegenwärtigen Stand des Strahlungsproblems," *Phys. Zs., 10,* 185–193 (1909) (Russian translation: [397], Vol. 3, pp. 164–179 (1966)).

[53] K. Longren and A. Scott (eds.), *Solitons in Action,* Academic Press, New York (1978) (Russian translation published by Mir, Moscow (1981)).

[54] R.K. Billough and P.J. Caudrey (eds.), *Solitons,* Springer, Berlin (1980) (Russian translation published by Mir, Moscow (1983)).

[55] A. Einstein, "Über die Entwicklung unserer Anshauungen über das Wesen und die Konstitution der Strahlung," *Phys. ZS., 10,* 817–825 (1909) (Russian translation: [397], *3,* pp. 181–195 (1966)).

[56] P. Speziali, ed., *Albert Einstein–Michele Besso Correspondance 1903–1955,* Hermann, Paris (1972) (Partial translation into Russian in: *Einshteinovskii Sbornik, 1974,* Nauka, Moscow (1976), pp. 5–112).

[57] C. Seelig, *Albert Einstein; eine dokumentarische Biographie,* Europa Verlag, Zürich (1954) (English translation: *Albert Einstein: A Documentary Biography,* Staples Press, London (1956). Russian translation published by Atomizdat, Moscow (1964)).

[58] G.A.F.W.L. Mie, *Lehrbuch der Elektrizität und des Magnetismus,* Enke, Stuttgart (1910) (Russian translation published by Matezis, Odessa (1914)).

[59] G. Mie, "Die Grundlagen einer Theorie der Materie," *Ann. Phys., 37,* 511–534 (1912).

[60] G. Mie, "Die Grundlagen einer Theorie der Materie," *Ann. Phys., 40,* 1–66 (1913).

[61] G. Mie, "Die Grundlagen einer Theorie der Materie," *Ann. Phys., 39,* 1–40 (1912); *40,* 1–66 (1912).

[62] H. Weyl. *Raum, Zeit, Materie,* Springer, Berlin (1918), 227 pp.

[63] W. Pauli, *The Theory of Relativity,* Pergamon Press, Oxford (1958) (translation of [10] with supplementary notes).

[64] H. Weyl, *Raum, Zeit, Materie,* 5th ed., Springer, Berlin (1923), 338 pp.

[65] D.D. Ivanenko and A.A. Sokolov, *Klassicheskaya Teoriya Polya: Novye Problemy (Classical Field Theory: New Problems)*, Gostekhteorizdat, Moscow–Leningrad (1949), 432 pp.

[66] A.A. Sokolov and D.D. Ivanenko, *Kvantovaya Teoriya Polya (Izbrannye Voprosy) (Quantum Field Theory (Selected Problems))*, Gostekhteorizdat, Moscow–Leningrad (1952), 780 pp.

[67] A.J.W. Sommerfeld, *Elektrodynamik*, Dieterich, Wiesbaden (1948) (English translation: *Electrodynamics*, Academic Press, New York (1952). Russian translation published by Izd-vo Inostr. Lit., Moscow (1958)).

[68] From the correspondence between Sommerfeld and Einstein in: A. Sommerfeld: *Paths to Understanding in Physics* [in Russian], Nauka, Moscow (1973), pp. 191–246. Russian translation of: A. Hermann, *Briefwechsel, 60 Breife aus dem goldenen Zeitalter der modern Physik von Albert Einstein und Arnold Sommerfeld*, Schwabe, Basel-Stuttgart (1968).

[69] L. Pyenson, "The Göttingen reception of Einstein general theory of relativity," Ph.D. Diss., John Hopkins University, Baltimore (1974).

[70] J. Ishiwara, "Zür Theorie der Gravitation," *Phys. Zs.*, *13*, 1189–1193 (1912).

[71] J. Ishiwara, "Die Grundlagen einer relativistischen und elektromagnetischen Gravitationstheorie," *Phys. Zs.*, *15*, 294–298, 506–510 (1914).

[72] G. Nordström, "Ueber die Möglichkeit das elektromagnetische Feld und das Gravitationsfeld zu vereinigen," *Phys. Zs.*, *15*, 504–506 (1914).

[73] G.T. Holton, *Thematic Origins of Scientific Thought: Kepler to Einstein*, Harvard University Press, Cambridge, Mass. (1973) (Russian translation published by Progress, Moscow (1981)).

[74] A. Einstein, "Die Feldgleichungen der Gravitation," *Sitzungsber. Preuss. Akad. Wiss.*, *48*, 2, 844–847 (1915) (Russian translation: [397], vol. 1, pp. 448–451).

[75] A. Einstein, "Zur allgemeinen Relativitätstheorie. (Nachtrag)," *Sitzungsber. Preuss. Akad. Wiss.*, *46*, 2, 799–801 (1915) (Russian translation: [397], vol. 1, pp. 435–438).

[76] A. Einstein, "Die Grundlage der allgemeinen Relativiätstheorie," *Ann. Phys. 49*, 769–822 (1916) (Russian translation: [397], vol. 1, pp. 452–504).

[77] A. Einstein, "Näherungweise Integration der Feldgleichungen der Gravitation," *Sitzungsber. Preuss. Akad. Wiss.*, *1*, 688–696 (1916) (Russian translation: [397], vol. 1, pp. 514–523).

[78] V.P. Vizgin, *Razvitie Vzaimosvyazi Printsipov Simmetrii s Zakonami Sokhraneniya v Klassicheskoi Fizike (Development of the Interconnections Between Symmetry Principles and Conservation Laws in Classical Physics)*, Nauka, Moscow (1972), 240 pp.

[79] D. Hilbert, "Die Grundlagen der Physik," *Nachrichten von der Kön. Ges. der Wissenschaften zu Göttingen, Math.–Phys. Kl. Heft*. 3, 395–407 (19150 (Russian translation in: *Variatsionnye Printsipy Mekhaniki (Variational Principles of Mechanics)*, ed. L.S. Polak, Fizmatgiz, Moscow (1959), pp. 589–598).

[80] D. Hilbert, "Mathematische Probleme," *Archiv. f. Math. u. Phys. 3. Reihe*. *1*, 44-63, 213–237 (1901) (*Gesammelte Abhandlungen*, Vol. 3, pp. 280–329) (Russian translation in: *Problemy Gil'berta (Hilbert's Problems)*, Nauka, Moscow (1969)).

[81] L. Pyenson, "Physics in the shadow of mathematics: The Göttingen electron-theory seminar of 1905," *Arch. Hist. Exact Sci., 21*, 55–89 (1979).

[82] E. Noether, "Invariante Variationsprobleme," *Nachrichten von der Kön. Ges. der Wissenschaften zu Göttingen, Math.–Phys. Kl.* Heft. 2, 235–258 (1918) (For Russian translation, see [79], pp. 611-630).

[83] A. Trautman, "Conservation laws in the general theory of relativity," in *Gravitation: Introduction to Current Research*, L. Witten, ed. J. Wiley, New York, London, 1962, pp. 169–198 (Russian translation in: *Einshteinovskii Sbornik, 1968*, Nauka, Moscow (1967), pp. 308–344).

[84] J. Earman and C. Glymour, "Einstein and Hilbert: Two months in the history of general relativity," *Arch. Hist. Exact Sci., 19*, 291–308 (1978).

[85] H. Weyl, "David Hilbert and his mathematical works," in: [106].

[86] H. Weyl, "Gravitation und Elektrizität," *Sitzungsber. Preuss. Akad. Wiss.*, 465–480 (1918) (English translation in: *The Principle of Relativity, A Collection of Original Memoirs by H. A. Lorentz, A. Einstein, H. Minkowski and H. Weyl*, Dover, New York (1952), pp. 200–216; Russian translation in: *Al'bert Éinshtein i Teoriya Gravitatsii (Albert Einstein and the Theory of Gravitation)*, Mir. Moscow (1979), pp. 513–527).

[87] D. Hilbert, "Die Grundlagen der Physik, 2 Mitt," *Nachrichten von der Kön. Ges. der Wissenschaften zu Göttingen, Math.–Phys. Kl.* (1917), pp. 53–76.

[88] F. Klein and D. Hilbert, "Zür Hilberts Note über die Grundlagen der Physik," in: F. Klein, *Gesammelte mathematische Abhandlungen*, Vol.1, Springer, Berlin (1921), Abh. XXXI.

[89] Russian translation of part of [254] published in: *Einshteinovskii Sbornik, 1971*, Nauka, Moscow (1972), pp. 7–54 (For English translation, see [254]).

[90] D. Hilbert, "Die Grundlagen der Physik (1924)," *Math. Annalen, 92*, 1–32 (1924); see also: D. Hilbert, *Gesammelte Abhandlungen*, Vol. 3, Springer, Berlin (1935), pp. 259–291.

[91] J. Mehra, *Einstein, Hilbert and the Theory of Gravitation*, Dordrecht (1974), 88 p.

[92] D. Hilbert, "Referat über die geometrischen Schriften und Abhandlungen H. Weyl's, erstattet der Physiko-Mathematischen Gesellschaft an der Universität Kasan," Izv. Fiz.-Mat. o-va pri Kazan. Un-ta, *11*, Ser. 3. 66–70, 1927).

[93] A. Einstein, "Spielen die Gravitationsfelder im Aufbau der materiellen Elementarteilchen eine wesentliche Rolle?" *Sitzungsber. Preuss. Akad. Wiss., 1*, 349–356 (1919) (for English translation, see [86]. Russian translation: [397], Vol. 1, pp. 664–671).

[94] D. Hilbert, *Grundlagen der Geometrie*, Teubner, Leipzig (1899), first part of: *Festschrift zur Feier. der Enthüllung des Gauss–Weber–Denkmals in Göttingen)* (English translation: *Foundations of Geometry*, OpenCourt, Chicago (1902). Russian translation published by Gostekhteorizdat, Moscow–Leningrad (1948)).

[95] H. Weyl, "Erkenntnis und Besinnung (ein Lebensrückblick)," in: H. Weyl, *Gesammelte Abhandlungen*, Vol. 4, Springer, Berlin (1968), pp.631–649.

[96] H. Weyl, "Zür Gravitationstheorie," *Ann. Phys. 54*, 117–145 (1919).

[97] L. Pyenson, "La réception de la relativité généralisée: Disciplinarité et institutionalisation en physique," *Rev. Hist. Sci., 28*, 61–73 (1975).

[98] L. Pyenson, "Mathematics, education and the Göttingen approach to physical reality, 1890–1914," *Europ. J. Interdiscipl. Stud., 2*, 91–127 (1979).

[99] H. Weyl, *Philosophie der Mathematik und Naturwissenschaft*, Oldenbourg-Verl., Munich (1927) (English translation: *Philosophy of Mathematics and Natural Science*, Princeton Press (1949)).

[100] M.H.A. Newman, "Herman Weyl," *J. London Math. Soc., 33*, 500–511 (1958) (Russian translation published in: *Usp. Mat. Nauk, 31*, 239–250 (1976)).

[101] H. Poincaré, *La science ei l'hypothèse*, Flammarion, Paris (1902) (English translation: *Schience and Hypothesis*, Walter Scott Publ. Co., London (1905). Russian translation published by Slovo, St. Pertersburg (1906)).

[102] F.A. Lange, *Geschichte des Materialismus*, Vols. 1 and 2, Baedeker, Iserlohn (1873) (Russian translation published in St. Petersburg 1881–1883 and 1899, 2nd edition).

[103] Quoted from: N.V. Motroshilova, *Printsipy i Protivorechiya Fenomenologicheskoi Filosofii (Principles and Contradictions of Phenomenological Philosophy)*, Vyssh. Shk., Moscow (1968), 128 pp.

[104] H. Weyl, "Ueber die neue Grundlagenkrise der Mathematik (1921)," in: H. Weyl, *Gesammelte Abhandlungen*, Vol. 2, Springer, Berlin (1968), pp. 143–180.

[105] H. Weyl, *O Filosofii Matematiki (On Philosophy of Mathematics)*, Gostekhteorizdat, Moscow-Leningrad (1934). (Russian translation of part of [99], [104], and Weyl's article "Die heitige Erkenntnislage in der Mathematik," 1925, Abhandlungen, Vol. 2, pp. 511–542).

[106] C. Reid, *Hilbert*, Springer, Berlin (1970). Russian translation published by Nauka, Moscow (1977).

[107] P.P. Gaidenko, *Filosifoya Fikhte i Sovremennost' (The Philosophy of Fichte and the Present)*, Mysl', Moscow (1979), 288 pp.

[108] B.U. Babushkin, "The transcendental–phenomenological foundation of philosophy and its results (review of western European literature of the 1960s and 1970s)," in: *Priroda Filosofskogo Znaniya (The Nature of Philosophical Knowledge)*, INION AN SSSR, Moscow (1977), Vol. 1, Part 2, pp. 16–53.

[109] P.P. Gaidenko, "Husserl," in: *Filosofskii Éntsiklopedicheskii Slovar' (Philosophical Encyclopedic Dictionary)*, Sov. Éntsiklopediya, Moscow (1983), pp. 132–133.

[110] V.N. Trostnikov, *Konstruktivnye Protsessy v Matematike (Constructive Processes in Mathematics)*, Nauka, Moscow (1975), 256 p.

[111] E. Reichenbächer, "Grundzüge zu einer Theorie der Elektrizität und der Gravitation," *Ann. Phys. 52*, 134–178 (1917).

[112] E. Reichenbächer, "Ueber die Nichintegrabilität der Streckenübertragung und die Weltfunktion in der Weylschen verallgemeinerten Relativitätstheorie, *Ann. Phys. 63*, 93–114 (1920).

[113] M. Abraham, "Neure Gravitationstheorien," *Jb. Radioaktiv. und Elekton., 11*, 470–520 (1914).

[114] G. Mie, "Die Grundlagen einer Theorie der Materie," *Ann. Phys.*, *37*, 511–534 (1912); *39*, 1–40 (1912); *40*, 1–66 (1913).

[115] J. Ishiwara, "Die Grundlagen einer relativistischen und elektro- magnetischen Gravitationstheorie," *Phys.Zs.*, *15*, 294–298, 506–510 (1914).

[116] G. Nordström, "Zür Theories der Gravitation vom Standpunkt des Relativitäts-prinzips," *Ann. Phys.*, *42*, 533–554 (1913).

[117] G. Nordström, "Ueber die Möglichkeit das electromagnetische Feld und das Gravitationsfeld zu vereinigen," *Phys. Zs.*, *15*, 504–506 (1914).

[118] E. Wiechert, "Die Gravitation als electromagnetische Erscheinung," *Ann Phys*, *63*, 301–381 (1920).

[119] T. Levi-Civita, "Nozione di parallelismo in uno varieta qualunque e conse-quente spezificazione geometrica della curvatura Riemanniana," *Rend. Circ. Mat. Palermo*, *42*, 173–205 (1917).

[120] G. Ricci and T. Levi-Civita, "Méthodes du calcul différentiel absolu et leurs applications," *Math. Ann.*, *54*, 125–201 (1901).

[121] G. Hessenberg, "Vektorielle Bergründung der Differentialgeometrie," *Math. Ann.*, *78*, 187–217 (1917).

[122] J. Schouten, "Die Direkte Analysis zur neuren Relativitätstheorie," *Verh. Akad. Wetensch. Amsterdam*, *12*, No. 6, 3–98 (1918).

[123] J. Schouten, *Ricci–Kalkül*, Springer, Berlin (1924).

[124] J. Schouten, "Erlanger Programm und Uebertragungslehre: Neue Gesichts-punkte zur Grundlegung der Geometrie," *Rend. Circ Mat. Palermo*, *50*, 142–169 (1926).

[125] H. Weyl, *Raum. Zeit. Materie*, 4th ed., Springer, Berlin (1921), 300 p.

[126] V.F. Kagan, "The geometrical ideas of Riemann and their further development," in: V.F. Kagan, *Ocherki po Geometrii (Essays on Geometry)*, Izd-vo MGU, Moscow (1963), pp. 437–516.

[127] H. Weyl, see [86].

[128] H. Weyl, *Space, Time, Matter*, Dover, New Yok (1952).

[129] C.W. Misner, K.S. Thorne, and J.A. Wheeler, *Gravitation*, W.H. Freeman, San Francisco (1973) (Russian translation published by Mir, Moscow (1977)).

[130] H. Weyl. "Reine Infinitesimalgeometrie," *Math. Zeitsr.*, *2*, 384–411 (1918), also in: H. Weyl, *Gesammelte Adhandlungen*, Vol. 2, Springer, Berlin (1968), pp. 1–28.

[131] H.Hönl, "Intensitäts – und Quantitätsgrössen," *Phys. Bl.*, *24*, 498–502 (1968).

[132] A. Einstein, Comment on Weyl's paper [86], published at the end of that paper (Not included in English translation) (Russian translation as in [86], pp. 525–526).

[133] H. Weyl, Response to Einstein's comment of [132], published at the end of [86] (Not included in English translation) (Russian translation as in [86], pp. 526–527).

[134] H. Weyl, "Gravitation und Elektrizität. Nachtrag Juni 1955," in: *Selecta Hermann Weyl*, Birkhäuser, Basel (1956).

[135] V.P. Vizgin, "Einstein and the problem of the construction of scientific theory
       (based on material of the general theory of relativty)," *Vopr. Filos.*, No. 10,
       56–64 (1979).

[136] St. Richter, *Wolfgang Pauli. Die Jahre 1918–1930. Skizzen zu einer wissen-
       schaftlichen Biographie*, Sauerländer, Aarau (1979), 112re

[137] W. Pauli, "Ueber die Energiekomponenten des Gravitationsfeldes," *Phys. Zs.,
       20*, 25–27 (1919).

[138] W. Pauli, "Zur Theorie der Gravitation und der Elektrizität von Hermann Weyl,"
       Phys. Zs., 20, 457–467 (1919).

[139] W. Pauli, "Mercurperihellbewegung und Strahlenablenkung in Weyls Gravita-
       tionstheorie," *Verh. Dt. Phys. Ges., 21*, 742–750 (1919).

[140] D.N. Trifonov, A.N. Krivomazov, and Yu.I. Lisnevskii, *Uchenie o Periodichnosti
       i Uchenie o Radioaktivnosti (Theory of Periodicity and Theory of Radioactivity)*,
       Atomizat, Moscow (1974), 248 pp.

[141] W. Pauli, *Wissenschaftlicher Briefwechsel mit Bohr, Einstein, Heisenberg u.A.*
       (eds. A. Herman, K. Meyenn, and V.F. Weisskopf) (Sources in the History of
       Mathematics and Physics, 2), Bd. 1, Springer, New York (1979).

[142] H. Weyl, "Die Relativitätstheorie auf der Naturforscherversammlung (1922),"
       in: H. Weyl, *Gesammelte Abhandlungen*, Vol. 2, Springer, Berlin (1968), pp.
       315–327.

[143] General Discussion on relativity theory at the 86th Naturforscherversammlung
       in Nauheim Sep. 19–25, 1920, in: *Phys. Zs., 21*, 666-668 (1920) (Russian trans-
       lation in: *Einshteinovskii Sbornik, 1971* (Einstein Collection, 1971), Nauka,
       Moscow (1972), pp. 374–378).

[144] Discussion on Weyl's paper "Elektrizität und Gravitation" at the 86th Natur-
       forscherversammlung in Nauheim Sept. 19–25, 1920, in" *Phys. Zs., 21*, 650–651
       (1920) (Russian translation in: *Einshteinovskii Sbornik, 1971* (Einstein Collec-
       tion, 1971), Nauka, Moscow (1972), pp. 371–373).

[145] N. Bohr, "Foreword," in:*Theoretical Physics in the Twentieth Century*, A Memo-
       rial Volume to Wolfgang Pauli (eds. M. Fierz and V.F. Weisskopf), Wiley, New
       York (Russian translation published by Izd-vo Inostr. Lit. (1962)).

[146] W. Heisenberg, "The development of quantum theory, 1918–1928," Russian
       translation in: *Priroda*, No. 9, 113–123 (1977).

[147] F. Hund, *Geschichte der Quantentheorie*, Bibliographisches Institute, Mann-
       heim (1967). English translation: *The History of Quantum Theory*, Harrap, Lon-
       don (1974). Russian translation published by Naukova Dumka, Kiev
       (1980).

[148] A.J.W. Sommerfeld, *Atombau und Spektrallinien*, Vieweg, Braunschweig
       (1919). English translation: *Atomic Structure and Spectral Lines*, Methuen, Lon-
       don (1932). Russian translation published by Gostekhizdat, Moscow (1956).

[149] W. Heisenberg, *Der Teil und das Ganze*, Munich (1969), 288 pp.

[150] W. Heisenberg, Russian translation of [149], in: *Problema Ob'ekta v Sovremen-
       noi Nauke (The Problem of the Object in Modern Science)*, INION AN SSSR,
       Moscow (1980), pp. 46–143.

[151] M. Born, *Vorlesungen über Atommechanik*, Vols. 1 and 2, Springer, Berlin (1925–1930). Russian translation vol. 1 published by GNTIU, Khar'kov (1934).

[152] M.J. Klein, "The first phase of the Bohr–Einstein dialogue," Hist. Stud. Phys. Sci., *2*, 1 (1970) (Russian translation in: *Einshteinovskii Sbornik, 1974*, (Einstein Collection, 1974), Nauka, Moscow (1976), pp. 115–155).

[153] S.I. Vavilov, "Translator's supplement," in: N. Bohr, *Tri Stat'i o Spektrakh i Stroenii Atomov (Three Papers of Spectra and the Structure of Atoms)*, translated by S.I. Vavilov, GIZ, Moscow, Petrograd (1923), pp. 148–155.

[154] A. Einstein, "Über ein den Elementarprozess der Lichtemission betreffendes Experiment," *Sitzungsber. Preuss. Akad. Wiss.* 882–883 (1921) (Russian translation: [398], Vol. 3, pp. 430–431).

[155] H.A. Kramers and H. Holst, *Bohrs Atomteori* (in Danish), Copenhagen (1922) (Russian translation published by GIZ, Moscow-Petrograd (1923)).

[156] M. Planck, "The atom of Bohr's theory" (Russian translation) in: *Novye Idei v Fizike*, No. 10, *Stroenie Atoma i Spektry*, Obrazovanie, Petrograd (1924), pp. 42–46.

[157] N. Bohr, *Drei Aufsätze über Spektren und Atombau*, Braunschweig (1922) (For Russian translation, see [153]).

[158] A.J.W. Sommerfeld, "Foundations of quantum theory and Bohr's model of the atom" (Russian translation), in: [68], pp. 8–14.

[159] A.E. Haas, *Das Naturbild der neuen Physik*, W. de Gruyter, Berlin (1924). Russian translation published by Izd.L.D. Frenkel', Moscow-Petrograd (1924).

[160] N. Bohr, "Der Bau der Atome und die physikalischen und chemischen Eigenschaften der Elemente," *Zeitschrift für Physik, 9*, 1–67 (1922) (Russian translation in Vol. 1 of Russian edition of selected works of Bohr published by Nauka, Moskow (1970)).

[161] M.A. El'yashevich, "From the origin of quantum concepts to the establishment of quantum mechanics," *Usp. Fiz. Nauk, 122*, 673–717 (1977) (English translation in: *Sov. Phys. Uspekhi, 20*, 656–682 (1977)).

[162] P. Forman, "Weimar culture, causality and quantum theory 1918–1927: Adaptation by German physicists and mathematicians to a hostile environment," *Hist. Stud. Phys. Sci., 3*, 1–115 (1971).

[163] S.I. Vavilov, see [153], translator's foreword, pp. 5–6.

[164] H. Weyl, "Eine neue Erweiterung der Relativitätstheorie," *Ann. Phys. 59*, 101–133 (1919).

[165] H. Weyl, "Ueber die physikalischen Sinn der erweiterten Relativitätstheorie," *Phys. Zs., 22*, 473–480 (1921).

[166] A. Einstein, *Äther und Relativitätstheorie*, Springer, Berlin (1920) (Russian translation: [397]. Vol. 1, pp. 682–689).

[167] H. Weyl, "Feld und Materie," *Ann. Phys., 65*, 541–563 (1921). In: H. Weyl, *Gesammelte Abhandlungen*, Vol. 2, Springer, Berlin (1968), p. 237–259.

[168] C. Kahn and F. Kahn, "Letters from Einstein to de Sitter on the nature of the Universe," *Nature, 257*, 451–454 (1975).

[169] A.V. Douglas, *The Life of Arthur Stanley Eddington*, Nelson, Edinburgh (1957).

[170] W. De Sitter, "On Einstein's theory of gravitation, and its astronomical consequences, Part 3," *Mon. Not. R. Astron. Soc.*, *78*, 3 (1917) (Russian translation
in: *Al'bert Éinshtein i Teoríya Gravitatsii (Albert Einstein and the Theory of
Gravitation)*, Mir, Moscow (1979), pp. 299–319).

[171] A.S. Eddington, *Report on the Relativity Theory of Gravitation*, Fleetmay Press,
London (1918).

[172] F.W. Dyson, A.S. Eddington, and C. Davidson, "Deflection of light by sun's
gravitational field. Total eclipse of May 29, 1919," *Roy. Soc., Phil. Trans.*, *220*,
291–333 (1920) (For Russian translation see [170], pp. 564–570).

[173] A.S. Eddington, *Space, Time and Gravitation, An Outline of the General Relativity Theory*, Cambridge University Press, Cambridge (1920) (Russian translation
published by Matezis, Odessa (1923)).

[174] A.S. Eddington, "A generalization of Weyl's theory of the electromagnetic and
gravitational field," *Proc. R. Soc. London A*, *99*, 104–122 (1921).

[175] A. Einstein, "Grundgedanken und Probleme der Relativitätstheorie," in: *Nobelstiftelsen, Les prix Nobel en 1921–1922*, Imprimerie Royale, Stockholm (1923)
(Russian translation: [397], Vol. 2, pp. 120–129).

[176] A. Einstein, "Zür allgemeinen Relativitätstheorie," *Sitzungsber. Preuss. Akad.
Wiss., Phys.-Math. Kl.*, 32–38 (1923) (Russian translation: [397], Vol. 2, pp.
134–141).

[177] V.V. Raman, "Kaluza Theodor Franz Eduard," in: *Dictionary of Scientific Biography* (ed. Ch.C. Gillispie), Ch. Scribner's Sons, New York (1973), Vol. 7,
pp. 211–212.

[178] T. Kaluza, "Zur Relativitätstheorie," *Phys. Zs.*, *11*, 977 (1910).

[179] T. Kaluza, Russian translation of [180] in [170], pp. 529–534.

[180] T. Kaluza, "Zum Unitätsproblem der Physik," *Sitzungber. Preuss. Akad. Wiss.*
966–972 (1921).

[181] H. Thirring, "Ueber die formale Analogie zwischen den elektromagnetischen
Grundgleichungen und den Einsteinschen Gravitationsgleichungen erster Nä-
herung," *Phys. Zs.*, *19*, 204–205 (1918).

[182] A.A. Onishchenko and A.D. Levkovich, "Yakob Grommer—Einstein's assistant
(Ya.P. Grommer in Belorussia)," (manuscript).

[183] A. Einstein and J. Grommer, "Beweis der Nichtexsistenz eines überall regulären
zentrisch symmetrischen Felds nach der Feld-Theorie von Th. Kaluza," *Scripta
Universitatis atque Bibliothecae Hierosolymitanarum, Mathematica et Physica
(Jerusalem)*, *1*, No. 7 (1923) (Russian translation [397], Vol. 2, pp. 130–133).

[184] A. Einstein, "Näherungsweise Integration der Feldgleichungen der Gravitation,"
Sitzungsber. Preuss. Akad. Wiss., *1*, 688–696 (1916) (Russian translation: [397],
Vol. 1, pp. 514–523).

[185] A. Einstein, "Besprechung: Hermann Weyl, Raum-Zeit-Materie," *Naturwissenschaften*, 6. Jahrgang, 373 (1918) (Russian translation: [397], Vol. 4, pp. 42–43).

[186] A. Einstein, "Motiv des Forschens," in: *Zu Max Plancks 60. Geburtstag: An-
sprachen in der Deutsche physikalische Gesellschaft*, Karlsruhe (1918), pp. 29–
32 (Russian translation: [397], Vol. 4, pp. 39–41).

[187] A. Einstein, "Spielen Gravitationsfelder im Aufbau der materiellen Elementarteilchen eine wesentliche Rolle?" *Sitzungsber. Preuss. Akad. Wiss.*, 349–356 (1918) (For English translation, see [86]).

[188] F. Herneck, *Albert Einstein; ein Leben für Wahrheit, Menschlichkeit und Frieden*, Der Morgen, Berlin (1963) (Russian translation published by Progress, Moscow (1966)).

[189] A. Einstein, Russian translation of [190].

[190] A. Einstein, "Geometrie und Erfahrung," *Sitzunsber Preuss. Akad. Wiss.* pp. 123–130 (1921).

[191] A. Einstein, "A brief outline of the development of the theory of relativity," *Nature, 106*, 782–784 (1921) (Russian translation: [397], Vol. 2, pp. 99–104).

[192] A. Einstein, Russian translation ([397], Vol. 2, p. 105) of [193].

[193] A. Einstein, "Ueber naheliegende Ergänzung des Fundamentes der allgemeinen Relativitätstheorie," *Sitzungsber. Preuss. Akad. Wiss.* pp. 261–264 (1921).

[194] E. Cartan, *Spaces of Affine, Projective, and Conformal Connection* (Russian translation ed. P. A. Shirokov), Izd. Kazan Un., Kazan (1962).

[195] R. Bach, "Zür Weylschen Relativitätstheorie und der Weylschen Erweiterung des Krümmungstensorbegriffes,: *Math. Zs., 9*, 110–135 (1921).

[196] A. Einstein, *The Meaning of Relativity*, Princeton University Press, Princeton, New York (1921) (Russian translation: [397], Vol. 2, pp. 5–82).

[197] A. Einstein, "Zür Theorie der Lichtfortpflanzung in dispergierenden Medien," *Sitzungsber. Preuss. Akad. Wiss. Phys.-Math., Kl.,* 18–22 (1922) (Russian translation: [397], Vol. 3, pp. 437–441).

[198] A. Einstein (with P. Bergmann), "Generalization of Kaluza's theory of electricity," *Ann. Math.*, Ser. 2, *39*, 683–701 (1938) (Russian translation: [397], Vol. 2, pp. 492–513).

[199] Yu.B. Rumer, "Unified field theory," in: *Fizicheskii Éntsiklopedicheskii Slovar' (Physics Encyclopedic Dictionary)*, Sov. Éntsiklopediya, Moscow (1962), Vol. 2, pp. 5–6.

[200] A.G. Webster, *The Dynamics of Particles and of Rigid, Elastic and Fluid Bodies* (Math. Wiss., Vol. 11), Teubner, Leipzig (1904).

[201] M.-L. Tonnelat, *Les principles de la théorie électromagnétique et de la relativité Masson*, Paris, (1959). Russian translation published by Izd-vo Instr. Lit., Moscow (1962)).

[202] J.A. Schouten, "Ueber die verschiedenen Arten der Uebertragung, die einer Differentialgeometrie zu Grunde gelegt werden können" *Math Zs., 13*, 56–81 (1922).

[203] J.A. Schouten and D.J. Struik, *Einführung in die neueren Methoden der Differentialgeometrie*, Noordhoff, Groningen (1938) (Russian translation published by GONTI, Moscow–Leningrad (1939), Vol. 1)).

[204] V.I. Rodichev, "Methodological aspects of unified field theory," in: *Éinshtein i Filosofskie Problemy Fiziki XX Veka (Einstein and Philosophical Problem of Physics of the 20th Century)* (ed. É.M. Chudinov), Nauka, Moscow (1979), pp. 408–440.

[205] V.F. Kagan, "The geometrical ideas of Riemann and their further development," in: [126] pp. 437–516.

[206] A. Einstein, "Vorwort," Japanese collected scientific papers of Albert Einstein Einstein (1922) (Russian translation: [397], Vol. 4, pp. 53–54).

[207] A. Einstein, "Über die gegenwärtige Krise in der theoretischen Physik," *Kaizo, 4*, No. 22, 1–8 (1922) (Russian translation:[397], Vol. 4, pp. 55–60).

[208] A. Einstein, "Zür allegemeinen Relativitätstheorie," *Sitzungsber. Preuss. Akad. Wiss.* pp. 32–38 (1923).

[209] A. Einstein, "Bemerkung zu meiner Arbeit 'Zur allegemeinen Relativitätstheorie'," *Sitzungsber. Preuss. Akad. Wiss., Phys.-Math. Kl.,* pp. 76–77 (1923) (Russian translation: [397], Vol. 2, pp. 142–144).

[210] A. Einstein, "Zur affinen Feldtheorie," *Sitzungsber. Preuss. Akad. Wiss., Phys.-Math. Kl.,* pp. 137–140 (1923) (Russian translation: [397], Vol.2, pp. 145–148).

[211] P. Ehrenfest and A.F. Ioffe, *Nauchnaya Perepiska (1907–1933) (Scientific Correspondence (1907–1933)),* Nauka, Leningrad (1973).

[212] A. Einstein, "The theory of the affine field," *Nature, 112*, 448–449 (1923) (Russian translation: [397], Vol. 2, pp. 149–153).

[213] A. Einstein, "Notiz zu der Bemerkung zu der Arbeit von A. Friedmann. 'Ueber die Krümmung des Raumes'," *Zeitschrift für Physik, 16*, 228 (1923) (Russian translation: [397], Vol. 2, p. 119).

[214] A. Einstein, "Grundgedanken und Probleme der Relativitätstheorie," in: *Nobelstiftelsen, Les prix Nobel en 1921–1922,* Imprimerie Royale, Stockholm (1923) (Russian translation: [397], Vol. 2, pp. 120–129).

[215] W. Gerlach, "Erinnerungen an Albert Einstein, 1908–1930," *Phys. Bl., 35*, 93–102 (1979).

[216] H.-J. Treder, "Albert Einstein an der Berliner Akademie der Wissenschaften," in: *Albert Einstein in Berlin 1913–1933: Darstellung und Dokumente,* Akad.-Verl., Berlin (1979), pp. 7–78.

[217] A. Einstein, "Über den Äther," *Schweiz. Naturforsch. Geselleschaft, Verhandlungen, 105*, 85–93 (1924) (Russian translation: [397], Vol. 2, pp. 154–160).

[218] V.Ya. Frenkel' and B.E. Yavelov, *Éinshtein—izobretatel' (Einstein the Inventor),* Nauka, Moscow (1981), 161 p.

[219] A. Einstein (with W.J. de Haas), "Experimenteller Nachweis der Ampèreschen Molekularströme," *Verhandl. Dtsch. Phys. Ges., 17*, 152–170 (1915) (Russian translation: [397], Vol. 3, pp. 363–379).

[220] P.M.S. Blackett, "The magnetic field of massive rotating bodies," *Nature, 159*, 658–666 (1947) (Russian translation published in *Usp. Fiz. Nauk, 38*, 52–76 (1947)).

[221] G.M. Idlis, "Force fields in space and some problems of the structure and evolution of galactic matter," *Izv. Astrofiz. In-ta., 4*, 3–159 (1957).

[222] N.P. Ben'kova, "Terrestrial magnetism," in: *Fizicheskii Éntsiklopedicheskii Slovat' (Physics Encyclopedic Dictionary),* Sov. Éntsiklopediya, Moscow (1962), Vol. 2, pp. 74–75.

[223] J. Stachel, "Einstein and the rigidly rotating disk," in: *General Relativity and Gravitation* (ed. A. Held), Plenum Press, New York (1980), Vol. 1, pp. 1–15.

[224] M.A. Markov, "Gaudeamus igitur juvenes dum sumus," in: *Sergei Ivanovich Vavilov: Ocherki i Vospominaniya (Sergei Ivanovich Vavilov: Essays and Reminiscences)*, Nauka, Moscow (1981), pp. 230–237.

[225] A. Einstein, Russian translation ([397], Vol. 3, pp. 456–462) of [226]).

[226] A. Einstein, "Bietet die Feldtheorie Möglichkeiten für die Lösung des Quantenproblems?" *Sitzungsber. Preuss. Akad. Wiss.* pp. 359–364 (1923).

[227] Continuation of [56] (Einstein–Besso Correspondence) published in: *Einshteinovskii Sbornik, 1975–1976*, Nauka, Moscow (1976), pp. 5–42.

[228] A. Einstein (with Grommer), "Allgemeine Relativitätstheorie und Bewegungsgesetz," *Sitzungsber. Preuss. Akad. Wiss., Phys.-Math. Kl.*, pp. 2–13 (1927) (Russian translation: [397], Vol. 2, pp. 198–210).

[229] A. Einstein, "Allgemeine Relativitätstheorie und Bewegungsgesetz," *Sitzungsber. Preuss. Akad. Wiss., Phys.-Math. Kl.*, pp. 235–245 (1927) (Russian translation: [397], Vol. 2, pp. 211–222).

[230] A. Einstein (with N. Rosen), "The particle problem in the general theory of relativity," *Phys. Rev., 48*, 73–77 (1935) (Russian translation: [397], Vol. 2, pp. 424–433).

[231] A. Einstein (with L. Infeld and B. Hoffmann), "Gravitational equations and problems of motion," *Ann. Math., 39*, 65–100 (1938) (Russian translation: [397], Vol. 2, pp. 450–491).

[232] A. Einstein, "On the generalized theory of gravitation," *Sci. Am., 182*, No. 4, 13–17 (1950) (Russian translation: [397], Vol. 2, pp. 719–731).

[233] W. Heisenberg, "Quantentheoretische Umdeutung kinematischen und mechanischen Beziehungen," *Zeitschrift für Physik, 33*, 879–893 (1925) (Russian translation in: *Usp. Fiz. Nauk, 122*, 574–585 (1977)).

[234] A. Pais, "Einstein and the quantum theory," *Rev. Mod. Phys., 51*, 863–914 (1979).

[235] M.A. El'yashevich, "From the origin of quantum concepts to the establishment of quantum mechanics," *Usp. Fiz. Nauk, 122*, 673–718 (1977) (English translation in: *Sov. Phys. Uspekhi, 20*, 656–682 (1977)).

[236] A. Einstein, "Quantentheorie des einatomigen idealen Gases. Zweite Abhandlung," *Sitzungsber. Preuss. Akad. Wiss., Phys.-Math., Kl.*, pp. 3–14 (1925) (Russian translation: [397], Vol. 3, pp. 489–502).

[237] E. Schrödinger, *Briefe zur Wellenmechanik* (correspondence with M. Planck, A. Einstein, H. Lorentz) Springer, Vienna (1963). Russian translation in selected works of Schrödinger published by Nauka, Moscow (1976), pp. 302–338.

[238] A. Einstein, "Eddingtons Theorie und Hamiltonsches Prinzip," Supplement to: A.S. Eddington, *Relativitätstheorie in mathematischer Behandlung*, Springer Verlag, Berlin (1925), pp. 366–371 (Russian translation: [397], Vol. 2, pp. 161–166).

[239] A.S. Eddington, "Einsteins neue Theorie," supplementary paper added to A. S. Eddington, *Relativitätstheorie in mathematischer Behandlung*, Springer, Berlin (1925) (Russian translation: A.S. Eddington, *Teoriya Otnositel'nosti*, Gostekhteorizdat, Moscow–Leningrad (1934)).

[240] A. Einstein, "Einheitliche Feldtheorie von Gravitation und Elektrizität," *Sitzungsber. Preuss. Akad. Wiss., Phys.-Math. Kl.*, pp. 414–419 (1925) (Russian translation: [397], Vol. 2, pp. 171–177).

[241] M. Ferraris, M. Francaviglia, and C. Reina, "Variational formulation of general relativity from 1915 to 1925: 'Palatini's method' discovered by Einstein in 1925," *Gen. Rel. and Grav., 14*, No. 1, 1–12 (1982).

[242] A. Palatini, "Deduzione invariantiva delle equazioni gravitazionali dal principio di Hamilton," *Rend. Circ. Mat. Palermo, 43*, 203–212 (1919).

[243] A. Einstein, "Elektron und allgemeine Relativitätstheorie," *Physica*, 5 Jaargang, 330–334 (1925) (Russian translation: [397], Vol. 2, pp. 167–170).

[244] H.-J. Treder, "Antimatter and the particle problem in Einstein's cosmology and field theory of elementary particles: (An historical essay on Einstein's work at the Akademie der Wissenschaften zu Berlin)," *Astron. Mach., 296*, 149–161 (1975).

[245] H.-J. Treder, *Grosse Physiker und ihre Probleme: Studien zur Geschichte der Physik*, Akad. Verl., Berlin (1983), pp. 209–213.

[246] A. Einstein, "Über die formale Beziehung des Riemannschen Krümmungstensors zu den Feldgleichungen der Gravitation," *Math. Ann.*, 99–103 (1926–7) (Russian translation: [397], Vol. 2, pp. 183–187).

[247] Russian translation of part of [254] published in: *Éinshteinovskii Sbornik, 1972*, Nauka, Moscow (1974), pp. 7–103 (For English translation, see [254]).

[248] V.A. Bazhanov, *Problema Polnoty Kvantovoi Teorii: Poisk Novykh Podkhodov (Filosofskii Aspekt) (The Problem of the Completeness of Quantum Theory; The Search for New Approaches (Philosophical Aspect))*, Izd-vo Kazan. Un-ta, Kazan' (1983), 104 p.

[249] A.F. Ioffe, "Albert Einstein," in: A.F. Ioffe, *O Fizike i Fizikakh (On Physics and Physicists)*, Nauka, Leningrad (1977), pp. 224–229.

[250] A. Einstein, "Vorschlag zu einem die Natur des elementaren Strahlungs-Emissionsprozesses betreffenden Experiment," *Naturwissenschaften, 14*, 300–301 (1926) (Russian translation: [397], Vol. 3, pp. 514–516).

[251] A. Einstein, "Über die Interferenzeigenschaften des durch Kanalstrahlen emittierten Lichtes," *Sitzungsber. Preuss. Akad. Wiss.*, pp. 334–340 (1926) (Russian translation: [397], Vol. 3, pp. 517–524).

[252] A. Einstein, "Theoretisches und Experimentelles zur Frage der Lichtentstehung," *Zs. Angew. Chemie, 40*, 546 (1927) (Russian translation: [397], Vol. 3, pp. 525–527).

[253] L.I. Mandel'shtam, "Lectures on selected problems of optics (1932–33)," in: L. I. Mandel'shtam, *Lektsii po Optike, Teorii Otnositel'nosti i Kvantovoi Mekhanike (Lectures on Optics, the Theory of Relativity, and Quantum Mechanics)*, Nauka, Moscow (1972), Lecures Nos. 1, 8, 9.

[254] A. Einstein, *Hedwig und Max Born, Briefwechsel, 1916–1955*, Nymphenburger Verl., Munich (1969), 330 p. (English translation: *The Born–Einstein Letters. Correspondence between Albert Einstein and Max and Hedwig Bron from 1916 to 1955 with Commentaries by Max Born*, Macmillan, London (1971)).

[255] A. Einstein (with B. Kaufmann), "A new form of the general relativistic field equations," *Ann. Math., 62*, 128–138 (1955) (Russian translation: [397], Vol. 2, 835–848).

[256] A. Einstein, "Relativistic theory of the non-symmetric field," in: *The Meaning of Relativity*, Fifth Edition, Princeton (1955) (Russian translation: [397], Vol. 2, pp. 849–873).

[257] M. Jammer, *The Conceptual Development of Quantum Mechanics*, McGraw-Hill, New York (1966).

[258] O. Klein, "Quantentheorie und fünfdimensionale Relativitätstheorie," *Zeitschrift für Physik, 37*, 895–906 (1926).

[259] A. Einstein, "Zu Kaluzas Theorie des Zusammenhäng von Gravitation und Elektrizität. II," *Sitzungsber. Preuss. Akad. Wiss., Phys.-Math. Kl.*, pp. 26–30 (1927) (Russian translation: [397], Vol. 2, pp. 193–197).

[260] A. Einstein, "Zu Kaluzas Theorie des Zusammenhängs von Gravitation und Elektrizität," *Sitzungsber. Preuss Akad. Wiss., Phys.-Math. Kl.*, pp. 23–25 (1927).

[261] A. Einstein, "Allgemeine Relativitätstheorie und Bewegungsgesetz," *Sitsungsber Preuss. Akad. Wiss., Phys.-Math. Kl.*, pp. 235–245 (1927) (Russian translation: [397], Vol. 2, pp. 211–222).

[262] L. Infeld and J. Plebánski, *Motion and Relativity*, Polish Academy of Sciences, Oxford (1960) (Russian translation published by Izd. Inostr. Lit., Moscow (1962)).

[263] L. Witten (ed.), *Gravitation*, J. Wiley, New York (1962).

[264] V.A. Fock, "Some applications of the idea of Lobachevskii's non-Euclidean geometry in physics," in: A. P. Kotel'nikov and V. A. Fock, *Nekotorye Primenenya Idei Lobachevskogo v Mekhanike i Fizike (Some Applications of Lobachevskii's Idea in Mechanics and Physics)*, Gostekhteorizdat, Moscow–Leningrad (1950), pp. 48–57.

[265] V.A. Fok, *Teoriya Prostranstva, Vremeni i Tyagoteniya*, Gostekhizdat, Moscow (1955) (English translation: V.A. Fock, *Theory of Space, Time, and Gravitation*, Pergamon Press, Oxford (1964)).

[266] V.A. Fock, "On the motion of finite masses in the general theory of relativity," in: *Al'bert Éinshtein i Teoriya Gravitatsii (Albert Einstein and the Theory of Gravitation)*, Mir, Moscow (1979), pp. 232–284.

[267] A. Einstein, "Newtons Mechanik und ihr Einfluss auf die Gestaltung der theoretischen Physik," *Naturwissenschaften 15*, 273–276 (1927) (Russian translation: [397], Vol. 4, pp. 82–88).

[268] A. Einstein, "Electrons et photons," *Rapports et discussions du cinquieme conseil de physique-Bruxelles du 24 au 29 octobre 1927 sous les auspices de l'Institut International de Physique Solvay*, Gautier-Villars et Cie, Paris (1928), pp. 252–256 (Russian translation: [397], Vol. 3, pp. 528–530).

[269] A. Einstein (with J. Grommer), "Beweis der Nichtexsistenz eines überall regulären zentrisch symmetrischen Felds nach der Feld-Theorie von Th. Kaluza," *Scripta Universitatis atque Bibliothecae Hierosolymitanarum, Mathematica et Physica (Jerusalem), 1*, No. 7 (1923) (Russian translation: [397], Vol. 2, pp. 130–133).

[270] V. Fock, "Ueber die invariante Form der Wellen- und der Bewegungsgleichungen für einen geladenen Massenpunkt," *Zeitschrift für Physik, 39*, 226–232 (1926).

[271] H. Mandel, "Zür Herleitung der Feldgleichungen in der allgemeinen Relativitätstheorie," *Zeitschrift für Physik, 39*, pp. 136–145 (1926).

[272] V.K. Frederiks and A.A. Fridman, *Osnovy Teorii Otnositel'nosti (Fundamental of the Theory of Relativity)*, Academia, Leningrad (1924), Pt. 1, 167 pp.

[273] V.K. Frederiks, "Schrödinger's theory and the general theory of relativity," in: *Osnovaniya Novoi Kvantovoi Mekhaniki (Fundamentals of the New Quantum Mechanics)*, (ed. A.F. Ioffe), GIZ, Moscow-Leningrad (1927), pp. 83–98.

[274] O. Klein, "The atomicity of electricity as a quantum theoretical law," *Nature, 118*, 516 (1926).

[275] F. London, "Die Theorie von Weyl und die Quantenmechanik," *Naturwissenschaften, 15*, 187 (1928).

[276] I.A. Voronov and Ya.I. Kogan, "Spontaneous compactification in Kaluza–Klein models and the Casimir effect," *Pis'ma Zh. Eksp. Teor. Fiz., 38*, 262–265 (1983).

[277] A. Einstein, Russian translation of [285], [397], Vol. 2, pp. 223–228.

[278] A. Einstein, "Neue Möglichkeit für eine einheitliche Feldtheorie von Gravitation und Elektrizität," *Sitzungsber. Preuss. Akad. Wiss., Phys.-Math. Kl.*, pp. 224–227 (1928) (Russian translation: [397], Vol. 2, pp. 229–233).

[279] D.D. Ivanenko, "Commentaries on Einstein's letter of July 11, 1929 to I.V. Obreimov (the publication of an unpublished letter of A. Einstein')," *Vopr. Istor. Estestvozn. Tekh.*, No. 67–68, 70–71 (1980).

[280] E. Cartan, *Riemannian Geometry in Orthogonal Frame* (Russian translation), Izd. MGU, Moscow (1960).

[281] R. Weitzenböck, "Neuere Arbeiten der algebraischen Invariantentheorie; Differentialinvarianten," in: *Encykl. Math. Wiss., 3* (3), 6 (1922).

[282] L.P. Eisenhart, *Riemannian Geometry*, Princeton (1966) (Russian translation published by Izd-vo Inostr. Lit. (1948), 316 pp.

[283] L.P. Eisenhart, *Non-Riemannian Geometry*, Amer. Math Soc. Colloq. Publ., New York (1927), Vol. 8.

[284] V.I. Rodichev, *Teoriya Tyagoteniya v Ortogonal'nom Repere (The Theory of Gravitation in Tetrad Form)*, Nauka, Moscow (1974), 184 pp.

[285] A. Einstein, "Riemann-Geometrie mit Aufrechterhaltung des Begriffes des Fernparallelismus," *Sitzungber. Preuss. Akad. Wiss.* 217–221 (1928).

[286] E. Cartan, "Sur la structure des groupes infinis de transformations," *Ann Sci. Ecole Norm Supér., 21*, 153–206 (1904); 22, 219–308 (1905).

[287] E. Cartan and J. A. Schouten, "On the geometry of the group manifold of simple and semi-simple groups," *Proc. Akad. Amsterdam, 29*, 803–815 (1926).

[288] E. Cartan, "La géometrie des groupes de transformations," *J. Math. Pures Appl.*, Ser. 9, 6, 1–119 (1927).

[289] E. Cartan, *Geometry of Lie Groups and Symmetric Spaces* (collection of papers translated into Russian), Izd.-vo Inostr. Lit., Moscow (1949).

[290] E. Cartan, "Notice historique sur la notion de parallelisme absolu," *Math. Ann., 102*, 698–706 (1930).

[291] A. Einstein, "Auf die Riemann-Metrik und den Fern-Parallelismus gegründete einheitliche Feldtheorie," *Math. Ann.*, *102*, 685–697 (1930) (Russian translation: [397], Vol. 2, pp. 307–320).

[291a] *Elie Cartan–Albert Einstein, Letters on Absolute Parallelism 1929–1932* (ed. R. Debever), Princeton University Press, Princeton (1979).

[292] P.K. Rashevskii, *Rimanova Geometriya i Tenzornyi Analiz (Riemannian Geometry and Tensor Analysis)*, Nauka, Moscow (1967), 664 pp.

[293] A. Einstein, "Zur einheitlichen Feldtheorie," *Sitzungsber. Preuss. Akad. Wiss., Phys.-Math. Kl.*, 2–7 (1929) (Russian translation:[397], Vol. 2, pp. 252–259).

[294] R. Weitzenböck, "Differentialinvarianten in der Einsteinschen Theorie des Fernparallelismus," *Sitzungsber. Preuss. Akad. Wiss.* 466 (1928).

[295] "News and Views", *Nature, 123*, 174–175 (1929).

[296] A. Einstein, "New field theory. I, II,"*Observatory, 52*, 82–87, 114–118 (1929) (Russian translation: [397], Vol. 2, pp. 260–264, 265–269).

[297] A.S. Eddington, "Einstein's field theory," *Nature, 123*, 280–281 (1929).

[298] A. Einstein, "Einheitliche Feldtheorie und Hamiltonsches Prinzip," *Sitzungsber Preuss. Akad. Wiss., Phys.-Math. Kl.*, 156–159 (1929) (Russian translation: [397], Vol. 2, pp. 270–274).

[299] A. Einstein, "Über den gegenwärtigen Stand der Feld-Theorie," *Festschrift Prof. A. Stodola Zum Geburtstag*, Füssli Verlag, Zurich and Leipzig (1929), pp. 126–132 (Russian translation: [397], Vol. 2, pp. 244–251).

[300] T. Levi-Civita, "A proposed modification of Einstein's field theory," *Nature, 123*, 678–679 (1929).

[301] E. Cartan, *Metod Podvizhnogo Repera, Teoriya Nepreryvnykh Grupp i Obobshchennye Prostranstva (The Tetrad Method, the Theory of Continuous Groups, and Generalized Spaces* (lectures read at the Institute of Mathematics and Mechanics, Moscow, June 16–20, 1930)), Gostekhteorizdat, Moscow–Leningrad (1933), 72 p.

[302] C. Lanczos, "Die neue Feldtheorie Einsteins," *Ergeb. Exakt. Naturwiss., 10*, 97 (1931).

[303] G.C. McVittie, "Solution with axial symmetry of Einstein's equations of teleparallelism," *Proc. Edinburgh Math. Soc., 2*, 140 (1931).

[304] H. Piaggio, "Einstein's and other unitary field theories; An explanation for the general reader," *Nature, 123*, 839–841, 877–879 (1929).

[305] N. Wiener and M.S. Vallarta, "Unified field theory with electricity and gravitation," *Nature, 123*, 317 (1929).

[306] N. Wiener, "Dirac equations and Einstein theory," *Nature, 123*, 950 (1929).

[307] I.E. Tamm, "On the connection between Einstein's unified field theory and quantum theory," in *Sobr. Nauch. Tr. (Collected Scientific Works)*, Nauka, Moscow (1975), Vol. 2, pp. 184–187.

[308] I.E. Tamm and M.A. Leontovich, "Comments on Einstein's unified field theory," *ibid*, p. 191–201.

[309] S.M. Rytov, "From ancient times," in: *Vospominaniya o I.E. Tamme (Recollections of I.E. Tamm)*, Nauka, Moscow (1981), pp. 185–188.

[310] V. Fock and D. Ivanenko, "Quantum geometry," *Nature, 123*, 838 (1929).

[311] H. Weyl. "Gravitation and electron (1929)," in H. Weyl, *Gesammelte Abhabdlungen*, Springer, Berlin (1968), Vol. 3, pp. 217–228.

[312] H. Weyl, "Gravitation and electron (1929)," in: H. Weyl, *Gesammelte Abhabdlungen*, Springer, Berlin (1968), Vol. 3, pp. 229–244.

[313] H. Weyl. "Elektron und Gravitation (1929)," in: H. Weyl, *Gesammelte Abhabdlungen*, Springer, Berlin (1968), Vol. 3, pp. 245–267.

[314] B. van der Waerden, "Exclusion principle and spin," in [145], pp. 189–244.

[315] Yu.S. Vladimirov, *Sistemy Otscheta v Teorii Gravitatsii (Frames of Reference in the Theory of Gravitation)*, Énergoizdat, Moscow (1982), 256 pp.

[316] A. Einstein, "Théorie unitaire de champ physique," *Ann. Inst. J. Poincaré 1*, 1–24 (1930) (Russian translation: [397], Vol. 2, pp. 286–306).

[317] A. Einstein, *Brief an Maurice Solovine*, Berlin, 1960. *Lettres à Maurice Solovine*, Paris, 1960 (Russian translation: [397], Vol. 4, pp. 547–575).

[318] A. Einstein (with W. Mayer), "Zwei strenge statische Lösungen der Feldgleichungen der einheitlichen Feldtheorie," *Sitzungsber. Preuss. Akad. Wiss., Phys.-Math. Kl.*, pp. 110–120 (1930) (Russian translation: [397], Vol. 2, pp. 329–341).

[319] A. Einstein, "Gravitational and electrical fields," *Science*, 74, 438–439 (Russian translation: [397], Vol. 2, pp. 347–348).

[320] A. Einstein, "Zur Theorie der Räume mit Riemann-Metrik und Fernparallelismus," *Sitzungsber. Preuss. Akad. Wiss., Phys.-Math. Kl.*, pp. 401–402 (1930) (Russian translation: [397], Vol. 2, pp. 342–343).

[321] A. Einstein (with W. Mayer), "Systematische Untersuchung über kompatible Feldgleichungen, welche in einem Riemannschen Räume mit Fernparallelismus gesetzt werden können," *Sitzungsber. Preuss. Akad. Wiss., Phys.-Math. Kl.*, pp. 257–265 (1931) (Russian translation: [397], Vol. 2, pp. 353–365).

[322] A. Einstein (with W. Mayer), "Semivektoren und Spinoren," *Sitzungsber. Preuss. Akad. Wiss., Phys.-Math. Kl.*, pp. 522–550 (1932) (Russian translation: [397], Vol. 3, pp. 535–567).

[323] A. Einstein (with W. Mayer), "Die Diracgleichungen für Semivektoren," *Proc. Acad. Wet. (Amsterdam)*, 36, 497–516 (1933) (Russian translation: [397], Vol. 3, pp. 568–590).

[324] A. Einstein (with W. Mayer), "Spaltung der natürlichsten Feldgleichungen für Semi-Vektoren in Spinor-Gleichungen vom Diracschen Typus," *Proc. Akad. Wet. (Amsterdam)*, 36, Pt. 2, 615–619 (1933) (Russian translation: [397], Vol. 3, pp. 591–596).

[325] A. Einstein (with W. Mayer), "Darstellung der Semi-Vektoren als gewöhnliche Vektoren von besonderem Differentiations Charakter," *Ann. Math.*, 35, 104–110 (1934) (Russian translation: [397], Vol. 3, pp. 596–603).

[326] E. Eyraud, *Les équations de la dynamique de l'éther*, Blanchard, Paris (1926).

[327] L. Infeld, "Zum Problem einer einheitlichen Feldtheorie von Elektrizität und Gravitation," *Zeitschrift für Physik, 50*, 137 (1928).

[328] L. Infeld, "Zur Feldtheorie von Elektrizität und Gravitation," *Phys. Zs.*, 29, 145 (1928).

[329] L. Infeld, "Les équations de Maxwell dans la théorie commune a la gravitation et la électricité," *C. R. Acad. Sci.*, 186, 1280 (1928).

[330] A. Einstein, "Der gegenwärtige Stand der Relativitätstheorie," *Die Quelle (Pedagogischer Führer)*, *82*, 440–442 (1932) (Russian translation: [397], Vol. 2, pp. 399–402).

[331] G. Ludwig, *Fortschritte der projektiven Relativitätstheorie*, Vieweg, Braunschweig (1951).

[332] O. Veblen and B. Hoffman, "Projective relativity," *Phys. Rev.*, *36*, 810–822 (1930).

[333] W. Pauli, "Ueber die Formulierung der Naturgesetze mit fünf homogenen Koordinaten," *Ann. Phys.*, *18*, 305–312 (1933).

[334] L.D. Landau and E.M. Lifshitz, *Teoriya polya (Field Theory)*, Gostekhteorizdat, Moscow–Leningrad (1941), 283 p.

[335] A. Einstein (with W. Mayer), "Einheitliche Theorie von Gravitation und Elektrizität," *Sitzungsber. Preuss. Akad. Wiss., Phys.-Math. Kl.*, pp. 541–557 (1931) (Russian translation: [397], Vol. 2, pp. 366–386).

[336] A. Einstein (with W. Mayer), "Einheitliche Theorie von Gravitation und Elektrizität. II." *Sitzungsber. Preuss. Akad. Wiss., Phys.-Math. Kl.*, 130–137 (1932) (Russian translation: [397], Vol. 2, pp. 387–395).

[337] V.F. Weisskopf, *Physics in the Twentieth Century: Selected Essays*, Cambridge, Mass. (1972) (Russian translation published by Atomizdat, Moscow (1977)).

[338] E. Schrödinger, "Diracsches Elektron im Schwerefeld. I," *Sitzungsber. Preuss. Akad. Wiss.*, 105 (1932).

[339] V. Bargmann, "Bemerkungen zur allgemein-relativistischen Fassung der Quantentheorie," *Sitzungsber. Preuss. Akad. Wiss.*, 346 (1932).

[340] L. Infeld and B.L. van der Waerden, "Die Wellengleichung des Elektrons in der allgemeinen Relativitätstheorie," *Sitzungsber. Preuss. Akad. Wiss.*, 380, 474 (1933).

[341] A. Einstein (with N. Rosen), "The particle problem in the general theory of relativity," *Phys. Rev.*, *48*, 73–77 (1935) (Russian translation: [397], Vol. 2, pp. 424–433).

[342] A. Einstein (with N. Rosen), "Two-body problem in general theory of relativity," *Phys. Rev.*, *49*, 404–405 (1936) (Russian translation: [397], Vol. 2, pp. 434-435).

[343] A. Einstein (with W. Pauli), "Non-existence of regular stationary solutions of relativistic field equation," *Ann. Math.*, *44*, 131–137 (1943) (Russian translation: [397], Vol. 2, pp. 560–567).

[344] A. Einstein, "Bivector fields," *Ann. Math.*, *45*, 15–23 (1944) (Russian translation: [397], Vol. 2, pp. 586-596).

[345] A. Einstein, "Relativistic theory of the non-symmetric field," in: *The Meaning of Relativity*. Fifth edition. Princeton, 1955. Supplement II (Russian translation: [397], Vol. 2, pp. 849–873).

[346] C.N. Yang and R.L. Mills, "Conservation of isotopic spin and isotopic gauge invariance," *Phys. Rev.*, *96*. 191–195 (1954) (Russian published in [358], pp. 28–37).

[347] H.-J. Treder, "Der heutige Stand der Geometrisierung der Physik und der Physikalisierung der Geometrie," *Sitzungsber. Preuss. Akad. Wiss, Berlin*, pp. 1–33 (1975).

[348] D. Hoffmann, *Erwin Schrödinger*, Teubner, Berlin (1984), 92 pp.

[349] E. Schrödinger, Autobiographical sketch, in: *Les prix Nobel en 1933*, M.C.G. Santesson, ed., Imprimerie Royale, Stockholm (1933), pp. 86–88 (Russian translation in [237], pp. 343–346).

[350] E. Schrödinger, "Die Energiekomponenten des Gravitationsfeld," *Phys. Zs., 19*, 4–7 (1918).

[351] T.S. Kuhn, J.L. Heilbron, P. Forman, and L. Allen, *Sources for History of Quantum Physics*, Philadelphia (1967), 176 pp.

[352] E. Schrödinger, "Über eine bemerkenswerte Eigenschaft der Quantenbahnen eines einzelnen Elektrons," *Zeitschrift für Physik, 12*, 13–23 (1922) (Russian translation in [237], pp. 1676–171).

[353] E. Schrödinger, "Quantisierung als Eigenwertproblem I–IV," *Ann. Phys., 79*, 361–376, 489–527 (1926); *80*, 437–490 (1926); *81*, 109–139 (1926) (Russian translation in [237], pp. 9–50, 75–138).

[354] V.V. Raman and P. Forman, "Why was it Schrödinger who developed de Broglie's ideas?" *Hist. Stud. Phys. Sci., 1*, 291–314 (1969).

[355] F. London, "Quantenmechanische Deutung der Theorie von Weyl," *Zeitschrift für Physik, 42*, 375–389 (1927).

[356] A.A. Slavnov and L.D. Faddeev, *Vvedenie v Kvantovuyu Teoriyu Kalibrovochnykh Polei*, Nauka, Moscow (1978), 240 p. (English translation: *Gauge Fields, Introduction to Quantum Theory* (Frontiers in Physics, Vol. 50), Addison-Wesley, Reading, Mass. (1980)).

[357] J.J. Sakurai "Theory of strong interactions," *Ann. Phys.* (USA), *11*, 1–48 (1960) (Russian translation in [358]).

[358] D.D. Ivanenko, ed., *Elementary Particles and Compensating Fields* (Russian translations), Mir, Moscow (1964).

[359] "Einstein and the physics of the future: discussion," in: *Some Strangeness in the Proportion: A Centennial Symposium to Celabrate the Achievements of Albert Einstein*, (ed. H. Woolf), Addison-Wesley (1980), pp. 491–506.

[360] J.J. Sakurai, "Vector theory of strong interaction," This report was given at the special session of the American Physical Society on November 24, 1960. The article was published in *Bull. Amer. Phys. Soc., 5*, 4–14, (1960). The text of this report is published as a preprint of E. Fermi's Institute of Nuclear Research (University of Chicago, EFINS-60-63). Russian translation in [358].

[361] E.P. Wigner, "Invariance in physical theory," in: *Symmetries and Reflections, Scientific Essays*, Indiana University Press (1967), pp. 3–13. Russian translation published by Mir, Moscow (1971).

[362] W. Pauli, "Relativistic field theories of elementary particles," *Rev. Mod. Phys., 13*, 203–232 (1941) (Russian translation published as book by Izd. Inostr. Lit., Moscow (1947)).

[363] J.C. Maxwell, *A Treatise on Electricity and Magnetism*, Vols. 1 and 2, Clarendon Press, Oxford (1892) (Russian translation in a volume of selected works published by Gostekhteorizdat, Moscow (1954), pp. 345–632)).

[364] M. Faraday, *Experimental Researched in Electricity*, Vols. 1, 2, 3, London (1849–1855) (Russian translations published by Izd. AN SSSR, Moscow (1947–1949)).

[365] T.P. Kravets, "Faraday and his *Experimental Researches in Electricity*," in: M. Faraday, *Éksperimental'nye Issledovaniya po Élektrichestvu*, Izd-vo AN SSSR, Moscow (1947), Vol. 1, pp. 733–780 (Russian translation of [366]).

[366] M. Faraday, see [364], Vol. 3.

[367] A.M. Bork, "Maxwell and the vector potential," *Isis, 58*, 210–222 (1967).

[368] J.C. Maxwell, "On Faraday's lines of force," in: *The Scientific Papers of James Clerk Maxwell*, ed. W. D. Niven, Vol. 1, University Press, Cambridge (1890), pp. 155–229 (Russian translation in [363], pp. 11–88)).

[369] J.C. Maxwell, "On physical lines of force," in [368], pp. 451–513 (Russian translation in [363], pp. 107–193)).

[370] J.C. Maxwell, "A dynamical theory of the electromagnetic field," in [368], pp. 526–597 (Russian translation in [363], pp. 251–341).

[371] J.C. Maxwell, "On a method of making a direct comparison of electrostatic with electromagnetic force; with a note on the electromagnetic theory of light," *Philos. Trans. R. Soc. London, 157*, 643–657 (1868).

[372] A.T. Grigor'yan and A.N. Vyal'tsev, *Genrikh Gerts (Heinrich Hertz)*, Nauka, Moscow (1968), 310 pp.

[373] B.G. Doran, "Origins and consolidation of field theory in nineteenth century Britain: From the mechanical to the electromagnetical view of nature," *Hist. Stud. Phys. Sci., 6*, 133–260 (1975).

[374] J. Larmor, *Aether and Matter*, University Press, Cambridge (1900).

[375] H.A. Lorentz, *The Theory of Electrons*, Teubner, Leipzig (1909) (Russian translation published by GITTL, Moscow (1956)).

[376] M. Abraham, *Theorie der Elektrizität* (fully revised by R. Becker), Vol. 1 and 2, Teubner, Leipzig (1930–1933) (English translation: *The Classical Theory of Electricity and Magnetism*, London (1932). Russian translation published by GONTI, Moscow–Leningrad (1939)).

[377] E. Wigner, *Symmetries and Reflection: Scientific Essays*, MIT Press (1970) (Russian translation published by Mir, Moscow, (1971), 320 pp.).

[378] Y. Aharonov and D. Bohm, "Significance of electromagnetic potentials in quantum theory," *Phys. Rev., 115*, 485–491 (1959).

[379] R.P. Feynman, R.B. Leighton, and M. Sands, *The Feynman Lectures on Physics*, Vol. 2, Addison-Wesley, Reading, Mass. (1964) (Russian translation published by Mir, Moscow (1966)).

[380] E.L. Feinberg, "On the 'special role' of the electromagnetic potentials in quantum mechanics," *Usp. Fiz. Nauk, 78*, 53–64 (1962) (English translation: Sov. Phys. Uspekhi, 5, 753–760 (1963)).

[381] H. Erlichson, "Aharonov–Bohm effect—quantum effects on charged particles in field-free regions," *Am. J. Phys., 38*, 162–173 (1970).

[382] K. Moriyasu, "The renaissance of gauge theories," *Contemp. Phys., 23*, 553–581 (1983).

[383] Yu.I. Manin, *Matematika i Fizika*, Znanie, Moscow (1979) (English translation: *Mathematics and Physics* (Progress in Physics, Vol. 3), Birkhäuser, Boston (1981)).

[384] H. Weyl, *Gruppentheorie und Quantenmechanik*, Leipzig (1928).

[385] V. Fock and D. Ivanenko, "Géometrie quantique linéare et déplacement parallèle," *C. R. Acad. Sci., 188*, 1470–1472 (1929).

[386] V. Fock and D. Ivanenko, "Ueber eine mögliche geometrische Deutung der relativistischen Quantentheorie," *Zeitschrift für Physik, 54*, 798–802 (1929).

[387] V.A. Fock, "Geometrization of Dirac's theory of the electron," in the Russian translation of [170], pp. 415–432.

[388] V.A. Fock, "Dirac's wave equation and Riemannian geometry," *ZhRFKhO Chast'. Fiz., 62*, 133-151 (1930).

[389] W. Pauli, "Prinzipien der Quantentheorie I," in: *Handbuch der Physik*, Vol. 24, Part 1 (eds. H. Geiger and K. Scheel), (1933) (Russian translation published by Gostekhteorizdat, Moscow–Leningrad (1947), English translation: *General Principles of Quantum Mechanics*, Springer, Berlin (1980)).

[390] W. Heisenberg and W. Pauli, "Quantentheorie der Wellenfelder, 2 mitt.," *Zeitschrift für Physik, 59*, 168–190 (1930) (Russian translation in: W. Pauli, *Trudy po Kvantovoi Teorii, Stat'i, 1928–1958*, Nauka, Moscow (1977), pp. 89–111)).

[391] P.A.M. Dirac, "Quantized singularities in the electromagnetic field," *Proc. R. Soc. London, 133*, 60–72 (1931) (Russian translation in: *Monopol' Diraka*, Mir, Moscow (1970), pp. 40–57)).

[392] H. Weyl, *Theory of Groups and Quantum Mechanics*, Dover, New York (1931).

[393] G. Wentzel, *Einführung in die Quantentheorie der Wellenfelder*, F. Deuticke, Vienna (1943) (English translation: *Quantum Theory of Fields*, New York (1949), Russian translation published by Gostekhizdat, Moscow–Leningrad (1947)).

[394] J. Schwinger, "The theory of quantized fields, I–III," *Phys. Rev., 82*, 914–927 (1951); *91*, 713–728 (1953); *91*, 728–740 (1953) (Russian translation publsihed as separate book by Izd-vo Inostr. Lit., Moscow (1956), 252 p.).

[395] E.S. Abers and B.W. Lee, "Gauge theories," *Phys. Rep. Phys. Lett. C (Netherlands), 9C*, 1–141 (1973) (Russian translation in: *Kvantovaya Teoriya Kalibrovochnykh Polei*, Mir, Moscow (1977), pp. 241–433).

[396] Yu.I. Manin, "Geometrical ideas in field theory," Introductory paper of editor of the translation of the collection: *Geometrical Ideas in Physics*, Mir, Moscow (1983), pp. 3–18.

[397] A. Éinshtein (Einstein), *Sobranie Nauchnykh Trudov (Collected Scientific Works in Russian translation)*, Vol. 1–4, Nauka, Moscow (1965–67).

[398] W.D. Niven (ed.), *The Scientific Works of James Clerk Maxwell*, Vol. 1, Cambridge University Press (1890).